Waves and Patterns in Chemical and Biological Media

Special Issues of *Physica D*

The titles in this series are paperback, readily accessible special issues of *Physica D*, produced by special agreement with Elsevier Science Publishers B.V.

Emergent Computation: Self-Organizing, Collective, and Cooperative Phenomena in Natural and Artificial Computing Networks, edited by Stephanie Forrest, 1991.

Lattice Gas Methods: Theory, Applications, and Hardware, edited by Gary D. Doolen, 1991.

Cellular Automata: Theory and Experiment, edited by Howard Gutowitz, 1991.

Waves and Patterns in Chemical and Biological Media, edited by Harry L. Swinney and Valentin I. Krinsky, 1992.

Waves and Patterns in Chemical and Biological Media

edited by
Harry L. Swinney
Valentin I. Krinsky

A Bradford Book
The MIT Press
Cambridge, Massachusetts
London, England

First MIT Press edition, 1992
© 1991 Elsevier Science Publishers, B.V., Amsterdam, the Netherlands

Reprinted from *Physica D*, Volume 49, Numbers 1–2, 1991. The MIT Press has exclusive license to sell this English-language book edition throughout the world.

Printed and bound in the Netherlands. This book is printed on acid-free paper.

Library of Congress Cataloging-in-Publication Data

Waves and patterns in chemical and biological media / edited by Harry L. Swinney, Valentin I. Krinsky. — 1st MIT Press ed.
     p.   cm. — (Special issues of physica D)
    Papers from a workshop organized by the USSR Academy of Sciences and others.
    Published also as v. 49, no. 1–2 of Physica D.
    "A Bradford book."
    Includes bibliographical references and index.
    ISBN 0-262-69150-7
    1. Waves—Congresses. 2. Transport theory—Congresses. 3. Excitation (Physiology)—Congresses. I. Swinney, H. L., 1939–   . II. Krinskiĭ, V. I. (Valentin Izrailevich) III. Akademiiâ nauk SSSR. IV. Series.
QC157.W39 1992
541.3'9—dc20
                                            91-23336
                                            CIP

MIT Press

0262691507

KRINSKY
WAVES PATTERNS CHEM BIO

Preface

This volume contains invited papers presented at an international research workshop, "Waves and Patterns in Biological and Chemical Excitable Media," held in Pushchino, USSR, in May 1990. The workshop was organized by the USSR Academy of Sciences, the Biological Research Center, Institute of Biological Physics of the USSR Academy of Sciences, and the Center for Nonlinear Dynamics of the University of Texas at Austin, USA.

Pushchino is a small town near Moscow, where the Biological Research Center of the USSR Academy of Sciences is situated. It was a great pleasure for us to be able to welcome there most of the pioneers of this field. A previous symposium on the same topic was held in Pushchino in 1983. Proceedings of that symposium were published in "Self-Organization, Autowaves and Structures Far from Equilibrium", V.I. Krinsky, ed., Synergetics, Vol. 28 (Springer, Berlin, Heidelberg, New York, 1984).

Experiments in this field are difficult, and theorists dominate. However, much of the theoretical work has been motivated by experiments, and in recent years digital imaging techniques and new reactor designs have led to the discovery of phenomena that were not anticipated theoretically. Also numerical simulations of reaction–diffusion systems have become more realistic as large fast computers become widely available. At this workshop we succeeded in having a good mix of theorists, experimentalists, and numericists from diverse disciplines, including chemistry, biology, physics, applied mathematics, and physiology.

The organisation of this workshop would not have been possible without the generous support of the sponsoring organizations: Dyna Electronics FRG–USSR; several institutes of the USSR Academy of Sciences – Institute of Applied Physics, Moscow Radio-technical Research Institute, Institute of Radio-electronics, Institute of Applied Mathematics, and the Physical Research Institute; the Institute of Atomic Energy; the Engineering Center for Physico-Chemical Biology; and the Physical Society of the USSR. Such unusually broad support reflects the interdisciplinary nature of the field.

Finally, we wish to thank the members of the Autowave Laboratory of the Institute of Biological Physics in Pushchino for their flawless organization of this meeting.

Harry Swinney
Valentin Krinsky

Physica 49 (1991) viii–ix
North-Holland

Contents

Physica D 49 (1991) 1–4
North-Holland

Chapter 1. Spiral, ring and scroll patterns: Experiments

Direct observation of vortex ring collapse in a chemically active medium

K.I. Agladze, R.A. Kocharyan and V.I. Krinsky

Institute of Biological Physics of the USSR Academy of Sciences, 142292 Pushchino, Moscow Region, USSR

We studied experimentally the behavior of vortex rings in a homogeneous layer of Belousov–Zhabotinsky (BZ) reagent in solution. It is observed that in a chemically active medium there is no spontaneous drift of vortex rings, only collapse. The diffusion coefficient for the autocatalytic variable ($HBrO_2$) estimated from the ring collapse velocity is $(1.9 \pm 0.2) \times 10^{-5}$ cm/s at 20°C.

The dynamics of three-dimensional vortex rings in excitable media was extensively studied in numerical experiments. Using the FitzHugh–Nagumo model, it was possible to study on a computer the dynamic properties of vortex rings, such as their spontaneous drift along the axis of the torus, shrinkage and collapse, and expansion [1, 2]. It was found subsequently that not all of the results based on diffusion of the fast variable only [3] are valid for chemically active media. Particularly, consideration of diffusion of the slow variable influences not only the magnitude but also the sign of the velocity of spontaneous drift of a vortex ring along its axis. In the equal diffusion case, only collapse of vortex ring is observed [4].

Experimental studies of 3D vortices in different chemically active media [5–7] have led to conflicting results. Keener and Tyson observed collapse of 3D vortices representing the so-called elongated sources [5], which randomly occur in thick layers of liquid-phase BZ reagent. These sources are vortex rings of irregular shape whose thread forms a complicated elongated oval, in contrast to rings whose thread is a circle. As measured in ref. [5], the thread shrinkage velocity for such a source turned out to be in good agreement with theoretical estimates for vortex rings of regular shape.

Winfree and co-workers [6] studied the behavior of vortex rings in gelled BZ reagent. The thread shrinkage velocity was found to be half as high as predicted theoretically. This could possibly be explained by the peculiar technique used by the authors to create vortex rings. The technique (layer wise spreading of the melted gel (layer by layer)) permitted obtaining vortex rings of the desired initial size and regular shape [6], but failed to ensure homogeneity of the medium. In inhomogeneous media, however, the vortex structure dynamics is greatly complicated [8]. Therefore, the problem is still open, and direct accurate studies are needed.

We studied vortex rings with clearly discernible structure as in refs. [6, 9], but our rings were created in layers of liquid reagent known to be uniform, as in refs. [5, 9]. The thickness of layers was 5–6 mm. The composition of the reagent was the following: 0.1 M $CHBr(COOH)_2$; 0.45 M H_2SO_4; 0.3 M $NaBrO_3$; 0.1 M $CH_2(COOH)_2$; 5 mM ferroin. We used a procedure combining the advantages of the abovementioned two methods and free of their limitations. By "rocking and rolling" a petri dish, we obtained a variety of 3D structures. From those, we chose structures shaped precisely as a regular vortex ring. Their percentage was at first very low, but it grew with experience, justifying labor and time expended

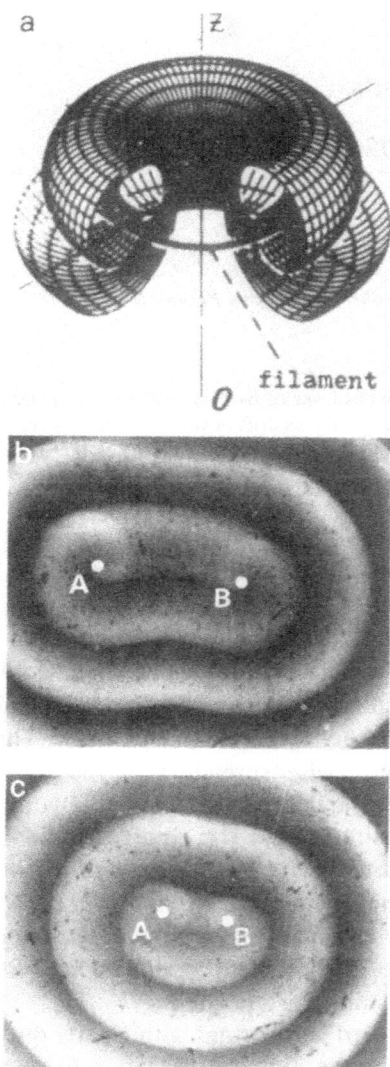

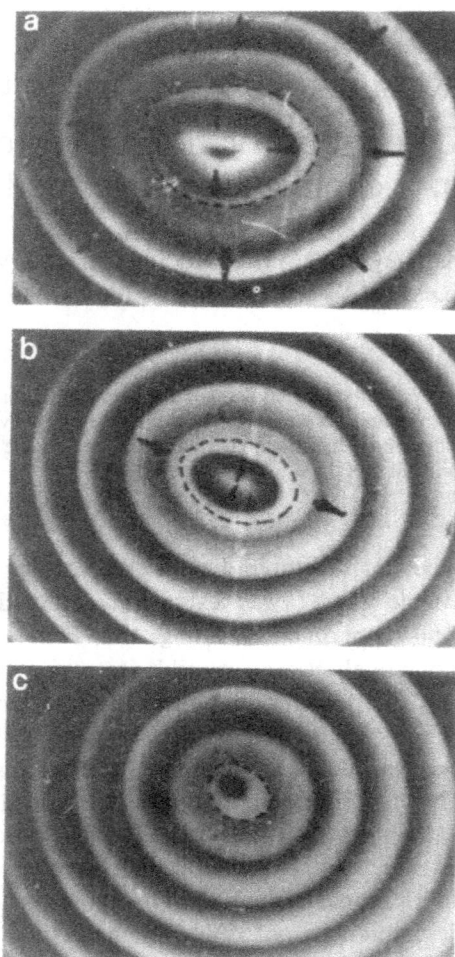

Fig. 1. Vortex ring. (a) Computer simulation of a vortex ring [2]. (b) A fragment of a vortex ring in a petri dish with BZ reagent. A and B are rotation centers of spiral waves in the plane of ring section by the solution surface, and the line from A to B is the ring thread. (c) The same ring fragment after 7 min; shrinkage of the ring thread is seen.

Fig. 2. Vortex ring with its thread running parallel to the solution surface. (a) Arrows show the direction of wave propagation; the dashed line is the ring thread. Inside the region bounded by the thread, waves collapse and outside it they propagate to the medium borders. (b, c) Progressive shrinkage of the ring thread. It is seen that the thread acquires with time a more regular ring shape.

for their selection: the results obtained with vortex rings of a varying diameter were wholly reproducible.

When the vortex thread was properly located with respect to the plane of photography the structure of 3D sources (fragments of vortex ring) was clearly observable; see fig. 1. In a number of cases the vortex thread plane was parallel to the petri dish plane. The wave pattern on the solution surface in such a case is shown in fig. 2. Inside the region bounded by the vortex thread, a wave moves to its center and collapses, and outside the region it propagates away from the center to the medium borders. At the same time, the ring thread slowly contracts. Examination of such

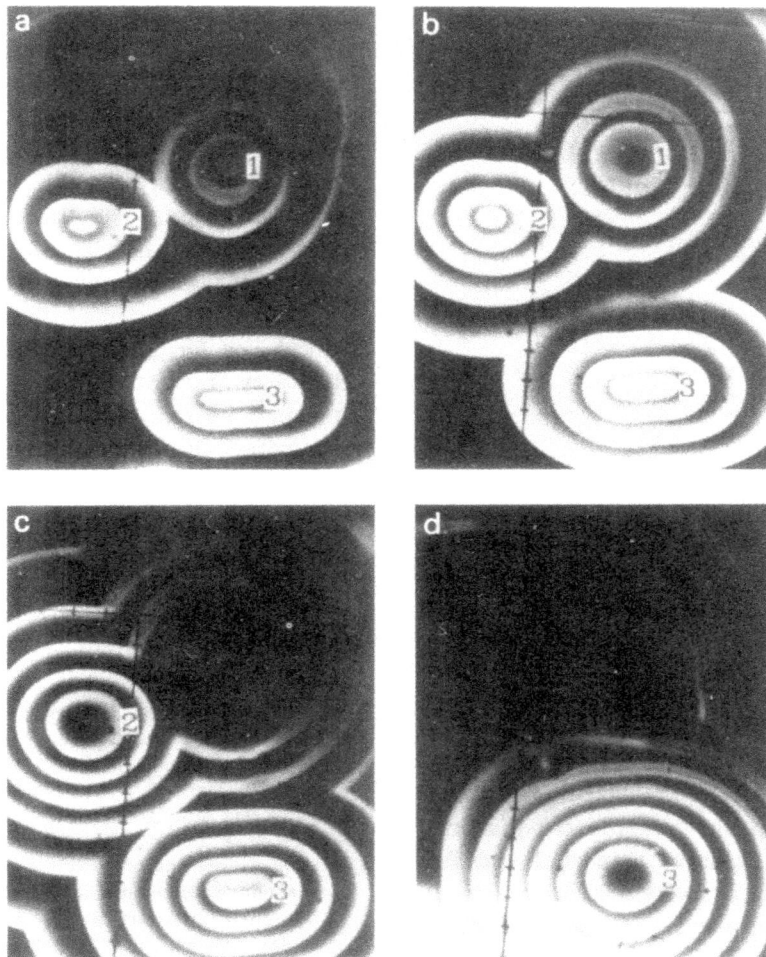

Fig. 3. Evolution of three-dimensional rings. Successive snapshots of the wave pattern created by the three sources (1, 2 and 3) in (a). The snapshots were taken against the coordinate grid with a 2 mm step. In (b), vortices 1 and 2 still persist and vortex ring 3 has disappeared; in its center, instead of waves, there appears a dark hole that expands with time (snapshots (b), (c), and (d)). The wave pattern that persists at a distance from the source center is due to propagation of waves emitted by the source earlier, before its decay. In (d) no trace of the wave structure of vortex ring 1 is seen. The instant of decay of vortex ring 2 is shown in (c); here again there appears a dark hole in its center that grows with time. It is seen that the greater the initial source size the longer its life time.

vortices permitted estimating the life time of vortex rings, depending on their initial size.

Fig. 3 presents a series of successive snapshots showing the evolution of three autowave sources of different size. It is seen that the source size diminishes with time and, finally, the source collapses. First the source with the smallest size collapses, (b, top, 1), then the next in size collapses (c, left, 2), and finally the largest one

collapses (d, bottom, 3).

No motion of vortex rings along the torus axis was observed experimentally (with an accuracy of 0.1 mm per 50 min). Note that in numerical experiments the effect of spontaneous vortex drift was obtained when diffusion of the fast variable only was considered [1]. In the case of equal diffusion coefficients for the fast and slow variable, the theory predicts the absence of such a

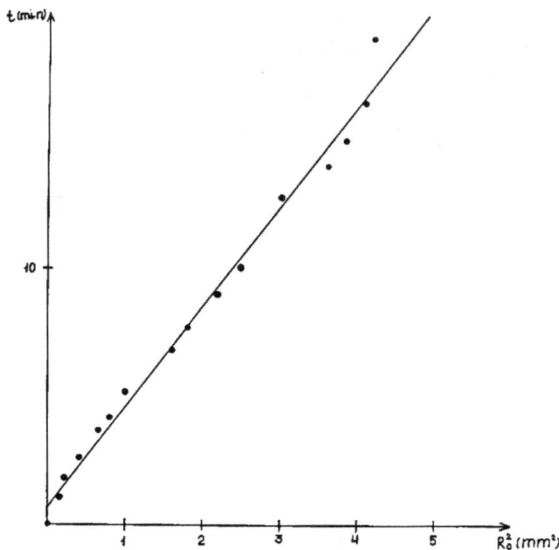

Fig. 4. Dependence of the life time τ on the vortex ring size. It is seen that the plot of τ against R_0^2 (R_0 being the initial ring radius) is a straight line.

drift [4]. As shown by Panfilov et al. [4], the velocity of ring shrinkage in an active medium with equal diffusion coefficients for all the components is independent of the shape of the non-linear function describing the medium and is only determined by the diffusion coefficient and the curvature radius for rings with a sufficiently large radius ($R = 2r$ or $3r$, where r is the core radius):

$$V = D/R, \tag{1}$$

where R is the thread radius and D is the diffusion coefficient.

Integrating (1), we obtain an estimate for the life time of a vortex ring:

$$\tau = R_0^2/2D, \tag{2}$$

where τ is the life time and R_0 is the initial ring radius.

Formula (1) describes rigorously only the initial stages of the evolution of a vortex ring, as long as $R \gg r$ and, therefore, it gives an approximate estimate for its life time.

The results of experimental measurements of the vortex ring life time are presented in fig. 4. It is evident that plotting the measured values against the initial radius squared gives practically a straight line. The diffusion coefficient, as determined from this plot, is $D = (1.9 \pm 0.2) \times 10^{-5}$ cm^2/s, coinciding with fair accuracy with that taken for HBrO$_2$ [10], $D = 1.8 \times 10^{-5}$ cm^2/s.

The approximate equality of the diffusion coefficients in the chemically active medium considered (for Br$^-$, HBrO$_2$ and Fe(phen)$_3$) is explained by the fact that $D \sim m^{-1/3}$, where m is molecular weight, and, as is known, low molecular compounds show close values of their diffusion coefficients.

References

[1] A.V. Panfilov and A.M. Pertsov, Vortex ring in a three-dimensional active medium of a reaction system with diffusion, Dokl. AN SSSR 274 (1984) 1500–1503.
[2] A.V. Panfilov and A.N. Rudenko, Two regimes of the scroll ring drift in the three-dimensional active media, Physica D 28 (1987) 215–218.
[3] A.M. Pertsov and A.V. Panfilov, Spiral waves in an active medium. Reverberator in a FitzHugh–Nagumo model, in: Active Processes in Systems with Diffusion (IPF, Gorki, 1981) pp. 77–84 [in Russian].
[4] A.V. Panfilov, A.N. Rudenko and V.I. Krinsky, Vortex rings in a three-dimensional medium with two-component diffusion, Biofizika 34 (1986) 850–854.
[5] J.P. Keener and J.J. Tyson, The motion of untwisted untorted scroll waves in Belousov–Zhabotinsky reagent, Science 239 (1988) 1284–1286.
[6] W. Jahnke, C. Henze and F.T. Winfree, Chemical vortex dynamics in three-dimensional excitable media, Nature 336 (1988) 662–665.
[7] A.V. Panfilov, R.R. Aliev and A.V. Mushinsky, An integral invariant for scroll rings in a reaction–diffusion system, Physica D 36 (1989) 181–188.
[8] A.N. Rudenko and A.V. Panfilov, Drift and interaction of vortices in a two-dimensional inhomogeneous active medium, Stud. Biophys. 99 (1983) 183–188.
[9] B. Welsh, J. Gomatam and A. Burgess, Three-dimensional waves in the Belousov–Zhabotinsky reaction, Nature 304 (1983) 611–614.
[10] P. Foerster, S. Müller and B. Hess, Curvature and propagation velocity of chemical waves, Science 241 (1988) 685–687.

Physica D 49 (1991) 5–12
North-Holland

Spatial patterns in a uniformly fed membrane reactor

G. Kshirsagar[1], Z. Noszticzius[2], W.D. McCormick and Harry L. Swinney[3]

Center for Nonlinear Dynamics and Department of Physics, The University of Texas, Austin, TX 78712, USA

A novel reactor, consisting of a thin membrane with each side exposed to fluid from a well-stirred flow reservoir, has been developed for the study of sustained chemical spatial patterns. This reactor bears some similarity to another continuously fed unstirred reactor with a uniform feed, the disk-shaped reactor developed by Tam et al. [J. Chem. Phys. 88 (1988) 3395], but the present reactor differs in two important aspects: (1) The effective residence time, which is determined by the thickness of the gel, can be very short (of the order of seconds). (2) The feed is from two reservoirs rather than one; this opens up new regions of chemical composition for study. Examples are presented of patterns obtained for the Belousov–Zhabotinsky reaction in the membrane reactor.

1. Introduction

Two decades ago Zhabotinsky [1], Winfree [2], and others conducted pioneering studies on the formation of spatial patterns in chemical systems. Those experiments have been followed by many others that have examined the patterns that form, usually in thin fluid layers, as a function of the initial chemical concentrations, temperature, and light intensity. Those studies have always concerned *transient* patterns – the patterns emerge, evolve continuously, and finally decay as the system approaches thermodynamic equilibrium. Recently a new generation of reactors has been developed that makes it possible to study *sustained* patterns: the chemical system is maintained in a well-defined nonequilibrium state by a chemical flux, just as in a CSTR (continuous flow stirred tank reactor), but in these new types of reactors there is no stirring in the region in which patterns can form. To emphasize the analogy and the contrast with stirred tank reactors, these new

types of reactors are sometimes called CFURs (continuous flow *un*stirred reactors). With these reactors it is now possible to study *bifurcations* in spatiotemporal patterns formed by reaction–diffusion systems. Control parameters can be increased and decreased through a particular transition to determine whether or not it is subcritical (hysteretic) or supercritical (nonhysteretic). Long-time series data can be obtained for a well-defined asymptotic state of the system and analyzed by the techniques of dynamical systems to obtain, for example, the generalized dimensions D_q, the spectrum of Lyapunov exponents, and the metric entropy [3]. The spatiotemporal character of a pattern can be deduced from space–time correlation functions and from the flow of information [4].

Two basic types of CFURs have been developed thus far. One type has an imposed chemical gradient, as in the two-dimensional ring-shaped reactor of Noszticzius et al. [5] and Kreisberg et al. [6], and in the effectively one-dimensional Couette reactor developed in Austin [7] and Bordeaux [8]. A second type of reactor has a uniform diffusive feed orthogonal to the line or plane in which a pattern forms. An example of this type of apparatus is the disk-shaped reactor of Tam et al. [9] and Skinner and Swinney [10]; the patterns form in a thin inert gel layer that is

[1] Present address: Puritan-Bennett Corporation, 10800 Pflumm Road, Lenexa, KS 66215, USA.
[2] Permanent address: Institute of Physics, Technical University of Budapest, 1251 Budapest, Hungary.
[3] To whom correspondence should be addressed. e-mail: swinney@chaos.utexas.edu

0167-2789/91/$03.50 © 1991 - Elsevier Science Publishers B.V. (North-Holland)

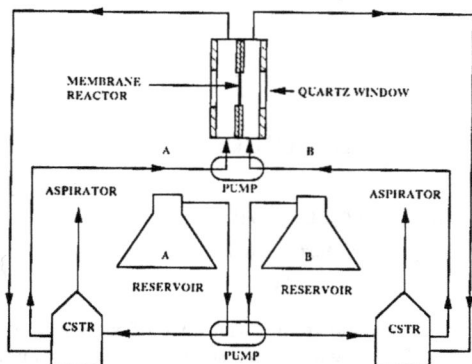

Fig. 1. Chemical patterns form in a thin gel-impregnated membrane placed between two compartments containing rapidly flowing reagents. In the present experiments reagents A and B are respectively the oxidant ($KBrO_3$ with or without H_2SO_4) and reductant of the BZ reaction (malonic acid and ferroin with or without H_2SO_4). The reactor body is constructed of plexiglas (polymethylmethacrylate). Circulating water from a thermostat maintains the temperature at $25.0 \pm 0.1°C$.

in diffusive contact with a CSTR through a thin capillary array.

This paper describes a new type of reactor of the second type: patterns form in a thin gel layer placed between *two* compartments that may respectively contain, for example, the oxidant and reductant of a reaction. This design offers considerable flexibility for studies of sustained patterns under a wide range of chemical compositions. We have examined the versatility of this reactor using the Belousov–Zhabotinsky (BZ) reaction.

2. Experimental system

Fig. 1 is a schematic diagram of the system. The reaction occurs in a polyacrylamide-impregnated PVC membrane, 0.75 mm thick by 25 mm in diameter. The two sides of the membrane are identical compartments (each 1.9 ml in volume) fed with chemicals from two external CSTRs (each 4.0 ml in volume), which in turn are fed from reservoirs A and B, respectively; see the diagram in fig. 2. The feeds are maintained with peristaltic pumps.

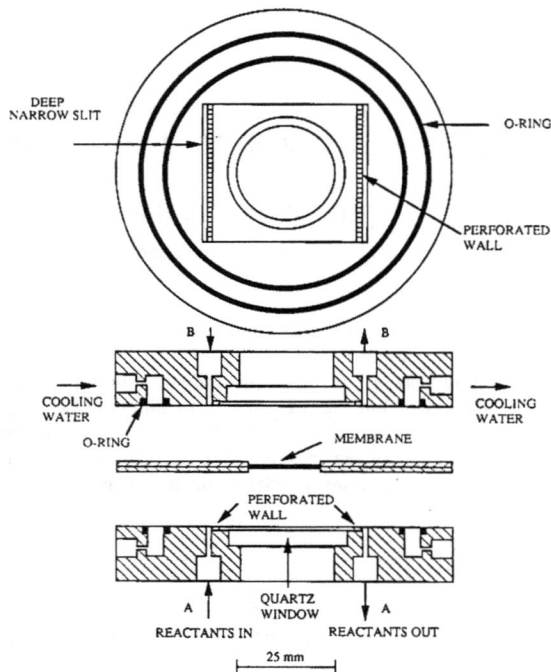

Fig. 2. Top view and exploded side view of the uniformly fed membrane reactor. The top view shows only the principal features while the side view gives more details. The circular membrane is held between two thin circular plexiglas plates that are compressed between the larger reactor components that contain the feed chemicals and have quartz windows for viewing the patterns. The perforated wall ensures uniform feed parallel to the membrane. The chemicals pass from the deep narrow slits through perforated walls into compartments (1 mm deep) in contact with the membrane.

The effective residence time of the reagents in the membrane reactor is $\sim d^2/\pi^2 D$, where D is the diffusion coefficient and d is the membrane thickness; this time is ~ 1 min for our membranes. The residence time of the reagents in the compartments in contact with the membrane must be short to insure uniform feed to the entire membrane. A compartment residence time of 18 s was found to yield reproducible patterns, but significantly longer residence times gave patterns that were clearly affected by nonuniformity of the feed. In the experiments the residence times of the two external CSTRs were varied in concert in the range from 17 to 68 min.

Most of our experiments were conducted with a PolySil[#1] membrane with a pore size of 0.5 μm and a porosity of 70–80%; the membrane is made by incorporating silica in a PVC matrix. To decrease the porosity we impregnated the membrane with a polyacrylamide gel, which was made from the following solutions: A, 20% acrylamide; B, 4% piperazine diacrylamide (this cross-linking agent, from Bio-Rad Laboratories, is superior for our purposes to the usual N,N′-methylene bisacrylamide); C, 30% triethanol amine; and D, 20% ammonium persulfate. The PolySil membrane was first soaked under vacuum (to remove air trapped in membrane pores) in a solution containing 5 ml of A, 25 drops of B, and 6 drops of C (1 drop equals about 0.03 ml). Then 3 drops of D, the initiator for gel formation, was added to the solution containing the membrane, and the solution was stirred rapidly. The membrane and solution were then placed for 30 min in a mold between plexiglas plates separated by a 0.65 mm spacer, after which the membrane was rinsed to wash out any unreacted gel solution. The gel was then placed in water overnight and it would swell to a thickness of about 0.75 mm.

Reservoir solutions for the BZ reaction were made using reagent grade chemicals as supplied, except for $KBrO_3$, which was doubly recrystallized before use. The solutions were passed through cellulose nitrate filters (0.45 μm pore size) and degassed before use. The control parameters in the experiments were $KBrO_3$ concentration in the reservoir and the residence time in the external CSTRs.

Patterns were viewed from the side of the membrane in contact with the compartment containing bromate, while the compartment on the other side of the membrane contained malonic acid and ferroin. Some experiments were conducted with the sulfuric acid on the bromate side,

but for most the acid solution was on the ferroin side. The ferroin concentration was observed to depend on the sulfuric acid concentration in the reservoir; therefore, we examined the stability and equilibrium composition of the ferroin as a function of acid concentration, as described in the appendix.

3. Results

We have varied the bromate concentration and the residence time of the two external CSTRs while other parameters were held fixed. Waves have been observed to originate only in the rim of the membrane. No pacemakers occur in the central region, probably because of the short residence time. Waves can be created in the central region, however, by perturbation with UV light. Pacemakers and sustained rotating waves (*pinwheels* [5, 6]) can form spontaneously in the rim of the membrane, where the residence time is long because the plexiglas plates that support the membrane prevent the direct feed of reagent to that region. Pinwheels can be seen in the rim in figs. 3a–3i. A wave in the bulk that originated from a pacemaker in the membrane rim can be seen in fig. 3a. The pacemakers and pinwheels in the membrane rim can be completely eliminated by blocking the entry of reagents into this region (e.g., by filling the membrane rim with plexiglas dissolved in chloroform before it is impregnated with gel). In this case waves can be triggered with UV light, as in fig. 3j–3k.

We conducted experiments with UV light as a perturbation, using a ring-shaped lamp with peak output at 350 nm and a half-width of 50 nm. When the medium had partially recovered from a previous wave, then only the edge of the membrane in contact with the waves of the pinwheel was excitable, as shown for example in fig. 3c. When the medium was excitable and had completely recovered from a previous perturbation, the whole membrane could be excited, as shown in fig. 3e. The clear 7-fold symmetry of figs. 3c

---

[#1] PolySil, a registered trademark of Polysciences, Inc., is a polyvinylchloride–silica-based membrane that is used for thin-layer chromatography. The silica in the membrane provides natural wetting characteristics, and it has greater thickness and mechanical strength than nitrocellulose membranes.

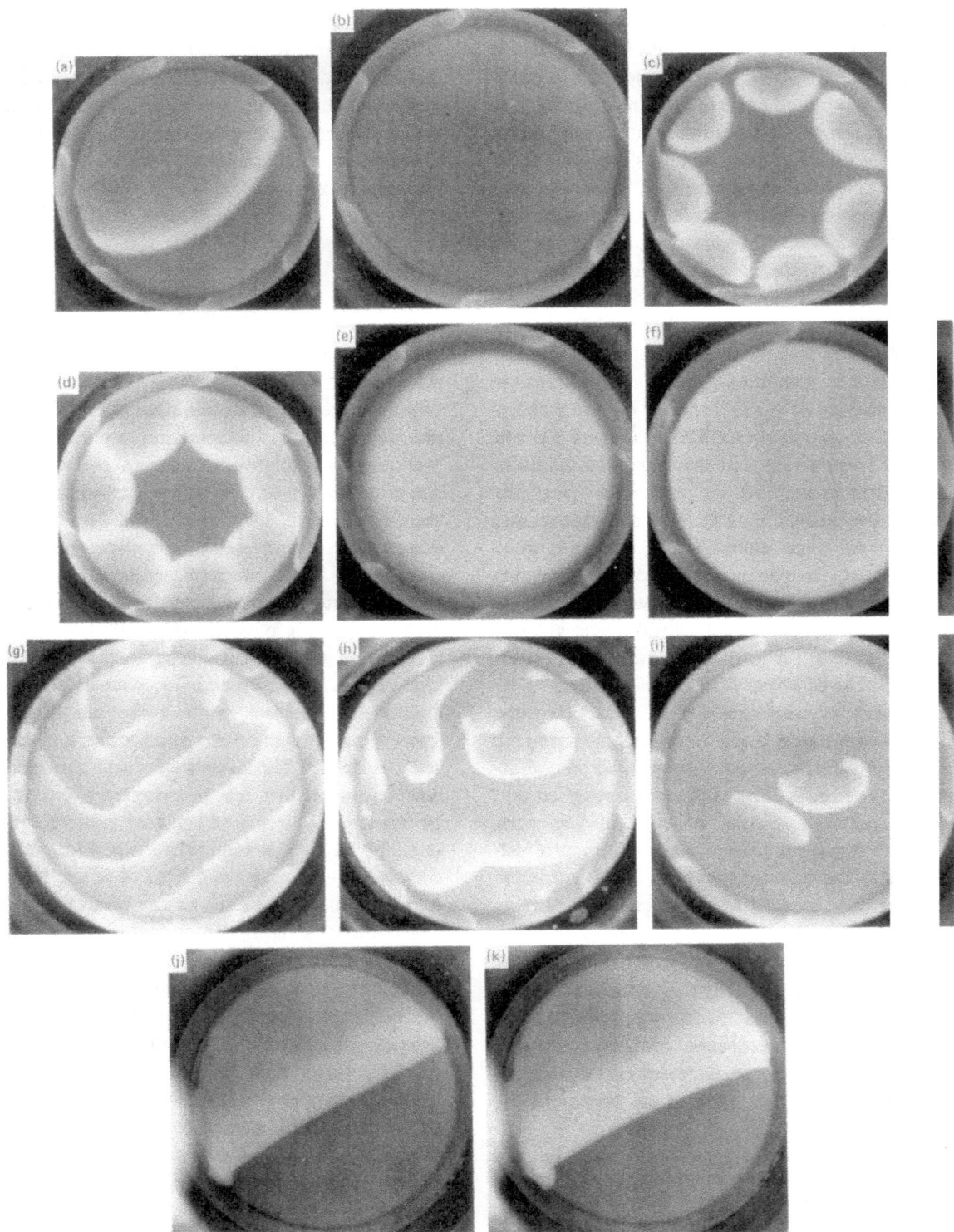

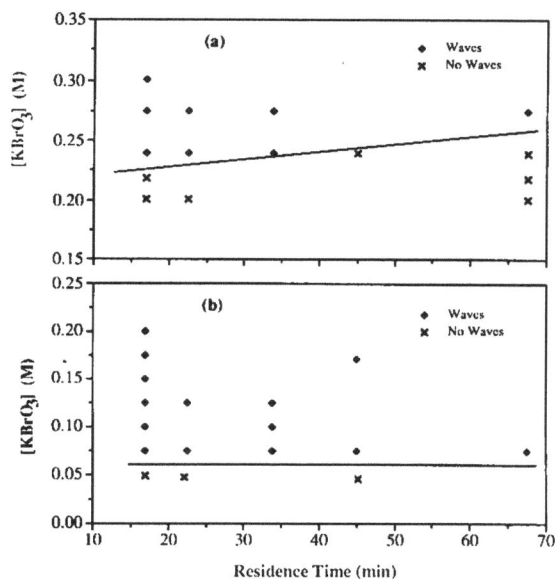

Fig. 4. Threshold for excitability determined with sulfuric acid (0.2 M) contained in the (a) ferroin–malonic acid reservoir and (b) bromate reservoir. The threshold for excitability is lower in (b) than in (a). Malonic acid concentration was 0.1 M; ferroin concentration was 0.00478 M in (a), 0.00600 M in (b).

and 3d, derived from the 7-wave pinwheel, and the rotational symmetry of figs. 3e and 3f attest to the uniformity of our reactor; thus this reactor should be ideal for studying the stability and lifetime of patterns imposed by exposure to light [11].

Fig. 4a is a phase diagram showing the threshold for excitability as a function of bromate concentration. Above the threshold waves spontaneously form in the unfed perimeter of the membrane where it is clamped between plexiglas plates. These waves propagate into the central portion of the membrane, as illustrated in fig. 3a. The wave front is sharp and the waves are thin when the sulfuric acid is fed with the malonic acid and ferroin, as in the experiments illustrated in figs. 3a–3i and the phase diagram in fig. 4a. The waves are thicker and the width increases with increasing bromate concentration when the sulfuric acid is fed with the bromate, as in figs. 3j–3k and in the phase diagram in fig. 4b. A similar effect was observed by Nagy-Ungvarai et al. [12] in experiments and simulations for the cerium-catalyzed BZ batch reaction. At a bromate concentration of 0.2 M or larger, three-dimensional effects and the effect of the finite size of the container become apparent. Simulations by Vasquez et al. [13] of a two-dimensional reaction–diffusion system with Oregonator kinetics support the observation that the width of the front increases with increasing bromate. Qualitatively, this is due to the higher bromate concentration, which leads both to a higher wave speed and to a longer time for the bromide concentration to build up and suppress the oxidation of ferroin.

◄ Fig. 3. Photographs of patterns formed in the membrane reactor: (a) A wave excited by a pacemaker in the rim of the gel. A spontaneously formed pinwheel exists in the rim, where the membrane holder blocks the direct feed of the reagents. (b) Quiescent state following the passage of the wave in (a); the system is in the refractory phase. (c) Effect of illumination with UV light for 5 min when the medium is in the refractory state after the passage of a wave; the 7-armed pinwheel generates a pattern with 7-fold symmetry. (d) Evolution of the system from (c) after the removal of the perturbation. (e) Effect of illumination with UV light for 2 min after the medium has sufficiently recovered from an excitation and is again fully excitable. (f) Evolution of the system from (e) after the removal of the perturbation. (g) At higher bromate concentration, 0.275 M rather than 0.240 M as in (a)–(f), the pinwheel excites waves that have different speeds in the rim and bulk, and this type of pattern forms. (h) Long-time evolution of the system in (g) to spirals (for same bromate concentration as in (g)). (i) Pattern formed from (h) after the residence time was reduced from 68 to 22 min (for same bromate concentration as in (g)). In (a)–(i), the sulfuric acid was contained in the ferroin–malonic acid reservoir. In the last two photographs, (j)–(k), the sulfuric acid was contained in the bromate reservoir, which results in thicker wave fronts: (j) Wave at 0.125 M bromate concentration. (k) A somewhat thicker wave obtains when the bromate concentration is increased to 0.175 M. Residence times: (a)–(f) 22 min, (g)–(h) 68 min, (i) 22 min, (j)–(k) 17 min. In all cases malonic acid concentration was 0.1 M and sulfuric acid concentration was 0.2 M; ferroin concentration was 0.00478 M in (a)–(i) and 0.00600 M in (j)–(k).

We also observe that leaching silica from the PolySil membrane (using 15% NaOH) leads to thicker waves. For the same chemistry the waves are considerably thicker if the gel-filled PolySil membrane is replaced by a gel that is supported by an Anopore filter[#2].

4. Discussion

The experiments described here successfully test and demonstrate the usefulness of the membrane reactor. Since the residence time for the reaction in the membrane is determined by the membrane thickness rather than by the residence time of the reagents in external CSTRs, the residence time for the membrane reactor can easily be made much shorter than in previous continuously fed unstirred reactors. Experiments can be conducted for a wide range of excitable and nonexcitable media with a flexibility in reagent composition that is not possible with other reactors developed for studying sustained two-dimensional patterns. For example, studies of chaotic spatiotemporal dynamics including phase turbulence [14, 15] and amplitude turbulence [14, 16] have been proposed for several years but have not been conducted in reaction–diffusion systems for lack of a suitable laboratory tool. The membrane reactor now provides such a tool.

Some measurements were made with membranes both thicker and thinner than the 0.75 mm thick membranes used for most of our study. In the thinnest membrane studied (0.2 mm thick), the patterns had a short wavelength (~ 0.2 mm) and poor contrast. The wave patterns appeared sharp and two-dimensional for membranes less than about 1 mm thick, but, of course, there are concentration gradients in the direction perpen-

dicular to the membrane, making all patterns in fact three-dimensional. Three-dimensional effects became apparent for thick membranes or high bromate concentration, and the visibility of the observed two-dimensional projection of the pattern rapidly decreased; no pattern was discernable for a 4 mm thick membrane, even for chemical compositions favoring excitability.

We have examined several other chemical compositions of the BZ reaction in addition to those described here. Most of the experiments were conducted with the sulfuric acid in the malonic acid and ferroin feed rather than the bromate feed because this leads to long gel life (~ 2 weeks) and to few CO_2 and CO bubbles. A malonic acid concentration of 0.1 M was found to be high enough to give good contrast but not so high that it led to significant bubble formation. Similarly, a ferroin concentration of 0.006 M was high enough to give good contrast but low enough so that bubble formation was not a problem. We have also obtained patterns with acetyl acetone rather than malonic acid as the organic substrate, a chemistry which has been shown to be nearly bubble free [17].

The membrane reactor should be particularly amenable to studies of dynamical behavior following a perturbation because the short residence times in the membrane lead to very rapid recovery. Future studies with this reactor should examine pattern formation for a range of chemical compositions and boundary conditions, particularly in inorganic oscillating reactions [18–20], which can have simpler mechanisms than the BZ system. Thus it should be possible to make quantitative comparisons of patterns observed in such experiments with the results of numerical simulations of reaction–diffusion model systems.

Acknowledgements

We thank Qi Ouyang and Mary Noszticzius for helpful discussions. This work is supported by BP Venture Research.

[#2]Anopore is a registered trademark for an inorganic membrane that is a high purity matrix with an extremely well-defined capillary structure It is also available as Anodisk, a membrane filter disk.

Appendix. Stability of ferroin under strongly acidic conditions

The concentration of ferroin obtained after mixing phenanthroline and ferrous ammonium sulfate decreases with increasing acid concentration. An excess of Fe(II) or phenanthroline can shift the complex equilibrium [21] and thereby stabilize ferroin at a higher concentration than that would be obtained from the stoichiometric ratio. For the typical acidic conditions of experiments on patterns in the BZ reaction there is essentially no free phenanthroline. Most of it is in the form of the 1:3 complex $Fe(phen)_3^{2+}$ (*ferroin*) or the 1:1 complex; the 1:2 complex is highly unstable and has a negligibly small concentration at equilibrium. There is also a relatively small amount of protonated phenanthroline present.

At low pH (2–5) the ferroin concentration can be computed within experimental accuracy from the stoichiometric amounts of phenanthroline and ferrous ammonium sulfate. Under the strongly acidic conditions of our experiments, however, the full set of chemical equilibria need to be considered in order to compute the ferroin concentration accurately. Ferroin concentrations determined independently from optical absorption measurements at 488 nm (extinction coefficient $= 10090 \pm 285$ M^{-1} cm^{-1} [22]) agree within 5% with those computed numerically from the equations for the complex equilibrium [21]. At acid concentrations greater than 0.2 M, a difference between experimental and calculated values arises mainly due to activity coefficient corrections, which then must be taken into account.

The calculated ratio of ferroin to monophenanthroline complex decreases from 118 to 31 in the range $0.005 < [H^+] < 0.5$ M, which is the typical acidity range for experiments on patterns in the BZ reaction. At $[H_2SO_4] = 0.2$ M ($[H^+] = 0.261$ M [23]), the effect of 50% excess iron is the same as 7.5% excess phenanthroline; however, for 50% excess iron there is significant amount of free Fe(II) and monophenanthroline complex (the ra-

tio of tris to mono complex is 21, compared to 62 for 7.5% excess phenanthroline).

When ferroin and malonic acid are together in a reservoir, the acidity is low (for malonic acid, $pK_1 = 2.86$, $pK_2 = 5.7$ [24]) and within the accuracy of the measurement (about 3%) the stoichiometric amount of ferroin is obtained; acidity prevents air oxidation of Fe(II) and the solutions remain stable. When ferroin, malonic acid, and sulfuric acid are in the same reservoir, the ferroin concentration is determined by the effective concentration of $[H^+]$, which is mainly due to the sulfuric acid alone since the dissociation constant of malonic acid is very low. At high acidity the solutions become less dark and the contrast of the patterns decreases. It is not clear that the mono complex of Fe(II) plays the same role in the BZ reaction as ferroin. Our experiments confirm the validity of the equilibrium data [21] at the acid concentrations we have used, and suggest that to increase the ratio of the tris to the mono complex of phenanthroline, an excess of phenanthroline rather than Fe(II) is preferred.

References

[1] A.M. Zhabotinsky, in: Oscillating Processes in Biological and Chemical Systems, ed. G.M. Frank (Science Publ., Moscow, 1967) p. 149;
A.N. Zaikin and A.M. Zhabotinsky, Nature 225 (1970) 535.

[2] A.T. Winfree, Science 175 (1972) 634.

[3] J.P. Eckmann and D. Ruelle, Rev. Mod. Phys. 57 (1985) 617.

[4] K. Kaneko, Physica D 37 (1989) 60;
J.A. Vastano and H.L. Swinney, Phys. Rev. Lett. 60 (1988) 1773.

[5] Z. Noszticzius, W. Horsthemke, W.D. McCormick, H.L. Swinney and W.Y. Tam, Nature 329 (1987) 619.

[6] N. Kreisberg, W.D. McCormick and H.L. Swinney, J. Chem. Phys. 91 (1989) 6532.

[7] W.Y. Tam, J.A. Vastano, H.L. Swinney and W. Horsthemke, Phys. Rev. Lett. 61 (1988) 2163;
W.Y. Tam and H.L. Swinney, Physica D 46 (1990) 10;
J.A. Vastano, T. Russo and H.L. Swinney, Physica D 46 (1990) 23.

[8] Q. Ouyang, J. Boissonade, J.C. Roux and P. De Kepper, Phys. Lett. A 134 (1989) 287;
P. De Kepper, Q. Ouyang, J. Boissonade, J.C. Roux,

Reaction Kinet. Catal. Lett. 42 (1990) 275;
Q. Ouyang, V. Castets, J. Boissonade, J.C. Roux, P. De Kepper and H.L. Swinney, J. Chem. Phys., submitted for publication.

[9] W.Y. Tam, W. Horsthemke, Z. Noszticzius and H.L. Swinney, J. Chem. Phys. 88 (1988) 3395.

[10] G. Skinner and H.L. Swinney, Physica D 48 (1991) 1.

[11] L. Kuhnert, K.I. Agladze and V.I. Krinsky, Nature 337 (1989) 244.

[12] Zs. Nagy-Ungvarai, S.C. Müller, J.J. Tyson and B. Hess, J. Phys. Chem. 93 (1989) 2760.

[13] D. Vasquez, W.D. McCormick, W. Horsthemke and Z. Noszticzius, Reaction Kinet. Catal. Lett. 42 (1990) 253.

[14] Y. Kuramoto, Prog. Theor. Phys. 71 (1984) 1182.

[15] Y. Pomeau, A. Pumir and P. Pelce, J. Stat. Phys. 37 (1984) 39.

[16] H.R. Brand, P.S. Lomdahl and A.C. Newell, Physica D 23 (1986) 345.

[17] Q. Ouyang, W.Y. Tam, P. De Kepper, W.D. McCormick, Z. Noszticzius and H.L. Swinney, J. Phys. Chem. 91 (1987) 2181.

[18] P. De Kepper, I.R. Epstein, K. Kustin and M. Orban, J. Phys. Chem. 86 (1982) 170.

[19] D.M. Weitz and I.R. Epstein, J. Phys. Chem. 88 (1984) 5300.

[20] I. Lengyel, G. Rabai and I.R. Epstein, J. Am. Chem. Soc. 112 (1990) 4660.

[21] W.W. Brandt, F.P. Dwyer and E.C. Gyarfas, Chem. Rev. 54 (1954) 959.

[22] P.M. Wood and J. Ross, J. Chem. Phys. 82 (1985) 1924.

[23] E.B. Robertson and H.B. Dunford, J. Am. Chem. Soc. 86 (1964) 5080.

[24] J.A. Dean, ed., Lange's Handbook of Chemistry, 12th Ed. (McGraw-Hill, New York, 1979) pp. 5–32.

Physica D 49 (1991) 13–20
North-Holland

Autowave propagation in heterogeneous active media

H. Linde and H. Engel

Institute of Physical Chemistry, Department of Theoretical Chemistry, Rudower Chaussee 5, Berlin, O-1199, Germany

New results obtained in self-completion of autowaves using a two-layer gelled Belousov–Zhabotinsky (BZ) medium with immobilized catalyst fed from above are presented. The stability of the collision level of the primary and of the secondary waves as well as different final outcomes of their interaction are discussed. Colliding spiral waves lead to characteristic cusp-shaped structures of the secondary waves with a high positive curvature at the edge of the cusp. Propagating through the gel the corresponding wave fronts can break at the edge of the cusp. These breaks generate small self-reproducing double spirals which possess a higher frequency than a single spiral wave. The spatiotemporal evolution of the small double spiral depends on the size of the break and on the excitability of the medium.

1. Introduction

Recently there have been studies of the propagation of autowaves in heterogeneous active Belousov–Zhabotinsky (BZ) media consisting of a reaction zone with immobilized catalyst covered by a feeding solution without catalyst. In the reaction zone the catalyst, usually ferroin, is fixed in a silicahydrogel matrix [1–3].

We recently reported about the tunneling of autowaves through gaps inside the gel where the catalyst is absent [2, 3]. Experimentally, generating wedge-shaped splits in the gel, and a critical gap width, I_{cr}, was determined. For $I < I_{cr}$, the autowave is able to overcome the gap whereas for $I > I_{cr}$ the wave will be stopped. Just below the critical value a pronounced damping of the wave amplitude as well as a retardation of the wave caused by the interaction with the gap were observed. Due to the delay of the front at the gap, the effective velocity of the wave is reduced compared to the propagation in a corresponding medium without gap. Moreover, in experiments with pulse trains the delay of a pulse at the gap may cause a frequency transformation similar to that observed by Pertsov et al. in diffraction experiments with autowaves [4]. The reason is that the refractory zone of the delayed pulse overlaps with the following pulse. Therefore, a pulse fol-

lowing behind the pulse delayed at the gap will be annihilated. Under these conditions only every second, third, etc. pulse will tunnel through the gap (cf. ref. [3]).

In this paper we will focus attention on some new results obtained in a so-called self-completion procedure of autowaves [1]. In our experiments we used a pseudo-two-layer gel system fed from above. The preparation of the gel system is described in section 2.

In the beginning waves propagate in the whole gel. At some moment the excitability of the upper part of the gel is decreased due to direct contact to oxygen. In this way the gel is divided into an active lower layer and an inhibited upper layer, where just now wave propagation becomes impossible. In the lower layer the waves continue to propagate. Let us call them the primary waves. The upper layer can be reactivated by covering the gel again with the feeding solution. Now the free ends of the primary waves at the boundary between the two layers become the filaments of spiral waves propagating into the upper layer inhibited previously. Section 3 describes how these secondary waves can interact with the primary waves.

In the last section we deal with the situation obtained if two secondary wave trains meet at some angle. From the experiments it follows that the secondary waves possibly fail to form a closed front and break. The appearance of the breaks may be connected with the emergence of small double spirals which possess a frequency higher than that of a single spiral. Thus these double spirals are able to eliminate all other wave sources in the system.

2. Experimental method; self-completion

In detail the procedure of immobilizing ferroin-type catalysts in silicahydrogel is described in refs. [3, 6]. We used silicahydrogel prepared by pouring 10 ml of water glass solution (alkali $\approx$ 2.94 M) into a mixture consisting of 6 ml Fe(phen)$_3$ (6.6×10^{-3} M) in water and 8.5 ml H$_2$SO$_4$ (1 M). The gel can be fed permanently by a feeding solution of NaBrO$_3$ (0.34 M), CH$_2$(COOH)$_2$ (0.13 M), KBr (0.067 M) and H$_2$SO$_4$ (0.34 M). In this way one obtains an open heterogeneous active medium suitable to study autowave propagation and pattern formation under quasistationary or even under stationary conditions. Convective or other disturbances of hydrodynamic origin cannot occur in the reaction zone and CO$_2$ bubbles very seldom form. Sometimes it may be useful to perform the experiments in a closed box filled with CO$_2$, N$_2$ or another inert gas to avoid or possibly to take advantage of the inhibitory influence of O$_2$. The experiments were carried out in a petri dish at room temperature.

Let us assume now that autowaves are propagating through the gel. After pouring off the feeding solution, the gel will come into direct contact with air. Near the boundary between feeding solution and gel the excitability drops due to the inhibitory effect of oxygen on the BZ reaction. The thickness of the inhibited layer of the reaction zone depends on the exposure time. The primary waves are damped out in this layer. Now, covering the gel with the feeding solution again, the layer inhibited previously will be reactivated rather quickly. Therefore, the primary wave

pattern starts to extend into the reactivated upper layer. Thus a pattern of secondary waves develops in the upper layer of the gel. Primary and secondary waves interact with each other in a different manner to produce different final states. This procedure was called *self-completion* of autowaves in refs. [1, 5].

We emphasize the difference between the experimental technique used in this work and the method described in ref. [1]. In ref. [1] the gel layer with the immobilized catalyst was covered with a solution containing all the reactants except for the catalyst. First the autowaves could propagate only within the gel. Then the soluble catalyst was added to the feeding solution and after mixing the waves entered the solution. Thus, the main difference from the method proposed in ref. [1] is that now the secondary waves propagate within a gel layer too.

3. Results of self-completion

Depending on the thickness of the two layers in the gel and on the wavelengths of primary and secondary waves we will distinguish four different cases of self-completion.

In the first case the thickness of the upper layer, h_2, is much less than the wavelengths of the primary waves, $h_2 \ll a_2$. The secondary waves tend to form spirals, which is however impossible because the upper layer is too thin to form complete spirals. Therefore, the fronts flatten and the final outcome of this behaviour represents a simple extension of the primary wave pattern into the reactivated upper layer. The situation of the first case is illustrated in fig. 1.

In the second case, given by $h_2 > a_2$, the open ends of the primary waves are able to form spirals. The spirals emanating from different primary wave fronts annihilate partly and form closed fronts traveling perpendicular to the primary waves until they disappear at the boundary between the gel and feeding solution (compare

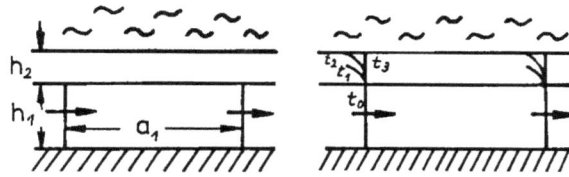

Fig. 1. Cross-section of the gel system before (left picture) and after (right picture) the upper layer was reactivated at time t_0, as described in section 2. Abbreviations used in this and in the following figures: h_1 and h_2 are the thickness of the lower and upper layers, respectively; a_1 and a_2 the wavelengths of the primary and secondary waves respectively. The arrows indicate the direction of the propagation of the primary waves. For clarity, after time steps denoted by t_1, t_2 and t_3 the wave fronts are removed virtually backwards compared to the direction of primary wave propagation. In other words, the fronts are drawn with respect to a coordinate system moving with the same velocity as the primary waves. In the figure the completion process is illustrated for the conditions of case 1, $h_2 \ll a_2$. The primary and the secondary waves form finally common plane fronts in the whole layer (side view!).

fig. 2). The secondary waves interact not only with each other but with the primary waves too. In general, the level of collision between both patterns will not remain stably located in space as self-completion proceeds. Usually the wavelengths of the primary waves is greater than that of the secondary waves, $a_1 > a_2$ (case 2). Therefore the collision level moves stepwise to the bottom of the petri dish. Finally the primary waves are extinguished completely. If $a_1 \gg a_2$, one completion cycle is sufficient to annihilate the primary pattern, as shown in fig. 3.

The opposite situation, $a_1 < a_2$ (case 3), is realized only under special conditions (see below).

The collision level moves to the top of the gel layer and finally the primary waves are extended into the whole gel layer (case 1 is recovered).

Concerning the stability of the position of the collision level, a special situation is given if the origin of the primary waves is a spiral center. Then we expect under homogeneous conditions (the completion procedure can be followed up over many cycles) that a_1 approximately equals a_2, so that the collision level should be stable in space. However, experimentally it was found that the collision level slowly moves stepwise to the bottom of the primary layer. This behaviour can be explained by the fact that the wavelength of

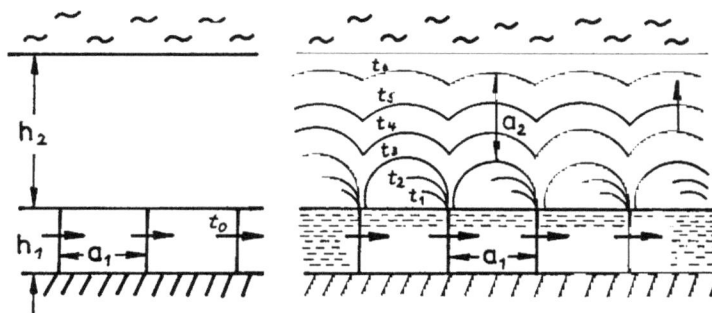

Fig. 2. Completion procedure in case 2, $h_2 > a_2$. The secondary waves form tunnel arch-shaped closed fronts traveling perpendicular to the primary waves. At the beginning the level of confrontation between the primary and the secondary waves is located near h_1. Later this collision level moves stepwise to the bottom (top) of the gel if $a_1 > a_2$ ($a_1 < a_2$); compare the dashed lines. In the picture one completion cycle is finished after the time t_3. The time intervals t_1 and t_4, t_2 and t_5 as well as t_3 and t_6 correspond to same stages of successive completion cycles.

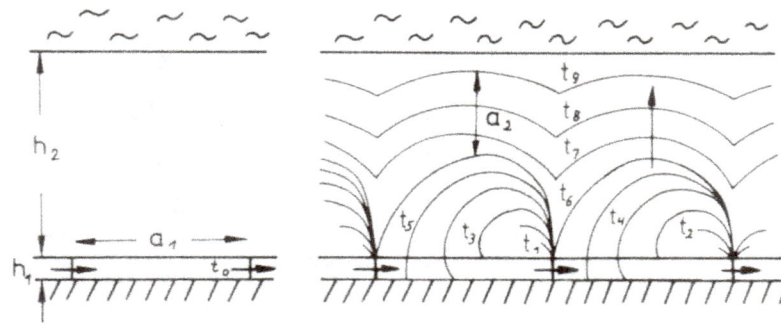

Fig. 3. Situation similar to fig. 2 but now $a_1 \gg a_2$. Therefore, the annihilation of the primary wave pattern is carried out during one completion cycle only.

the primary waves already has approached its asymptotic value denoted by a_1 whereas the section corresponding to the first half convolution of the secondary spiral is still less than the asymptotic value a_2.

The last case of self-completion is given by the conditions of case 2 and the additional inequality $h_1 > a_2$. Then, after eliminating the primary pattern the secondary waves form a system of scroll waves as described in refs. [1, 7].

The behaviour explained under case 2 can be recognized from the photographs in figs. 4a and 4b. The annihilation of the primary waves proceeds stepwise, starting from the periphery of the petri dish and spreading to the centre. This was caused by a variable thickness of the lower layer.

The completing behaviour of a big double spiral is presented in fig. 5. Conditions different from the abovementioned cases are valid in several areas of the gel. From fig. 5b it is evident

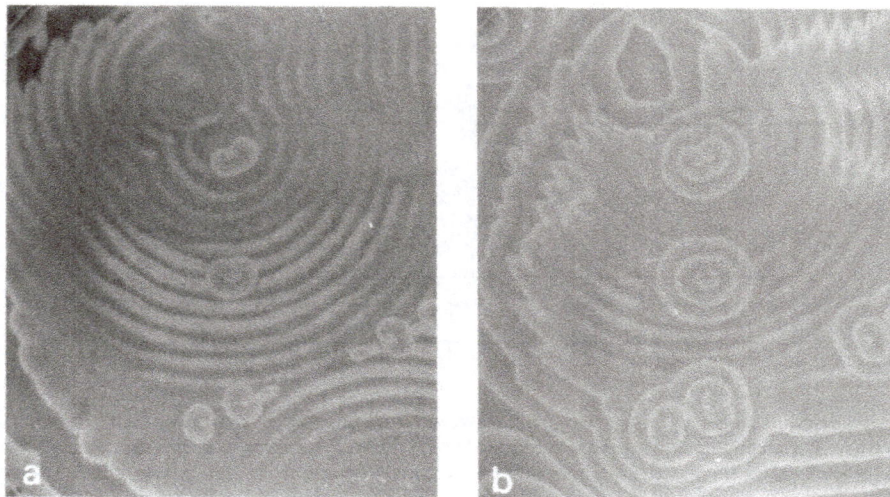

Fig. 4. Breaks of overcritical size have generated small double spirals which grow because their frequency is higher than that of the single spirals. Small breaks smooth out and do not leave any traces. The left picture (a) corresponds to 5 completion cycles, whereas on the right picture (b) 12 completion cycles are finished.

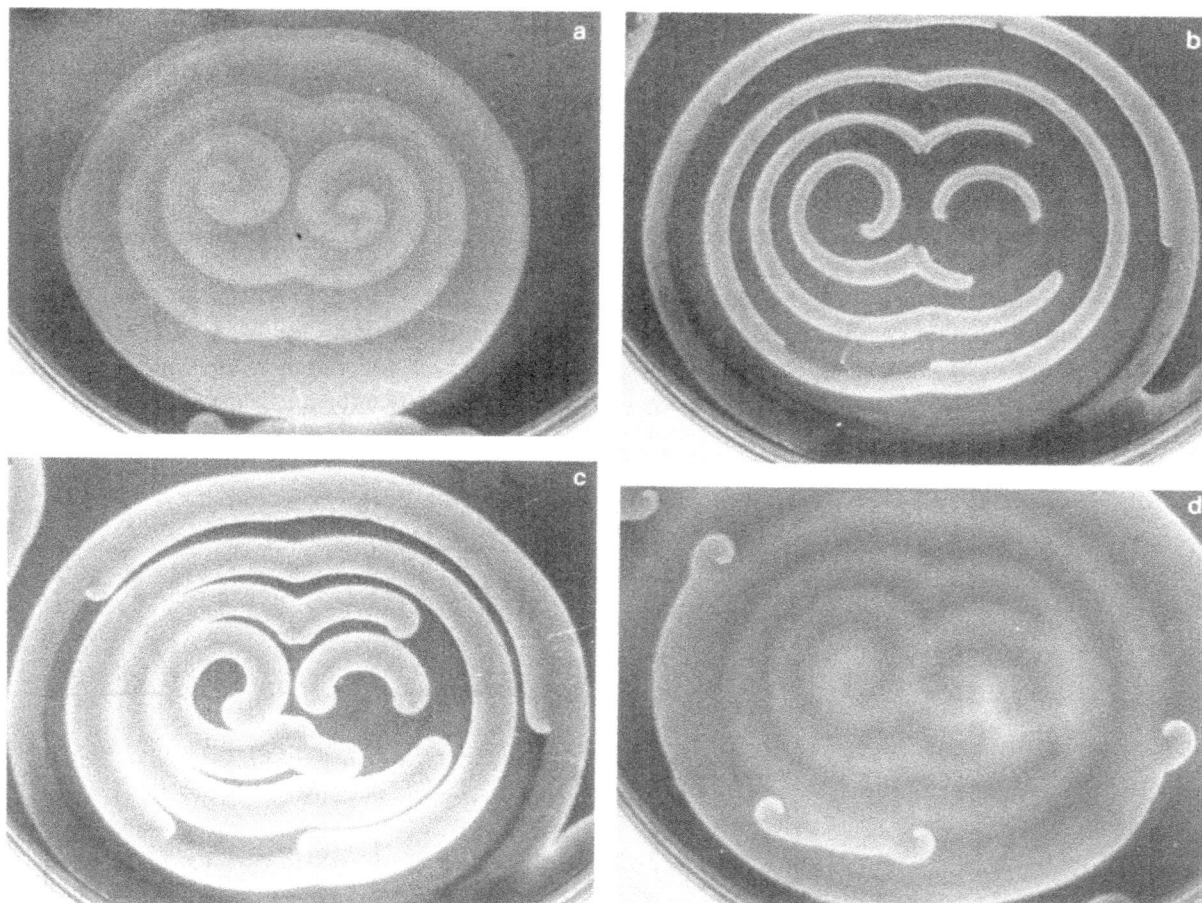

Fig. 5. Different stages of the completion procedure of a big double spiral: (a) Start of formation of secondary waves. (b) The secondary waves develop according to case 2. An exception are parts of the two external wave rings in the lower part of the photograph. Here the conditions of case 1 hold. In addition, it is recognizable that breaks appear in the cusp-shaped regions of the secondary wave fronts. (c) The secondary waves continue to propagate, moving against the primary waves. (d) The secondary and the primary wave fronts collide with each other. The primary waves are annihilated in one completion cycle because of $a_2 \gg a_1$. Only the nonannihilated lower parts of the primary wave rings survive and start to form spirals at their open ends.

Fig. 6. Formation of breaks in the secondary waves during the completion process according to case 2.

that the secondary waves emanating from the two external ring fronts are partly flattened (compare the lower part in fig. 5b), as explained in the first case (fig. 1). Therefore, in this part of the picture the primary waves are not annihilated. The rest of the medium behaves according to the scenario of case 2 with $a_1 \gg a_2$. The open ends of the primary waves form spirals and the emerging secondary waves annihilate the primary waves within one completion cycle only (fig. 5c). Only the two pieces of the nonannihilated primary

waves in the lower part of the petri dish remain finally and form spirals at their open ends (fig. 5d).

4. Breaks in secondary wave fronts. The evolution of small double spirals

Fig. 5 indicates that the secondary waves moving backwards compared to the primary waves tend to break spontaneously at divergent corners.

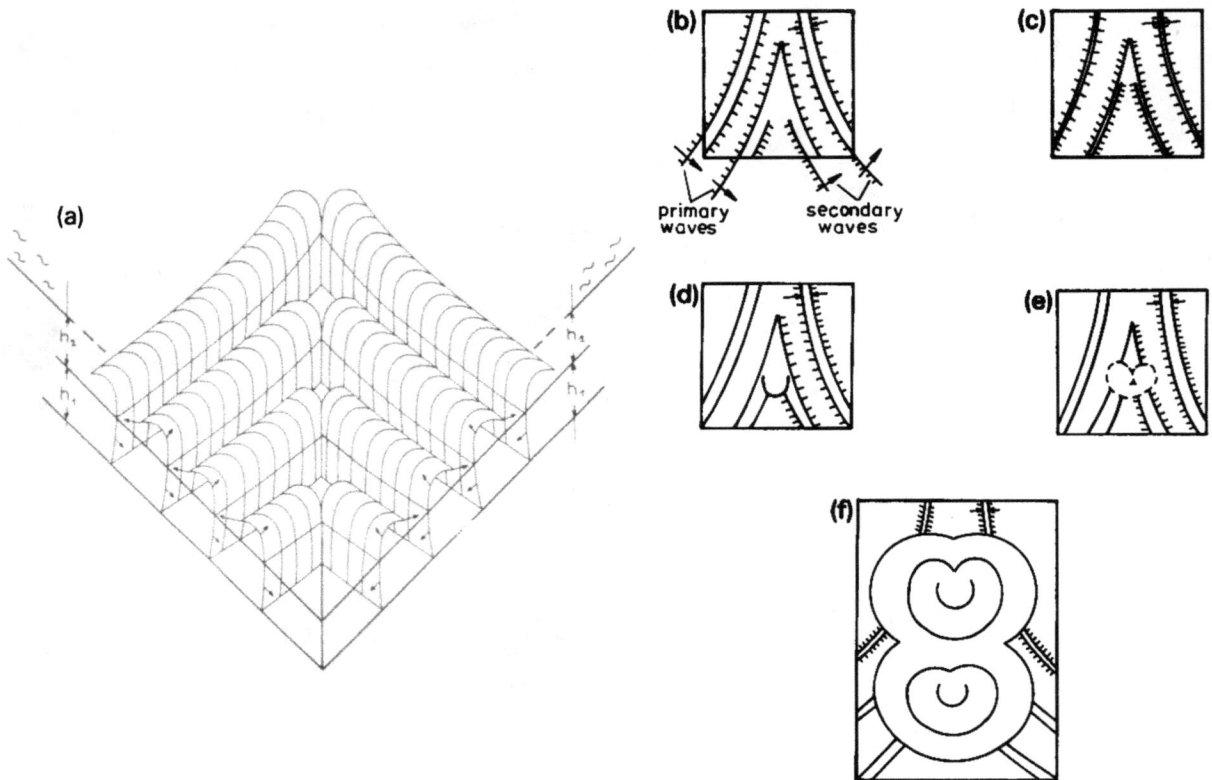

Fig. 7. (a) Perspective plot to illustrate the completion process if two primary wavetrains emanating from two single spiral centers collide with a certain angle between their direction of propagation. The picture corresponds to a completion stage shortly before the secondary spiral waves reach the collision level near h_1. (b)–(f) Sector of (a) in subsequent time steps (view from the top). Solid lines are wave fronts, hatching lines refractery zones, and arrows give the direction of propagation. (b) The secondary wave is broken at the cusp (divergent corner). (c) After collision between secondary waves a nonannihilated segment will remain in the upper layer near the cusp. (d) The nonannihilated segment generates a small double spiral which interacts as with the primary as well with the secondary waves. (e) The small double spiral reproduces itself with a higher frequency than that of a single spiral. Therefore, it succeeds in growing up and expanding against secondary and primary waves. (f) Single or even series of breaks can generate high-frequency leading centers in the form of small self-reproducing two-dimensional double spirals (compare fig. 6 also).

The further evolution of such breaks will be discussed in the present section (compare also with fig. 6).

As follows from fig. 5b and 5c, breaks which are very small in size will be smoothed and will disappear. A more complicated situation is depicted schematically in a perspective plot in fig. 7. Here the primary waves emanate from two leading centers. Partly they annihilate each other, but the remaining fronts form cusp-shaped convergent corners (compare refs. [8, 9]). As can be seen in fig. 7, the secondary waves form a cusp structure with a high positive curvature of the front at the cusp and a moderate negative curvature at the flank sides. Clearly this front will break in the immediate vicinity of the cusp if the curvature exceeds the critical value for wave propagation (fig. 7b, view from the top).

Now assume the conditions of case 2 of the preceding section are valid, i.e., the secondary wave fronts collide and annihilate each other partly as shown in fig. 2. However, now two trains of secondary waves meet each other at a certain angle, and because of the breaks in the secondary waves in the cusp regions no annihilation takes place during the completion procedure (fig. 7c). The segment of the front remaining after the annihilation becomes the origin of a small double spiral interacting both with the secondary as well as with the primary waves.

It was found that this double spiral exhibits at least two interesting properties. First, it possesses a very high frequency and thus is able to displace all other sources of autowaves in the system. For example, in the same time interval the primary waves succeeded in carrying out 11 cycles of self-completion, while the small double spiral shown in fig. 4b realized 13 cycles of self-reproduction. Therefore, three wave fronts around the double spiral center are left, as seen in fig. 4b. Second, the distance between the cores of the spiral centers grows in the course of time. Simultaneously the frequency decreases until it reaches the characteristic value for a single spiral in the medium considered. During the growth process

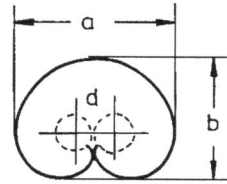

Fig. 8. Aspect ratio of small double spirals $\kappa = a/b$. The distance between the cores is denoted by d.

of the double spiral the aspect ratio $\kappa = a/b$ was found to change from $\kappa \approx 1.22$ at the moment of appearance to $\kappa \approx 1.36$ at $t_0 + 35$ min (compare fig. 8).

For wave propagation in the presence of obstacles the evolution of breaks in a wave front (or of remaining front pieces) depends very sensitively on the size of the breaks (respective pieces) and on the excitability of the medium [10, 11]. This will be discussed in greater detail in a forthcoming paper [12]. Here we only briefly summarize preliminary observations.

Three critical values, I_1, I_2, I_3, for the size of the segment of the front determine the final outcome of its evolution. For $I < I_1$ it disappears in the course of time. If the relation $I_1 \leq I < I_2$ holds, then the piece closes in a circle and a concentric leading center appears. If I is greater than I_2 but less than or equal to I_3 the piece of the front transforms into a double spiral. At the beginning its frequency is higher than that of a single spiral in the same medium. In the course of time the distance between the two spiral cores decreases. Simultaneously the frequency increases and finally one gets a concentric leading center, as in the preceding case. For $I > I_3$ a high-frequency double spiral occurs too. However, the distance between the cores increases with time while the frequency decreases until it reaches the value of a single spiral in the medium.

Finally, let us return to the distinct cases of section 3. It is possible to generate a small break in the front of a primary wave by local extinction using an oxygen beam. Then a small double spiral with smaller wavelength develops in the primary layer. If now the layer h_2 is reactivated, the self-completion proceeds according to case 3 ($h_2 > a_2$, $a_2 > a_1$), i.e., the collision level moves to the boundary between gel and feed solution.

References

[1] K.I. Agladze, V.I. Krinsky, A.V. Panfilov, H. Linde and L. Kuhnert, Physica D 39 (1989) 38.

[2] O.A. Mornev, H. Engel, J. Enderlein, H. Linde and V.I. Krinsky, in: Springer Series in Synergetics, to be published.

[3] H. Linde and Ch. Zirkel, Z. Phys. Chem. (Leipzig) (1990), to appear.

[4] A.M. Pertsov, E.A. Ermakova and E.E. Shnol, preprint, Pushchino (1989).

[5] A.V. Panfilov and A.M. Pertsov, Dokl. Akad. Nauk SSSR 274 (1984) 1500.

[6] L. Kuhnert, T. Yamaguchi, Zs. Nagy-Ungvarai, S.C. Müller and B. Hess, in: Proceedings of the Advanced Research Workshop on Nonlinear Wave Processes in Excitable Media, Leeds, Sept. 11–15, 1989.

[7] W. Jahnke, Ch. Henze and A.T. Winfree, Letter to Nature 336 (1988) 662.

[8] P. Foerster, S.C. Müller and B. Hess, Science 241 (1988) 685; Proc. Natl. Acad. Sci. USA 86 (1989) 6831.

[9] Ch. Zülicke, Ph.D. Thesis, Humboldt University, Berlin (1990).

[10] A.M. Pertsov, A.V. Panfilov and F.U. Medvedeva, Biofizika 28 (1983) 100.

[11] A. Engel and W. Ebeling, Phys. Lett. A 122 (1987) 20.

[12] M. Braune, H. Engel, H. Linde and Ch. Zülicke, in preparation.

Physica D 49 (1991) 21–32
North-Holland

Chemical waves in inhomogeneous excitable media

Jerzy Maselko[1] and Kenneth Showalter[2]

Department of Chemistry, West Virginia University, Morgantown, WV 26506-6045, USA

Propagating chemical waves are typically studied in homogeneous, excitable reaction mixtures. Chemical waves in an inhomogeneous excitable medium are examined in this paper. Cation exchange beads, loaded with ferroin, are bathed in Belousov–Zhabotinsky reaction mixtures containing no catalyst. Spiral waves are spontaneously initiated above a critical bromate concentration, which is dependent on the size of the ferroin-loaded beads. At high bromate concentrations, irregular patterns are formed due to an overcrowding of spirals. An upper limit in the number of individual waves is exhibited, which is independent of the bead size. Regular and irregular patterns are analyzed by calculating spatial correlation functions from digital images.

1. Introduction

For over two decades, chemical waves in excitable Belousov–Zhabotinsky (BZ) reaction mixtures [1] have been studied to provide insights into the behavior of excitable media in biological systems. Target and spiral patterns in two-dimensional thin films and scroll waves in three-dimensional bulk solutions have been experimentally and theoretically characterized [2–6]. Some dynamical features of the BZ system have been suggested as analogues for behavior found in heart muscle and other biological media. A number of important features of excitable biological media, however, cannot be explained in terms of homogeneous descriptions.

Spiral waves in homogeneous excitable media owe their existence to special and restrictive initial conditions. This requirement precludes the possibility of their spontaneous occurrence, behavior that is believed to be of importance in cardiac fibrillation. Spiral waves appear spontaneously in inhomogeneous media. Spontaneous change in wave populations is another feature not observed in homogeneous media. The number of spiral centers remains constant in such systems because the rotational frequency is established by the composition and temperature of the medium. In inhomogeneous media, the rotational frequency of a spiral is determined by the local environment, and the natural dispersion of frequencies gives rise to a decrease in the population of lower-frequency spirals. The number of spirals may also increase in such media due to inhomogeneity-induced wave breaks.

The consequences of inhomogeneity in excitable media have been addressed in relation to neuromuscular wave activity in heart tissue. Krinsky [7] examined the effects of inhomogeneity in excitable media over 20 years ago and more recently studied with Agladze and Pertsov [8] BZ reaction mixtures in which Bénard-like convection gives rise to wave breaks and subsequent spiral wave formation. Cohen and co-workers [9] have developed a description of spontaneous spiral formation in terms of a cellular automaton with a dispersion of refractory periods. A recent study by Chee et al. [10] examined the effects of inhomogeneous perturbations on a discrete, coupled-map Brusselator.

We have initiated studies of inhomogeneous, excitable chemical systems in which the inhomo-

---

[1] On leave from the Institute of Inorganic Chemistry and Metallurgy of Rare Elements, Technical University, Wroclaw, Poland.
[2] To whom correspondence should be addressed.

geneity is well defined and controllable. The system considered here is a Belousov–Zhabotinsky reaction in which the ferroin catalyst is immobilized on cation exchange resin beads. In a typical experiment, bead size varies over a two- to three-fold range, providing cellular character and inhomogeneity to the medium. Our initial study [11] showed that spiral waves are spontaneously initiated in the medium at concentrations of bromate or hydrogen ion above a critical level. At high bromate concentrations, an overcrowding of spiral waves results in patterns that are irregular in space but regular in time.

In this paper, we report on further studies of this inhomogeneous, excitable BZ system. Wave populations, measured as a function of reactant concentrations and as a function of time, are characterized in terms of spatial correlation functions. Behavior dependence on bead size is described and the local coupling and entrainment of neighboring beads is examined. Simple cellular automata are presented in which inhomogeneous excitability and refractory period result in behavior similar to the experimental behavior.

2. Experimental procedure

A thermostated petri dish maintained at 25.0°C was utilized in all the experiments. Analytical grade cation exchange resin was loaded with a desired amount of ferroin by stirring the beads in a ferroin solution of known concentration until the solution became colorless. The loaded resin was then collected and added to a reaction mixture containing sulfuric acid, potassium bromate, and malonic acid. Solution compositions and ferroin loading are listed in table 1; resin bead sizes and corresponding mesh ranges are listed in table 2.

Agitated reaction mixtures containing ferroin loaded beads were poured into the petri dish and the beads were allowed to settle onto the bottom of the dish. In a typical experiment, a 0.5 mm layer of beads was covered with a 1.6 cm layer

Table 1
System composition[a].

[KBrO$_3$]	0.05–0.50 M
[H$_2$SO$_4$]	0.25 M
[CH$_2$(COOH)$_2$]	0.025 M
[ferroin]	1.0×10^{-5} mol/g resin

[a]200.0 ml solution mixed with 10.0 g loaded resin.

Table 2
Cation exchange resin[a].

Size	Mesh	Bead diameter (μm)
1	< 400	38–75
2	200–400	75–150
3	100–200	106–250
4	50–100	180–425
5	20–50	300–1180

[a]Bio-Rad analytical grade 50W-X4.

of reaction mixture. In some experiments, a monolayer of beads was studied, which required adjusting the amount of resin to attain reasonably regular packing.

3. Results

Experiments were carried out to examine spontaneous wave initiation, displacement of low-frequency spirals by high-frequency spirals, spontaneous wave breaks, and the local interaction of neighboring beads. Digital imaging techniques, described elsewhere [11], were used to examine the spatiotemporal evolution of the system.

3.1. Wave initiation

The dependence of spontaneous wave initiation on bromate concentration was examined in two series of experiments. Similar experiments were reported earlier [11]; however, this behavior is reexamined because the earlier description was incomplete. Shown in fig. 1 is the number of initiation sites as a function of bromate concentration for bead sizes 3 and 4 in table 2. In each series of experiments, the number of initiation

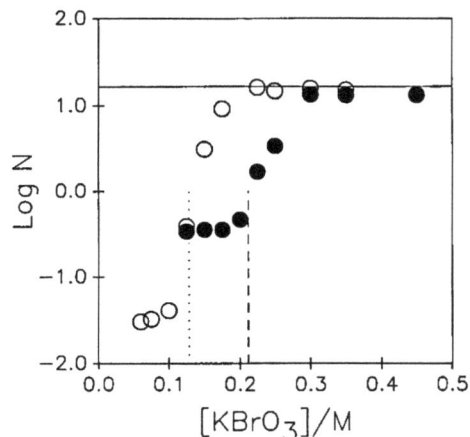

Fig. 1. Number of wave initiation sites N as a function of bromate concentration. Series of experiments with bead size 3 in table 2 are shown by filled circles and series with bead size 4 by open circles. The loaded resin beads are bathed in a 1.6 cm layer of reaction mixture maintained at 25.0°C; solution composition in table 1. Dotted and dashed vertical lines indicate calculated critical bromate concentrations for wave nucleation based on largest beads in each series. The horizontal solid line shows calculated maximum packing of counterrotating spiral pairs.

sites N was determined by counting the sites 2.0 min after the appearance of the first spontaneous initiation. The figure shows that there is a relatively small number of spontaneous initiations in each series at low bromate concentrations. As bromate concentration is increased, a critical concentration is reached where the number of sites increases rapidly. The critical concentration is lower for the larger beads (size 4) and higher for the smaller beads (size 3). At higher concentrations, the number of initiation sites levels off in each series to approximately the same value.

As reported earlier [11], there is a correlation between bead size and the critical bromate concentration at which the number of initiation sites rapidly increases. Above the critical concentration, an increasing number of beads are capable of serving as wave initiation sites. Each bead is a potential initiation site due to the acidic character of the cation exchange resin; however, it is effective only if it is larger than some critical nucleation size determined by the medium com-

position. According to arguments by Tyson and Keener [6, 12], this critical nucleation size arises from wave front curvature. Wave velocity and curvature in a three-dimensional medium are related by the eikonal equation,

$$v = v_c + 2DK, \qquad (1)$$

where v is the normal velocity of a wave with curvature $K = 1/r$, defined here as positive for concave and negative for convex fronts, v_c is the velocity of the corresponding planar wave, and D is the diffusion coefficient. The critical radius for nucleation is given by this equation when $v = 0$; at smaller values of r the wave collapses, and at larger values it propagates outward with a velocity determined by the curvature. As bromate concentration is increased, the largest beads in the mesh range will be the first to serve as initiation sites.

The critical bromate concentration when these beads become initiation sites can be estimated from the eikonal equation and an empirical velocity equation determined earlier [11],

$$v_c \ (\text{mm min}^{-1})$$
$$= -1.56 + 15.1([H_2SO_4][BrO_3^-])^{1/2} \ (\text{M}).$$
$$(2)$$

This velocity equation was measured for waves propagating in thin layers of ferroin-loaded beads and differs from the velocity equation for homogeneous BZ reaction mixtures [13]. The difference is presumably because the diffusion of the catalyst is negligible in the bead system, although it may be due to other factors such as the effective bromate and hydrogen ion concentration of the medium.

The critical bromate concentration calculated from eqs. (1) and (2) for the larger beads is 0.131 M; for the smaller beads, the critical concentration is 0.212 M. These concentrations, shown by the dotted and dashed lines in fig. 1, compare well with the rise in nucleation sites for each series. It should be noted that the correlation between bead size and spontaneous wave initiations in our previous study was made using a velocity equation for thin films of homogeneous solution [13] rather than eq. (2).

Fig. 1 shows that an upper limit to the number of initiation sites is exhibited at high bromate concentrations and that its value seems to be independent of the bead size. Insight into this upper limit, which is reminiscent of the upper limit of complexity proposed by Kopell and Ruelle [14], can be obtained by considering the maximum possible number of counterrotating spiral pairs in a given area. Accurate measurements by Müller et al. [15] have shown that BZ spiral waves are well described by spirals defined by the involute of a circle. If we assume that two counterrotating spirals can exist such that their cores are adjacent in the limit of maximum packing, the area occupied by the spiral pair can be readily calculated. The area of a circle enclosing one wavelength of a spiral is $\pi(2\pi r + r)^2$, where r is the radius of the spiral core. A circle of area $\pi(2\pi r + 2r)^2$ encloses the outermost regions of the two circles corresponding to each spiral of the counterrotating pair. The horizontal line at $\log N = 1.22$ in fig. 1 was calculated by assuming hexagonal closest packing of these circles. An average wavelength (1.0 mm) at $[KBrO_3] = 0.45$ M, corresponding to the last point in fig. 1, was used to determine this value; other points in the plateau give similar saturation values, as wavelength varies only slightly in this region. Fig. 1 shows that the upper limit is relatively independent of the medium coarseness, since both bead size ranges exhibit essentially the same plateau.

3.2. Spatial correlation function

As the concentration of bromate is increased and the number of wave initiation sites increases, the resulting pattern becomes increasingly disordered. Shown in fig. 2 are difference images of a regular pattern arising from asymmetric counterrotating spirals, a pattern of intermediate disorder with many spirals and wave segments, and a

Fig. 2. (A) Difference image of spiral pattern 20 min after mixing reactants. Solution composition in table 1 with $[KBrO_3] = 0.3$ M; bead size 1 in table 2. Image shows intensity difference between two images collected with delay of ≈ 10 s. Field of view 1.15×0.78 cm. (B) Difference image of pattern containing closely packed spirals 8 h after mixing reactants. Bead size 3, $[KBrO_3] = 0.3$ M, and delay of 12.8 s between primary images. Field of view 1.66×1.16 cm. (C) Difference image of irregular pattern 85 min after mixing reactants. Bead size 4, $[KBrO_3] = 0.5$ M, and delay of 12.8 s between primary images. Field of view 1.15×0.78 cm.

highly disordered pattern made up of short wave segments. The spatial correlation function,

$$A(j,t) = \frac{1}{N} \sum_{i=1}^{N} \delta I(i,t)\, \delta I(i+j,t), \qquad (3)$$

$$\delta I(i+j,t) = I(i+j,t) - I(t), \qquad (4)$$

$$\delta I(i,t) = I(i,t) - I(t), \qquad (5)$$

$$I(t) = \frac{1}{N} \sum_{i=1}^{N} I(i,t), \qquad (6)$$

defined here for a one-dimensional grid, provides a measure of order or disorder in a particular pattern. The intensity (gray level) at point i is $I(i,t)$, and $I(t)$ is the average intensity of the image. The difference between the intensity at point i and the average intensity is $\delta I(i,t)$, and $\delta I(i+j,t)$ is the difference between the intensity at point $i+j$ and the average intensity. The spatial correlation function $A(j,t)$ is determined for every point i as a function of j. For a two-dimensional array of pixels, $A(j,t)$ is calculated in an analogous manner, except now j defines the radius of a circle of points around point i. In order to smooth the circle in the discrete array, an annulus three pixels wide around point i at radius $j \pm 1$ is defined. The average intensity in the annulus is determined and $\delta I(i+j,t)$ is the difference between this average and the overall average $I(t)$.

Fig. 3 shows $A(j,t)$ calculated from the corresponding images in fig. 2. In each case, the image was smoothed using a moving average and then converted to binary by assigning 1 and 0 to pixels in dark and light areas in order to normalize the calculated values of $A(j,t)$. A grid of 200×200 was used for the calculation in fig. 3a; for figs. 3b and 3c, grids of 100×100 were used. Fig. 3a shows the regularity anticipated for the pattern generated by the counterrotating spirals. The strong correlations correspond to successive waves of wavelength 1.2 mm. Fig. 3b, corresponding to the intermediate pattern, shows a relatively strong correlation at a wavelength of about 0.9 mm.

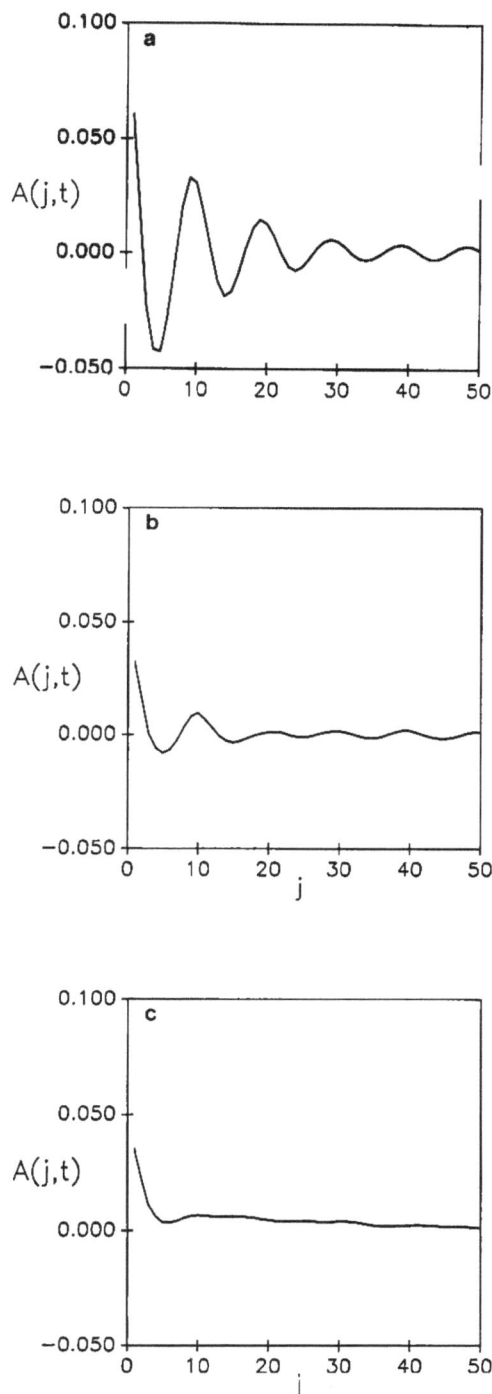

Fig. 3. Spatial correlation function $A(j,t)$ for images in fig. 2. Distance per pixel number j in (a), (b), and (c) is 0.116, 0.090, 0.058 mm.

Examination of the image in fig. 2B shows that beyond one wavelength, waves with random orientations may occur. This orientational disorder is responsible for the weaker correlations at successive wavelengths. Fig. 3c shows the spatial correlation function corresponding to the very irregular image. There is now little correlation because the wave segments do not develop into spirals, and a well-defined wavelength is not exhibited.

The power spectra of the spatial correlation functions show the trend from order to disorder even more dramatically. Fig. 4a shows the corresponding spectrum from fig. 3a. Although the data set is not sufficiently large to give a smooth curve, it is clear that the pattern in fig. 2A is characterized by a wavelength of 1.2 ± 0.1 mm ($j = 50/k$). Fig. 4b, corresponding to fig. 3b, shows a considerably reduced intensity; however, a wavelength of 0.9 ± 0.09 mm ($j = 50/k$) is still well defined. The very weak maximum in fig. 4c indicates that a well-defined wavelength is not exhibited in the irregular pattern of fig. 2C, although it is clear that the broken wave segments cannot be arbitrarily close. (Some of the spread in the power spectra results from an image aspect ratio of ≈ 0.9, which was not included in the calculations.)

3.3. Wave population

The number of individual waves per unit area typically decreases over long periods of time. The initiations per unit area shown in fig. 1 represent initial values, determined 2.0 min after the appearance of the first initiation. A subsequent increase in the number of individual wave segments may occur due to wave breaks, but at long times a decline in population is always exhibited. Shown in fig. 5 is the number of wave segments over a 23 h period in a system prepared with bead size 3 in table 2. Although there is some scatter in the values, an approximately linear decay can be discerned. The behavior can be understood in terms of differing spiral frequen-

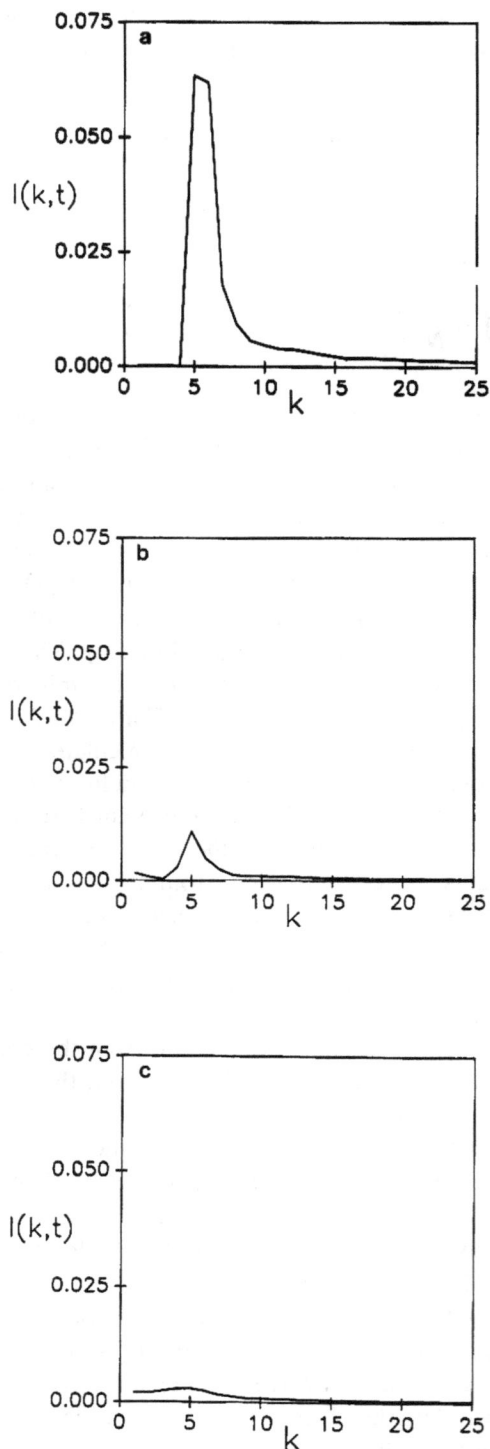

Fig. 4. Power spectra of spatial correlation functions in fig. 3.

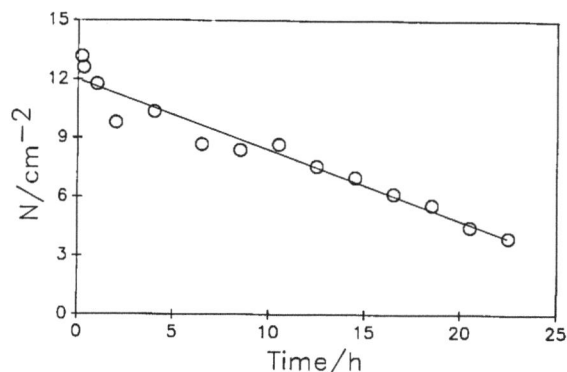

Fig. 5. Number of wave segments N per cm^2 as a function of time. Solution composition in table 1 with [KBrO$_3$] = 0.3 M; bead size 3 in table 2.

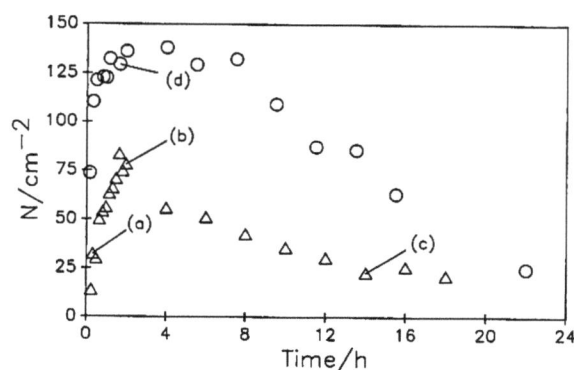

Fig. 6. Number of wave segments N per cm^2 as a function of time. Circles show series with [KBrO$_3$] = 0.5 M; triangles show series with [KBrO$_3$] = 0.3 M. Solution composition in table 1; bead size 4 in table 2.

cies at different sites in the inhomogeneous medium. The local environment at a spiral core determines the rotational frequency, and differences in the bead layer due to packing and bead size distribution lead to small differences in the spiral frequencies. As demonstrated by Krinsky and Agladze [16], a higher-frequency source serves to move a lower-frequency source as a dislocation to the boundary of the medium. An example of this behavior is shown in fig. 2C: the spiral remanent to the right of the central spiral is subsequently transformed into a dislocation and moved to the right as it is overtaken by waves from the higher-frequency source.

Systems prepared with larger beads exhibit an increase in the number of wave segments on a shorter time scale before the population decline at long times. Shown in fig. 6 are measurements of the number of wave segments as a function of time for systems prepared with bead size 4 in table 2 and [KBrO$_3$] = 0.3 and 0.5 M. In both series, the number of wave segments increases to a maximum before subsequently decreasing.

The spatial correlation function provides a measure of the temporal evolution of the pattern. Shown in fig. 7 are plots of $A(j, t)$ as a function of the pixel number j calculated from images taken at 20, 110, and 840 min (points (a), (b), and (c) in fig. 6). Wavelengths calculated from the first maximum in figs. 7a–7c are 1.16 ± 0.06, 1.04 ±

0.06, and 1.28 ± 0.06 mm. We see that the wavelength first contracts as the number of wave segments increases and then expands as the number decreases.

Wave segments resulting from spontaneous wave breaks are qualitatively different from those appearing during the wave initiation process. At early times, wave segments tend to form counter-rotating spirals, thereby giving rise to an ordered pattern. In experiments with larger beads, waves tend to break at inhomogeneities in the medium. The resulting broken waves do not form well defined spirals. Fig. 2C shows an example of the irregular pattern arising from these waves, which corresponds to point (d) in fig. 6.

3.4. Entrainment and Wenckebach cycles

Examination of the local spatiotemporal behavior in a monolayer of loaded beads provides insights into the overall spatiotemporal behavior of the medium. Shown in fig. 8 is gray level as a function of time for two adjacent beads in a layer of beads supporting propagating waves. Time series (a) and (b) differ in period by about a factor of two, with the average period 127.7 and 64.0 s, respectively. The time series with the shorter period is representative of most beads in the layer, while a few beads exhibit the 1:2 frequency

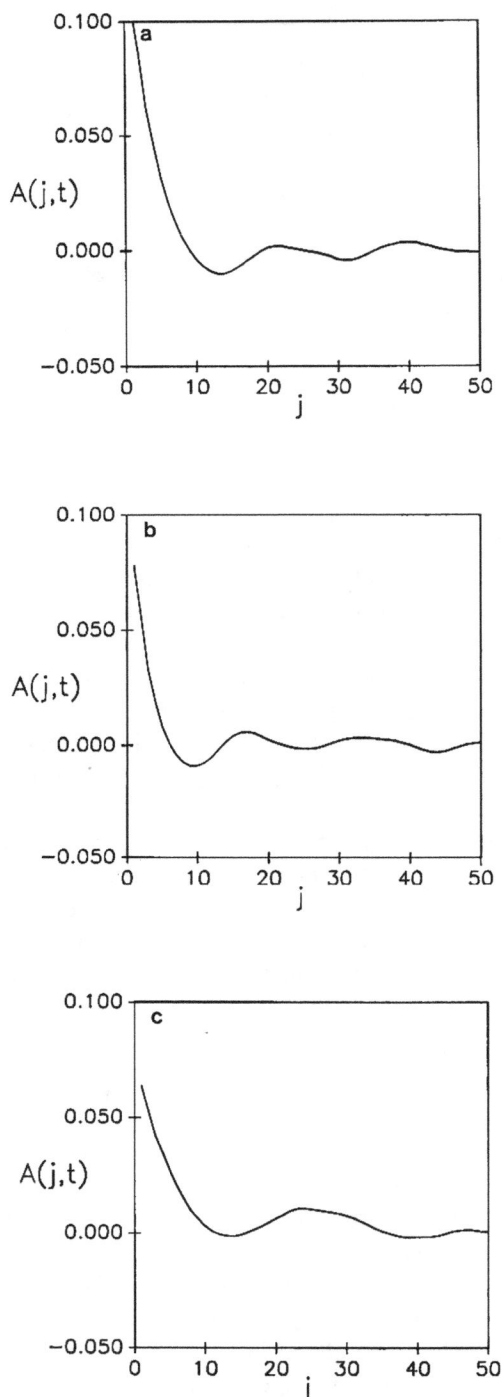

Fig. 7. Spatial correlation function $A(j,t)$ for images corresponding to points (a), (b), and (c) in fig. 6. Distance per pixel number j is 0.058 mm.

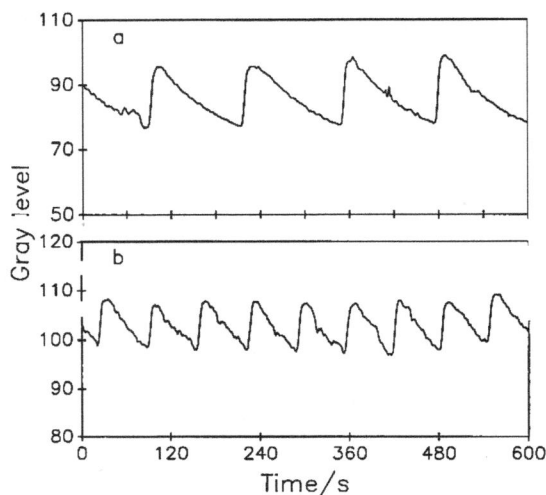

Fig. 8. Gray level as a function of time for two adjacent beads in monolayer supporting wave behavior. Values represent average of 30 measurements of intensity (440 nm) of a 5×5 pixel grid (60×60 μm) each second. The average period is 127.7 s in (a) and 64.0 s in (b). Solution composition in table 1 with $[KBrO_3] = 0.3$ M; bead size 4 in table 2.

locking. Examination of individual beads indicates that the longer-period oscillation occurs on beads without direct contact to neighboring beads due to packing defects; however, only a few examples of the behavior were found. Similar behavior is shown in fig. 9, with the typical bead exhibiting an oscillatory period of 40.4 s in (b) and the atypical bead in (a) exhibiting 1:5 frequency locking with a period of 203.4 s.

Shown in fig. 10 is another type of coupling between adjacent beads in a layer of beads supporting wave activity. Time series (a) exhibits Wenckebach-like cycles [17], where the regular oscillations are occasionally interrupted. The fourth and seventh oscillations are 67.8 s in duration, in contrast to the preceding and following oscillations with periods of ≈ 43 s. (The two oscillations between the Wenckebach cycles are 47.5 s in duration.) The period of the Wenckebach cycles is 1.5 times the period of the typical bead in (b) of 45.2 s. Examination of the Wenckebach beads revealed that they were in direct contact with neighboring beads, but with fewer nearest neighbors than the typical beads.

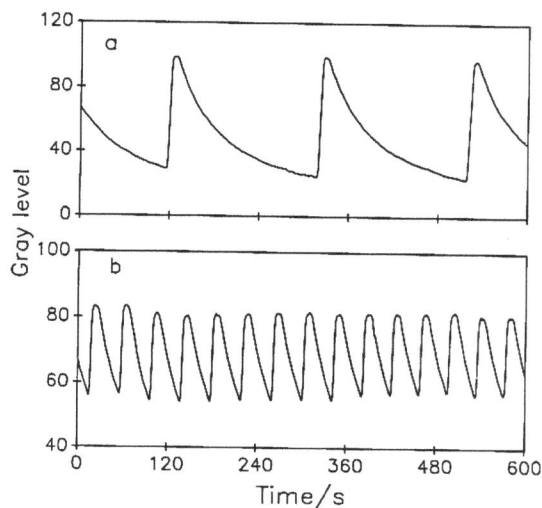

Fig. 9. Gray level as a function of time for two adjacent beads in monolayer supporting wave behavior. Values represent average of 30 measurements of intensity (440 nm) of a 5 × 5 pixel grid (55 × 55 μm) each second. The average period is 203.4 s in (a) and 40.4 s in (b). Solution composition in table 1 with [KBrO$_3$] = 0.3 M; bead size 5 in table 2.

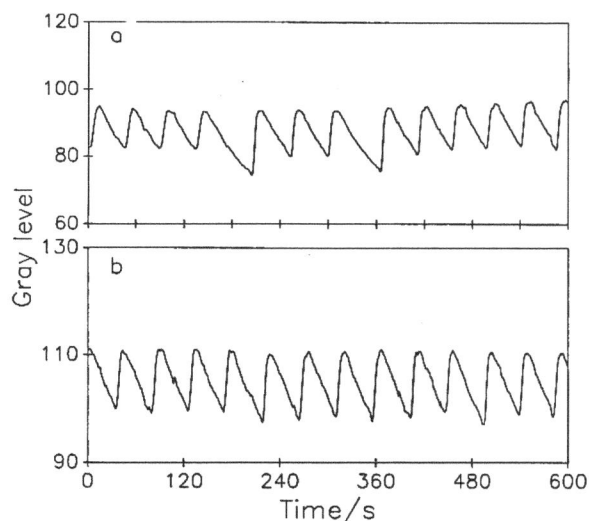

Fig. 10. Gray level as a function of time for two adjacent beads in monolayer supporting wave behavior. Values represent average of 30 measurements of intensity (440 nm) of a 5 × 5 pixel grid (60 × 60 μm) each second. Period of Wenckebach cycles in (a) is 67.8 s; period before and after cycles is ≈ 43 s. Average period in (b) is 45.3 s. Solution composition in table 1 with [KBrO$_3$] = 0.30 M; bead size 4 in table 2.

4. Discussion

The immobilized-catalyst Belousov–Zhabotinsky (BZ) reaction described here is an attractive model system for inhomogeneous excitable media. Each resin mesh range includes beads that vary in diameter between two- and three-fold. Because strongly acidic cation exchange resin carries a net negative charge due to fixed sulfonic acid groups, negative ions such as bromate or bromide do not significantly penetrate the polystyrene matrix of the resin. Wave activity is therefore confined primarily to the surface of the ferroin-loaded beads. Studies of spiral waves on a single ferroin-loaded bead have clearly demonstrated that the wave propagates on the surface without significantly penetrating the resin matrix [18].

Two features of the immobilized-catalyst BZ system are particularly noteworthy. The first is the spontaneous appearance of spiral waves. It is clear that inhomogeneity plays a critical role in the spiral formation of this system; what is not clear is whether the critical inhomogeneity is in excitability, refractory period, or intercellular (bead) coupling. Prior theoretical work [7–9] has focused on refractory period inhomogeneity as the primary source of spontaneous spiral formation; however, the effects of inhomogeneity in coupling and excitability are difficult to distinguish from refractory period inhomogeneity. The second notable feature involves changes in wave populations. This behavior includes both increases and decreases in the number of distinct wave segments. Population declines are straightforward: high-frequency spirals displace low-frequency spirals. The key to this behavior is that spirals in inhomogeneous media exhibit rotational frequencies dependent on the local environment. Population increases are less transparent; however, it is clear that medium inhomogeneities lead to spontaneous wave breaks.

Insights into these features can be gained from simple cellular automata [19, 20]. Fig. 11 shows three sequences for an automaton with inhomo-

Fig. 11. Five-state cellular automaton (see fig. 12) with inhomogeneous initial seeding. Each grid is 65 × 65 cells and each cell has eight nearest neighbors. Top: iteration 1, 3, and 20 with seeding probability of 1 in 125 for states 4 and 5. Middle: iteration 1, 3, and 35 with seeding probability of 1 in 114 for states 4 and 5. Bottom: iteration 1, 6, and 31 with seeding probability of 1 in 25 for states 4 and 5.

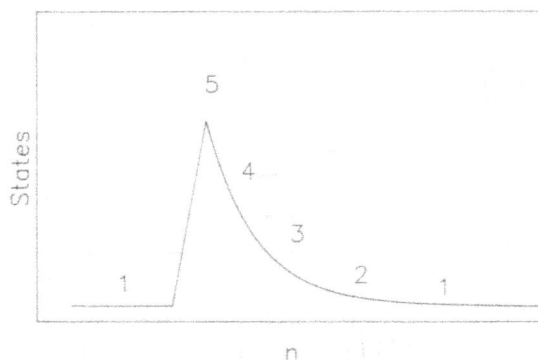

Fig. 12. Five states of cellular automaton simulating autocatalyst in chemical wave. State 1 is excitable, state 5 is excited, and states 4, 3, and 2 are refractory.

geneous initial conditions. The rule for this model, shown in fig. 12, reflects the autocatalyst concentration in a propagating chemical wave, such as bromous acid in the BZ reaction. There are five states: state 1 is excitable, state 5 is excited, and states 4, 3, and 2 are decreasingly refractory. Any state above the excitable state decays to that state ($n \rightarrow n - 1$ for $n > 1$). A cell in state 1 is excited if it has a neighbor in state 5:

$$1 \overset{5}{\rightarrow} 5.$$

The first sequence in fig. 11 shows a grid in which the initial probability of states 5 and 4 in each cell is 1 in 125, while the probability of the excitable state is 123 in 125. Waves are initiated in the first frame, which expand out in the second frame to either annihilate with each other or to move to the boundary of the medium. The last frame shows the medium returned to the excitable state. The second series shows a seeding probability of 1 in 114 for states 5 and 4. We see that most of the waves annihilate as before, but now one

counterrotating spiral pair is formed. The spirals serve as a periodic source which eventually entrains the entire medium, much like the experimental behavior shown in fig. 2A. The third series shows a seeding probability of 1 in 25 for states 5 and 4. In this case, a number of counterrotating spirals are formed which serve as periodic sources. The result is a persistent irregular pattern.

Although the automaton is a highly idealized model of the experimental system, its behavior suggests that spontaneous spiral formation and the accompanying irregular patterns result from wave initiations at sites with differing excitabilities and refractory periods. As bromate or hydrogen ion concentration is increased, the number of beads capable of serving as excitation sites increases. The largest beads appear in the tail of the size distribution, and only a small fraction of the beads at the critical bromate concentration are initially excited. The model in fig. 11 is based on seeding two different states, one in the excited state 5 and the other in the refractory state 4. Similar behavior can be obtained when the states 5 through 2 are used for the seeding, but subtle differences occur. The seeding in fig. 11 is especially interesting because the number of individual waves approaches an upper limit as in fig. 1 as the seeding probability is increased. Instead of exhibiting a plateau as in experiment, however,

Fig. 13. Cellular automaton with inhomogeneous initial seeding and spatially nonuniform rule. Seeding probability of 1 in 10 for states 4 and 5; probability of inhomogeneity 1 in 320 for each cell, which becomes a 5 × 5 grid by incorporation of neighbors. Sequence shows iteration 1, 7, and 20 in top row and iteration 60, 100, and 300 in bottom row.

the number of individual waves decreases at still higher seeding probabilities. Nevertheless, the upper limit in the number of individual waves helps explain the experimental behavior. The plateau in fig. 1 does not represent the maximum number of spirals possible per unit area; rather, it is the maximum number of spirals that can be spontaneously initiated (by this particular mechanism) per unit area. The simple model in fig. 11 is consistent with the experimental result that a minimum area defined by one wavelength is necessary for the birth of a spiral. The formation of a spiral pair in the automaton requires a spatial loop, formed initially by a cell in state 5 adjacent to a cell in state 4. In this sense, the upper limit in fig. 1 is only indirectly related to the upper limit of complexity as proposed by Kopell and Ruelle [14], which defines the maximum number of spirals possible per unit area.

Shown in fig. 13 is a similar automaton, except this model contains nonuniformities in addition to inhomogeneous initial seeding. States 5 and 4 are initially seeded with a probability of 1 in 10, and, in addition, the grid contains 5 × 5 inhomogeneities which appear with a probability of 1 in 320 (inhomogeneities may overlap). In an inhomogeneity, a cell in either the 1 or 2 state is excited by an adjacent cell in the 5 state. The rest of the medium has a similar rule except that in

order for a cell in the 2 state to be excited, it must be adjacent to two cells in the 5 state:

inhomogeneity: $1, 2 \xrightarrow{5} 5$,

medium: $\quad 1 \xrightarrow{5} 5, \quad 2 \xrightarrow{5,5} 5$.

The top row shows the evolution at iteration 1, 7, and 20, and the bottom row shows iteration 60, 100, and 300. We see that at iteration 7, a number of counterrotating spiral pairs are formed as in fig. 11; however, by iteration 60 it is evident that one source exhibits a higher frequency. The rule for the inhomogeneity results in an effectively shorter refractory period than the rest of the medium; therefore, a spiral that forms in such a region exhibits a shorter period. The sequence shows that the higher frequency spiral source displaces lower frequency spirals and eventually entrains the medium, much like the behavior observed in experiment (cf. figs. 2A, 2B). The remaining discontinuities at iteration 300 result primarily from the medium inhomogeneity; however, some memory of lower-frequency spirals seems to be retained. Other calculations with no inhomogeneities except that necessary to generate one higher-frequency spiral source result in uniform entrainment on displacement of lower frequency spirals to the boundary of the medium. This displacement is responsible for the decrease in the number of individual waves observed in experiment, as shown in figs. 5 and 6. The spiral rotational frequency is determined by the local environment, and the simple model in fig. 13 suggests that a nonuniform distribution of refractory period plays a critical role in the decline in wave populations at long times.

Fig. 14 shows an automaton that demonstrates spontaneous wave breaks, with the rules given by

inhomogeneity: $1 \xrightarrow{5} 5$,

medium: $\quad 1, 2 \xrightarrow{5} 5$.

Now the inhomogeneity has a longer effective refractory period than the rest of the medium. The top row shows the behavior of counterrotat-

Fig. 14. Cellular automaton with spatially nonuniform rule demonstrating spontaneous wave breaks. Top row shows inhomogeneity probabilities of 0,1 in 30, and 1 in 5, where the inhomogeneity is one cell. Bottom row shows inhomogeneity probability of 1 in 30 for 2×2, 3×3, and 4×4 cell area. Each calculation was initialized with a wave segment in order to generate a counterrotating spiral pair and each grid shows iteration 65.

ing spirals in which there is a probability of 0, 1 in 30, and 1 in 5 for the appearance of an inhomogeneity at each pixel. The bottom row shows behavior for a probability of 1 in 30 for inhomogeneities that are 2×2, 3×3, and 4×4 pixels in area. The wave breaks shown in the second frame of the bottom row are much like those observed in experiment. The resulting wave segments tend to be so crowded that spiral formation does not occur, in both model and experiment. This type of behavior is responsible for the transient increase in the number of individual wave segments shown in fig. 6.

5. Conclusion

The immobilized-catalyst BZ reaction is an attractive model system for study of inhomogeneous excitable media. The inhomogeneity is reasonably well defined and can be systematically varied. Our study suggests that the spontaneous appearance of spiral waves results from both inhomogeneous excitability and nonuniform refractory period, while increases and decreases in wave populations result primarily from a dispersion of refractory periods. Inhomogeneous coupling may also play a role, as demonstrated in the mono-

layer experiments where frequency locking and Wenckebach cycles are exhibited. The spatiotemporal behavior of this inhomogeneous excitable medium may provide insights into excitable media of biological systems.

Acknowledgements

We thank the National Science Foundation (Grant CHE-8920664) and NATO (Scientific Affairs Division, Grant 014/89) for financial support of this work. K.S. thanks Raymond Kapral for informative discussions on spatial correlation functions.

References

[1] A.N. Zaikin and A.M. Zhabotinsky, Nature 225 (1970) 535.
[2] A.M. Zhabotinsky and A.N. Zaikin, J. Theor. Biol. 40 (1973) 45.
[3] A.T. Winfree, The Geometry of Biological Time (Springer, Berlin, 1980).
[4] R.J. Field and M. Burger, eds., Oscillations and Traveling Waves in Chemical Systems (Wiley, New York, 1985).
[5] J. Ross, S.C. Müller and C. Vidal, Science 240 (1988) 460.
[6] J.J. Tyson and J.P. Keener, Physica D 32 (1988) 327.
[7] V.I. Krinsky, Biofizika 11 (1966) 676.
[8] K.I. Agladze, V.I. Krinsky and A.M. Pertsov, Nature 308 (1984) 834.
[9] D.T. Kaplan, J.M. Smith, B.E.H. Saxberg and R.J. Cohen, Math. Biosci. 90 (1988) 19.
[10] M.-N. Chee, R. Kapral and S.G. Whittington, J. Chem. Phys. 92 (1990) 7315.
[11] J. Maselko, J.S. Reckley and K. Showalter, J. Phys. Chem. 93 (1989) 2774.
[12] J.J. Tyson, J. Chim. Phys. 84 (1987) 1359.
[13] R.J. Field and R.M. Noyes, J. Am. Chem. Soc. 96 (1974) 2001.
[14] N. Kopell and D. Ruelle, SIAM J. Appl. Math. 46 (1986) 68.
[15] S.C. Müller, T. Plesser and B. Hess, Physica D 24 (1987) 87.
[16] V.I. Krinsky and K.I. Agladze, Physica D 8 (1983) 50.
[17] K.F. Wenckebach, Z. Klin. Med. 36 (1899) 181; 37 (1899) 475.
[18] J. Maselko and K. Showalter, Nature 339 (1989) 609.
[19] S. Wolfram, Nature 311 (1984) 419.
[20] J.M. Greenberg and S.P. Hastings, SIAM J. Appl. Math. 34 (1978) 515.

Physica D 49 (1991) 33–39
North-Holland

Control of dynamic pattern formation in the Belousov–Zhabotinsky reaction

Zsuzsanna Nagy-Ungvarai and Benno Hess

Max-Planck-Institut für Ernährungsphysiologie, Rheinlanddamm 201, W-4600 Dortmund 1, Germany

A new electrochemical method using specially prepared microelectrodes is presented for pattern detection in the Belousov–Zhabotinsky reaction. This method combined with 2D spectrophotometric data allows to relate the dynamic spatial distribution of different components to each other proving the validity of the FKN mechanism in thin layers. In addition, our results add to the understanding of the dispersion relation.

1. Introduction

The knowledge of the dynamic spatial distribution of the components of the Belousov–Zhabotinsky (BZ) reaction is a prerequisite for the understanding of the reaction–diffusion mechanism controlling its macroscopic patterns. In homogeneous stirred solutions of the BZ reaction many variables can be detected. Zhabotinsky monitored the periodical changes of the concentration of the oxidized and reduced form of the catalyst spectrophotometrically and the redox potential by a Pt electrode immersed into the solution [1]. The bromide ion concentration can be followed measuring the potential of a bromide-ion-selective electrode [2]. Several further variables like CO_2 [3], bromomalonic acid and other organic reaction products [4], CO [5], elementary Br_2 [6], BrO_2 radicals [7] as well as the heat evolution during the reaction [8] have been experimentally detected, too.

On the contrary, in distributed BZ systems, experimental work has been restricted to visual and photographic [9] and later 2D spectrophotometric [10] detection of the spatial concentration distribution of the metal ion catalyst, only. Theoretical description of the chemical waves in the BZ reaction based on the known kinetics of the elementary reactions leads to a set of three ordinary differential equations describing the behavior of the three variables $HBrO_2$, Br^- and a catalyst [11]. In order to verify this mathematical model knowledge of additional variables is needed. We now detected, for the first time, redox potential and bromide ion concentration in thin layers of the BZ reaction using specially prepared microelectrodes [12]. The three independent measurements enable us to relate different variables at different initial reagent concentrations in different BZ systems such as ferroin-, $Ru(bipy)_2^{2+}$-, Ce(III)- and Mn(II)-catalyzed reactions.

2. Experimental procedure

The experiments were carried out in a usual set-up [13] in non-stirred BZ reaction solutions of the following composition: 0.2–0.36 M BrO_3^-, 0.03–0.2 M malonic acid (MA), 0.21–0.51 M H_2SO_4, 0.001–0.01 M catalyst ($Fe(phen)_3SO_4$, $Ru(bipy)_3SO_4$, $Ce(SO_4)_2$ and $Mn(SO_4) \cdot 4H_2O$) and 0.09 M bromomalonic acid (BrMA). The solutions were placed in an optically flat petri dish giving a layer thickness of 0.6 mm. Circular waves were triggered by immersing a Ag wire (100 μm in diameter) into the solution. The Pt and Br^--sensitive microelectrodes for the measurement of redox potential and $[Br^-]$ in the BZ solution layer were prepared for every measurement by a method as described elsewhere [12]. Both electrodes were immersed into the solution layer by a micromanipulator in a depth of 0.3–0.4 mm and in a distance of 2 cm from the wave initiation point. The calomel reference electrode was connected to the reaction solution with a 0.1 M Mg–acetate salt bridge. The redox potential and [Br⁻] measurements were performed by a home-built electrochemical analyzer. 2D spectrophotometric measurements were performed as described before [10]. The velocity of each wave front was determined during the evaluation of time series of 2D spectrophotometric data by measuring the time interval necessary for a selected distance. All electrochemical experiments were carried out at $25.0 + 0.5°C$, spectrophotometric ones at $25.0 \pm 0.1°C$.

3. Results and discussion

3.1. Signal and specificity of the electrodes

The Pt electrode, used in these experiments, is an electrolytically edged pure Pt wire, while the bromide-selective electrode is a specially brominated silver layer coating a Pt supporting material, in both cases with a diameter of 0.5–5 μm. Their simultaneously recorded signal is shown in fig. 1. The Pt electrodes yield a signal in the usual potential range as can also be measured in homogeneous stirred solutions (900–1100 mV). The signal of the Br⁻-sensitive microelectrodes is, however, not fully specific. Their calibration curve made in pure KBr solution is practically identical with those of commercially available Br⁻-sensitive macroelectrodes. Addition of 0.1 m $KBrO_3$, 0.1 M H_2SO_4 or 0.1 M malonic acid solutions in separate experiments causes only little change in the calibration curve of the bromide electrodes; however, their signal in the reacting BZ solution layers always lies some hundred millivolts higher. The difference is possibly caused by the overall effect of the whole reacting BZ solution which might slightly damage the Br⁻-sensitive AgBr

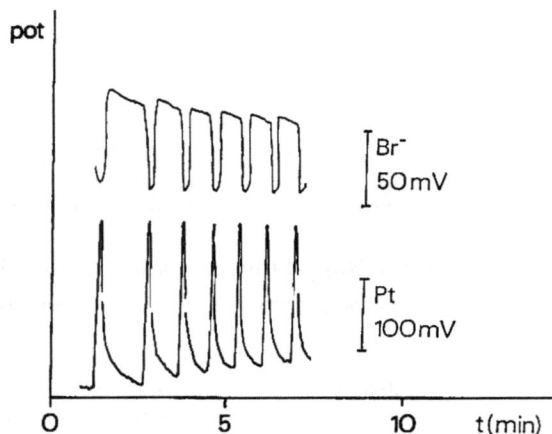

Fig. 1. Redox potential and Br⁻ curves in a thin layer of a BZ reaction solution with initial concentrations of 0.3 M $NaBrO_3$, 0.41 M H_2SO_4, 0.03 M malonic acid, 0.09 M bromomalonic acid and 0.006 M ferroin. The waves were initiated by placing a Ag wire into the solution with an initiation frequency of 0.033 s^{-1} for 2 s. The plot marked Br⁻ is the mirror image of the recorded potential curve (which equals to the potential of the reference electrode against the bromide electrode); this way, increasing traces of the potential plot indicate increasing concentrations of Br⁻.

layer in submicroscopic regions, where the silver coating acts as a simple Ag electrode [14]. Therefore, the signal of the Br⁻-selective electrodes could not be calibrated in terms of [Br⁻]. However, the recorded potential curves multiplied by −1 without any further logarithmic transformation, as shown throughout this paper, yield a clear qualitative picture with respect to an increase in [Br⁻].

Because of the corrosive chemical properties of the BZ solution, the Br⁻ electrodes lose their Br⁻-sensitivity usually after 3–5 measurements in a manner that their potential changes shown in a wave front decrease abruptly or their signal becomes suddenly identical with that of the Pt electrode. For this reason we used freshly prepared electrodes in each experiment.

Mechanical objects have a remarkable effect on wave fronts. Broad oxidized fronts may pass through small obstacles without any change, but narrow wave fronts passing through relatively large obstacles can be affected or completely disrupted [15]. Also, Ag$^+$ ions getting solved into the solution from the electrode can initiate waves and disturb the patterns observed. Both phenomena are curvature-related. According to the curvature–velocity relationship [11] there is a critical curvature which limits these effects. There exists a critical diameter necessary for wave propagation (in the case applied here 10–30 μm). Since the size of our working electrodes is much smaller than the latter value, the electrodes used in our measurements do not influence the wave initiation and propagation.

3.2. Conversion of electrode signals into spatial profiles

The electrochemical method described above yields the temporal change of the redox potential and the [Br$^-$] at a given position of the BZ solution layer (fig. 1). The data can readily be converted into spatial profiles with respect to the following considerations. Oxidized trigger waves in a BZ solution layer propagate with a velocity determined by reaction–diffusion properties of the given solution [13]. The results of our measurements are represented in table 1, summarizing the velocity v of the initiated wave fronts. In addition, as a control, we have shown by 2D spectrophotometric measurements, that the shape of every wave front remains constant during propagation (unpublished experiments). Therefore, temporal signals of the measured properties at a discrete place in the solution layer $(\delta c/\delta t)_x$ can directly be converted into spatial profiles $(\delta c/\delta x)_t$ using the equation

$$(\delta c/\delta t)_x = v(\delta c/\delta x)_t.$$

A profile converted from fig. 1 can be seen in fig. 2, showing the corresponding spatial distribution of the components. Because the method applied here is a dynamic one, a state of the patterns $t = vd$ time prior to recording is presented, where d is the distance of the electrode from the detected pattern at the beginning of the

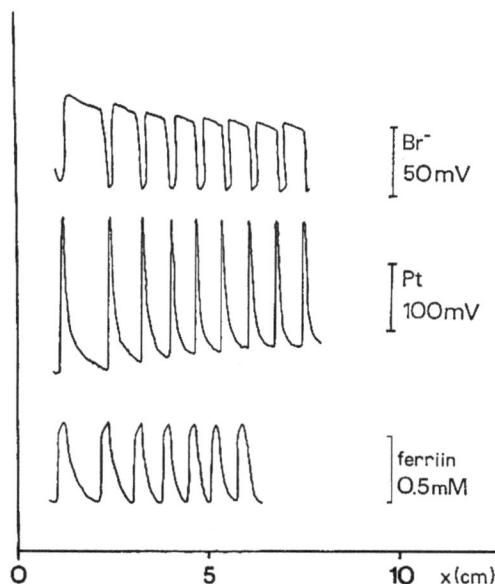

Fig. 2. Spatial profiles of redox potential, Br$^-$ (derived from fig. 1) and ferriin concentration (measured independently). The initial conditions are the same as given in fig. 1.

measurement and v the velocity of the respective wave front.

The third trace of fig. 2 is a ferriin concentration profile obtained independently by 2D spectrophotometric determination of the catalyst ferroin [10] in the same system. The waves were initiated both in the electrochemical and in the spectrophotometrical measurement by the same initiation procedure. The starting point of the curves was chosen in all three cases in the same distance from the place of wave initiation. As concentric waves have a radial symmetry the direction of the x axes does not change the form of the profiles. As can be seen from fig. 2 the calculated distances of the wave fronts in the potential curves are in good agreement with the directly measured wavelength of the ferriin concentration profile. The small difference in the position of the Br$^-$, Pt and ferriin curves is possibly caused by the temperature difference of the two sorts of measurements. While the solution layer was thermostated to $25.0 \pm 0.1°C$ during the spectrophotometric procedure (determination of [ferriin] and velocity) the temperature during the electrochemical measurements might have been slightly different (25.0 ± 0.5°C).

3.3. Validity of the Field–Körös–Noyes mechanism

In the BZ reaction malonic acid is oxidized by bromate in acid solution in the presence of a catalyst. The mechanism of the reaction was developed by Field, Körös and Noyes [2]. The main steps of the mechanism are as follows. In the presence of Br^- ion bromate reacts with bromide while bromous acid attains a low steady state concentration (process A). In the same system, when the $[Br^-]$ falls below a critical value, bromate oxidizes the reduced form of the catalyst creating a high steady state concentration of bromous acid as this latter reaction is autocatalytic in $HBrO_2$ (process B). The oxidized form of the catalyst reproduces Br^- ions (process C), therefore the whole cycle can be repeated again.

Inspection of fig. 2 illustrates clearly that in distributed layers we see a wave form of the redox potential and Br^- dynamics similar to what have been observed in batch reacting systems [2]. The $[Br^-]$ falls in a wave front where the autocatalytic reaction takes place, corresponding to the increase in ferriin concentration as predicted (process B). In the back of the wave, the $[Br^-]$ reaches a higher value (process C) inhibiting the autocatalytic reaction (process A). When the $[Br^-]$ reaches a critical level, the cycle is repeated. These results can be regarded as first experimental proof for the validity of the FKN mechanism in thin layers [2].

Fig. 3 shows redox potential and Br^- kinetics in BZ solution layers with the same initial concentrations using different catalysts, such as ferroin, $Ru(bipy)_3^{2+}$, cerium and manganese. The figure illustrates the differences in period between the four catalysts which can be understood on the basis of the FKN mechanism. The wavelength is decreasing in the series manganese < cerium < $Ru(bipy)_3^{2+}$ < ferroin, because regeneration of the oxidized form of the catalyst is enhanced with decreasing redox potential. Also, the velocity of the trigger waves was found to be lower in case of the high-redox-potential metal ion catalysts Ce(III) and Mn(II) compared to the low-redox-potential complexes ferroin and $Ru(bipy)_3^{2+}$ (table 1). Lower velocity and higher wavelength also correlates with lower excitability of the system. It should be added that the same result is obtained by changing the initial conditions, especially the concentration of H^+.

An important result of the FKN mechanism is the prediction of the dependence of critical bromide ion concentration on the initial reagent concentration. In the ferroin-catalyzed reaction we measured the redox potential and Br^- kinetics at different initial concentrations of the reactants. As the potential data of the Br^--selective electrodes are uncalibrated, it is not possible to calculate quantitative data on the critical bromide concentration. However, though uncalibrated, the potential values corresponding to the critical $[Br^-]$ show a characteristic dependence on the initial reagent concentrations. This allows the qualitative statement that the critical concentration of Br^- only depends on the initial $[BrO_3^-]$ and $[H_2SO_4]$, and it does not depend on the initial concentration of the catalyst and malonic acid.

3.4. Dispersion relation

Studying consecutive wave fronts and low- and high-frequency waves, the dynamics of the $[Br^-]$ data are relevant to the understanding of the dispersion relation [16]. In figs. 2 and 3 the decreasing distance between the first four successive wave fronts clearly indicates the effect of the dispersion relation: for any initiation frequency, especially for higher ones, the velocity of the successive wave fronts decreases until it reaches a lower limit, usually at the 4–6 wave (table 1). This is because the relaxation of the medium in the wake of the previous wave front is not complete. Thus, the concentration of the catalyst and other variables, especially Br^-, cannot be completely

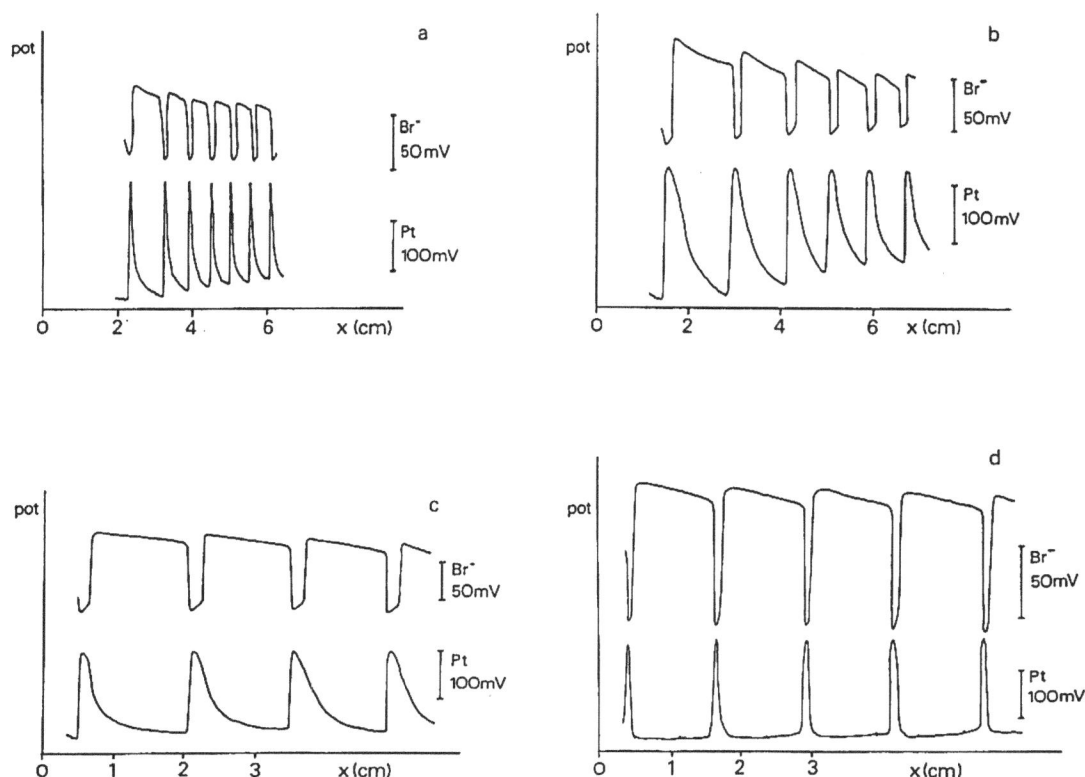

Fig. 3. Redox potential and Br⁻ curves as function of spatial coordinate x in thin layers of BZ reaction solutions with four catalyst: (a) ferroin, (b) Ru(bipy)$_3^{2+}$, (c) Ce(IV), (d) Mn(II). The initial conditions are the same as given in fig. 1.

Table 1
Velocity of the measured chemical waves.

Cat.	[BrO$_3^-$] (M)	[H$_2$SO$_4$] (M)	[MA] (M)	[BrMA] (M)	[Cat.] (M)	Wave No.	Velocity (μm/s)
ferroin	0.3	0.41	0.03	0.09	0.006	1	147
ferroin	0.3	0.41	0.03	0.09	0.006	2	105
ferroin	0.3	0.41	0.03	0.09	0.006	3	91
ferroin	0.3	0.41	0.03	0.09	0.006	4	82
Ru(bipy)$_3^{2+}$	0.3	0.41	0.03	0.09	0.006	1	123
Ce(IV)	0.3	0.41	0.03	0.09	0.006	1	85
Mn(II)	0.3	0.41	0.03	0.09	0.006	1	89
ferroin	0.27	0.41	0.03	0.09	0.006	1	139

recovered before the next front arrives. The faster the initiation frequency, the larger the effect. We studied this effect under conditions under which a spiral wave, having the highest frequency in a given medium, was overgrowing a low-frequency concentric wave. Fig. 4 illustrates the spatial propagation in a potential versus space plane. The electrodes were placed near the concentric wave monitoring the potential changes in its wave fronts. The sudden change of period following the fifth wave front indicates the time when the high-frequency spiral fronts arrived at the elec-

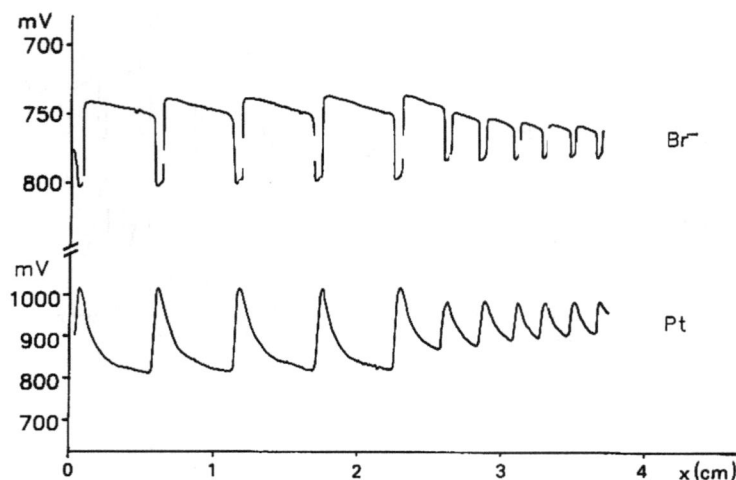

Fig. 4. Redox potential and Br⁻ curves for isolated concentric and spiral waves at initial concentrations of 0.27 M NaBrO₃, 0.41 M H₂SO₄, 0.03 M malonic acid, 0.09 M bromomalonic acid and 0.006 M ferroin.

trodes. The amplitude of both redox potential and Br⁻ traces is somewhat higher for concentric waves than for spiral waves indicating the incomplete recovery in the high-frequency waves.

4. Conclusion

The present work provides, for the first time, experimental data on the two-dimensional concentration distribution of three variables in distributed BZ solutions. The Pt electrode is most possibly measuring a mixed potential responding not only to the ratio of oxidized and reduced form of the catalyst but also to some oxybromine compounds [2]. A more accurate synchronization of the data on redox potential and catalyst concentration might allow to find out the difference between these two signals. The electrochemical method can easily be extended to the detection of chemical waves in thin layers of other batch oscillating solutions. The electrochemical detection has particular importance in systems which do not show any extinction changes and therefore identification of possible patterns is limited. Further modification of the preparation of the Br⁻-sensitive electrodes should also yield precise [Br⁻]

data serving as experimental basis for model calculations.

Acknowledgements

We thank H. Baumgärtl for the preparation of the microelectrodes, J. Ungvarai and S.C. Müller for critical discussions. The technical assistance of I. Beyer, Ch. Riemer and W. Zimelka is thankfully acknowledged.

References

[1] A.M. Zhabotinsky, Periodic liquid-phase oxidation reactions (rs), Dokl. Akad. Nauk. SSSR 157 (1964) 392.
[2] R.J. Field, E. Körös and R.M. Noyes, Oscillations in chemical systems. II. Thorough analysis of temporal oscillation in the bromate–cerium–malonic acid system, J. Am. Chem. Soc. 94 (1972) 8649.
[3] H. Degn, Effect of bromine derivatives of malonic acid on the oscillating reaction of malonic acid, cerium ions and bromate, Nature Phys. Sci. 243 (1973) 18.
[4] L. Bornmann, H. Busse and B. Hess, Oscillatory oxidation of malonic acid by bromate. 1. Thin-layer chromatography, Z. Naturforsch. 28b (1973) 93;
L. Bornmann, H. Busse, B. Hess, R. Riepe and C. Hesse, Oscillatory oxidation of malonic acid by bromate 2. Analysis of products by gas chromatography and mass spectrometry, Z. Naturforsch. 28b (1973) 824;

L. Bornmann, H. Busse and B. Hess, Oscillatory oxidation of malonic acid by bromate 3. CO_2 and BrO_3^- titration, Z. Naturforsch. 28c (1973) 514.

[5] Z. Noszticzius, Periodic carbon monoxide evolution in an oscillating reaction, J. Phys. Chem. 81 (1977) 185.

[6] U. Franck and W. Geiseler, Periodic reaction of malonic acid with potassium bromate in the presence of cerium ions (ge), Naturwissenschaften 58 (1971) 52.

[7] H.-D. Försterling, H. Schreiber and W. Zittlau, Detection of BrO_2^- in the Belousov–Zhabotinskii system (ge), Z. Naturforsch. 33a (1978) 1552.

[8] E. Körös, M. Orbán and Zs. Nagy, Calorimetric studies on the Belousov–Zhabotinskii oscillatory chemical reaction, Acta Chim. Acad. Sci. Hung. 100 (1979) 449.

[9] A.N. Zaikin and A.M. Zhabotinsky, Concentration wave propagation in two-dimensional liquid-phase self-oscillating system, Nature 225 (1970) 535.

[10] S.C. Müller, Th. Plesser and B. Hess, Two-dimensional spectrophotometry of spiral wave propagation in the Belousov–Zhabotinskii reaction 1. Experiments and digital data representation, Physica D 24 (1987) 71–86.

[11] J.P. Keener and J.J. Tyson, Spiral waves in the Belousov–Zhabotinskii reaction, Physica D 21 (1986) 307.

[12] Zs. Nagy-Ungvarai, H. Baumgärtl and B. Hess, Electrochemical detection of pattern formation in the Belousov–Zhabotinskii reaction, Chem. Phys. Lett. 168 (1990) 539.

[13] Zs. Nagy-Ungvarai, J.J. Tyson and B. Hess, Experimental study of the chemical waves in the cerium-catalyzed Belousov–Zhabotinskii reaction, 1. Velocity of trigger waves, J. Phys. Chem. 93 (1989) 707.

[14] Gy. Farsang, private communication.

[15] Zs. Nagy-Ungvarai, S.C. Müller and B. Hess, Spatial patterns in the Briggs–Rauscher reaction, Chem. Phys. Lett. 156 (1989) 433.

[16] J.D. Dockery, J.P. Keener and J.J. Tyson, Dispersion of traveling waves in the Belousov–Zhabotinskii reaction, Physica D 30 (1988) 177.

Physica D 49 (1991) 40–46
North-Holland

Front geometries of chemical waves under anisotropic conditions

Tomohiko Yamaguchi[1] and Stefan C. Müller

Max-Planck-Institut für Ernährungsphysiologie, Rheinlanddamm 201, W-4600 Dortmund 1, Germany

An experimental study of chemical waves in the Belousov–Zhabotinsky reaction under anisotropic conditions of substrate concentrations extending over the system was carried out by use of ferroin-immobilized silica gels as reacting matrices. Unidirectionally moving wave trains (similar to the so-called "chemical pinwheels") followed by the spontaneous fragmentation of wave fronts were observed in rectangular two-dimensional gels. A helicoidal chemical wave (or twisted scroll wave), which was predicted by numerical calculation, was detected in a micro-cylindrical gel system. A space–time representation of the chemical helix was constructed and its structural features are discussed qualitatively.

1. Introduction

Pattern formation and wave propagation in the Belousov–Zhabotinsky (BZ) reaction is well studied experimentally under isotropic conditions, but investigations under anisotropic conditions are still rare. Theoretical work suggests some interesting features of chemical waves under anisotropic conditions, which have not yet been observed experimentally:

(i) uniaxial anisotropy in two-dimensional (2D) reaction–diffusion systems produces an instability of the Turing type [1, 2],

(ii) in three-dimensional (3D) systems under mild uniaxial anisotropic conditions, a scroll BZ wave changes its shape into a twisted one [3, 4].

It is clear that macroscopic concentration gradients easily produce anisotropic conditions in the system of concern, but hydrodynamic effects have to be eliminated at the same time. Gel systems prove to be satisfactory for this purpose. An annular 2D polyacrylamide gel in which the catalyst ferroin was solubilized was the first for examining sustained chemical patterns [5]. For

our investigations we used two different geometries of silica gels in which ferroin was immobilized: 2D slab gels and microcylindrical gels.

Although Turing-type structures have not yet been observed in our 2D systems, the following two interesting patterns were recorded under concentration gradients extending over the systems: wave trains similar to the "chemical pinwheels" reported in ref. [5], and spontaneous fragmentation of wave fronts perpendicular to the direction of the anisotropy. Furthermore, in a micro-cylindrical gel a distinct 3D structure of a twisted scroll as mentioned above was observed and its helicoidal characters were analyzed qualitatively.

2. Experimental

Two types of ferroin-immobilized silica gels were prepared by mixing solutions of 15% (w/v) sodium silicate, 25 mM ferroin sulfate, water, and 1 M H_2SO_4 with a ratio of $10:2:1:2.2$ [6], which is a modification of the preparation given in refs. [7, 8].

(1) Two-dimensional gels: slab gels of 12 mm $\times$ 16 mm (thickness 0.45 mm) were prepared in situ in the cavity of a glass cell (see upper part

---

[1]Permanent address (to which correspondence should be sent): National Chemical Laboratory for Industry, Higashi 1-1, Tsukuba, Ibaraki 305, Japan.

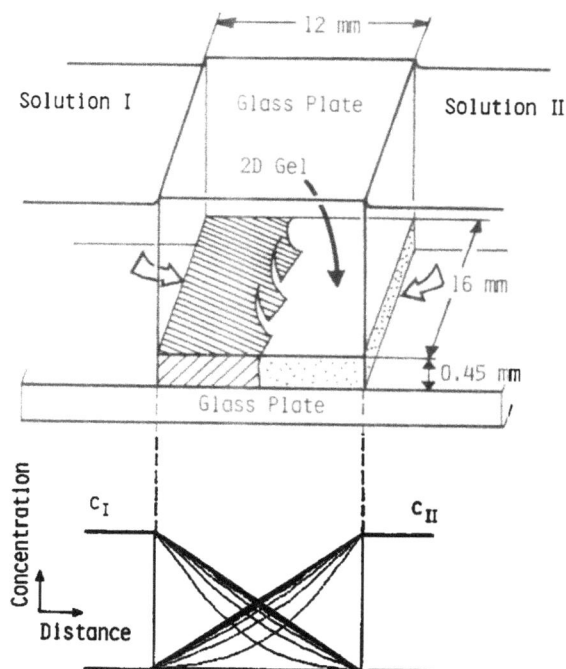

Fig. 1. Upper part: schematic representation of slab gel system (12 × 16 mm, gel thickness 0.45 mm). The open arrows indicate contact areas (16 mm × 0.45 mm) between the outer solutions I, II and the gel layer. The concentrations of solutions I and II (c_I and c_{II}, respectively) are specified in the text. Lower part: schematic representation of concentration profiles of malonic acid (c_I, for example) and bromate (c_{II}, for example) in the gel. The concave lines show the temporal evolution of concentration gradients at 1 h intervals, assuming the diffusion coefficients to be 2×10^{-5} cm² s⁻¹.

of fig. 1). Two lateral boundaries of the thin gel layer (indicated in fig. 1 with the white arrows) were open to the aqueous reacting solutions I and II, which contained at least one of the three components required in the BZ reactions (H_2SO_4, $NaBrO_3$ and malonic acid; the concentration of each solute was always 0.33 M). All other boundaries were covered by glass plates. The reactants penetrated from the two narrow open boundaries into the slab gel matrix and produced 2D patterns of chemical waves.

(2) Micro-cylindrical gels: these gel systems were prepared in situ in capillary glass tubes (inner diameter 1.1–1.2 mm, 5 mm in length). The tubes, which were open at both ends, were

then immersed into a reacting solution containing all the three reactants (H_2SO_4, $NaBrO_3$ and malonic acid) at 0.33 M concentration.

Chemical patterns evolving in these gels were monitored by transmitted green light and recorded with a time-lapse video recorder (Sony EVT 801 CE). In case of the micro-cylindrical gel experiments the recorded video images of BZ waves are equivalent to 2D projections of the 3D chemical waves. A strip of the digitized video images (2 × 460 pixels) was extracted parallel to the cylindrical axis from subsequent frames at 0.36 s intervals, and 250 serial strips (corresponding to a total time of 90 s) were arrayed to compose a space–time representation. This procedure transforms the 2D projection of 3D structures of the chemical waves into 1D space and time structures and enables one to visualize the time evolution of the systems. This processing was done on a personal computer (Siemens PCD) equipped with an image digitizer (Data Translation-2851, 515 × 512 pixels).

3. Results and discussion

3.1. 2D chemical waves

3.1.1. Chemical pinwheels

In general, when the reducing agent (malonic acid) and the oxidating agents (bromate and protons) penetrating from the opposite open boundaries of the 2D systems meet in the center region of the gel, there appears a train of chemical waves traveling *perpendicularly* to the direction of the concentration anisotropy extending over the system.

Usually, the chemical waves originate from the two covered boundaries of the rectangular 2D gel layer and annihilate upon collision at the center. Fig. 2A (96 min after the gel was in contact with the reacting solutions) shows such a beginning stage of chemical wave propagation and collision along the center line of the slab gel. In this top

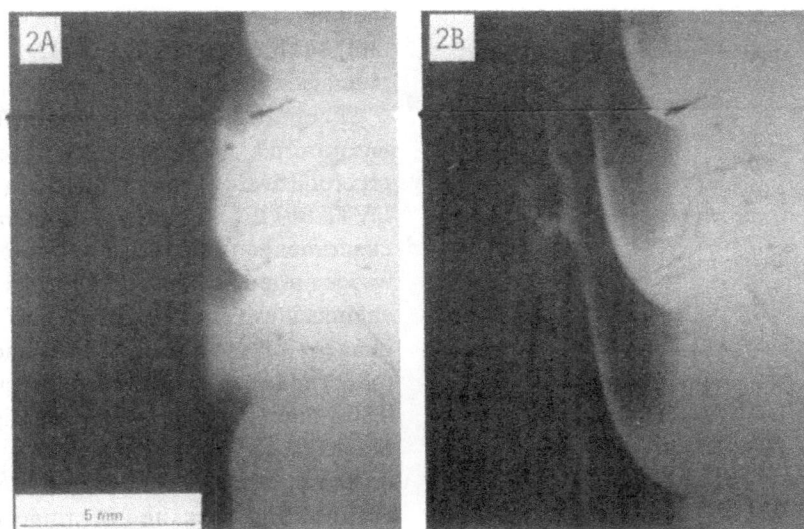

Fig. 2. (A) Top view of the rectangular ferroin-immobilized 2D silica gel, 96 min after the gel was placed in contact with the aqueous reacting solutions I and II. Black and white colors indicate reduced and oxidized states, respectively. Solution I (left) contains malonic acid and H_2SO_4, both with a concentration of 0.33 M, solution II (right) contains $NaBrO_3$ and H_2SO_4, both with a concentration of 0.33 M. (B) Same system after 149 min.

view of the gel layer the covered boundaries are located at the top and the bottom of the figure, where the solutions I and II penetrate from the left and the right, respectively. The white and the black parts correspond to oxidized and reduced areas in the gel system, respectively. The solution I (left) contained 0.33 M of malonic acid and H_2SO_4, and the solution II (right) contained 0.33 M of $NaBrO_3$ and H_2SO_4. The point of wave collision gradually shifted downward along the vertical center line, and finally there appeared an uni-directional wave train traveling downward as shown in fig. 2B ($t = 149$ min). This chemical pattern resembles the so-called "chemical pinwheels" observed in an annular gel system [5].

Under the current experimental setup the concentration profiles of the reactants in the gel change with time in accordance with their penetration processes after contact with the reacting solutions is established. Consequently the degree of anisotropy extending over the gel system changes with time, and finally reaches a stationary state after a certain interval of time. This time

evolution of the anisotropy can be estimated by a numerical calculation of the concentration profiles of malonic acid and bromate on the basis of a simple diffusion process, in which the value of their diffusion coefficients are assumed to be 2×10^{-5} cm^2 s^{-1}. These values are the same as in an aqueous solution [9], and their validity in gels is supported by the observation that the velocity of a trigger wave (which is proportional to the square root of the diffusion coefficient of $HBrO_2$) is found to be equal in gels and in aqueous solutions within an error of a few percent [6]. As indicated in fig. 1 (lower part; the time evolution of concentration profiles of the reactants in the gel layer is shown by the concave lines at 1 h intervals), the concentration change of bromate and malonic acid caused by this process is almost negligible after 4 to 5 h. In this experimental setup, of course, there is the concentration gradient of protons, which also may affect the production of macroscopic concentration anisotropy. The diffusion of protons is, however, faster than that of bromate and malonic acid by a factor of 10.

The concentration profile of protons in the investigated gels is therefore assumed to be equilibrated in a first approximation.

A characteristic feature of the present system is that the synchronized wave train occurs *spontaneously* after several collisions of the waves. In this rectangular geometry of the system there exist two pacemakers at the opposite covered ends of the gel. The origin of these pacemakers is attributed to the distortion of the concentration profiles at these boundaries, where the diffusion of the reactants is slightly delayed as compared to the diffusion at the middle part of the gel. This is supported by the slightly curved shape of the base line of the oxidative front in the gel (fig. 2A). After several collisions of the waves, the whole active domain of the system is finally dominated by the wave trains originating from the pacemaker whose frequency is higher than that of the other end of the system. In our observations, the frequency ratio between the pacemakers at the upper and the lower boundaries of fig. 2A

was 7:6, and only the upper one was present in the end as the source of uni-directional wave trains shown in fig. 2B.

3.1.2. Spontaneous fragmentation of wave fronts

Similar chemical patterns are observed also when the solution I contains $NaBrO_3$ instead of H_2SO_4; we investigated the case when solution I contained 0.33 M of $NaBrO_3$ and malonic acid while solution II contained 0.33 M of $NaBrO_3$ and H_2SO_4. Quite interestingly, under these conditions, a fragmentation of wave fronts takes place spontaneously during the later stages of the experiment, as shown in fig. 3.

230 min after the gel was in contact with the reacting solutions I and II, the wave front approaching the highly reductive domain (the dark part of the gel in fig. 3A) was no more smooth but became distorted. While the irregular shape of the first wave front gradually increased, the following second wave front remained to be smooth (fig. 3B, 100 s after fig. 3A). In the picture taken

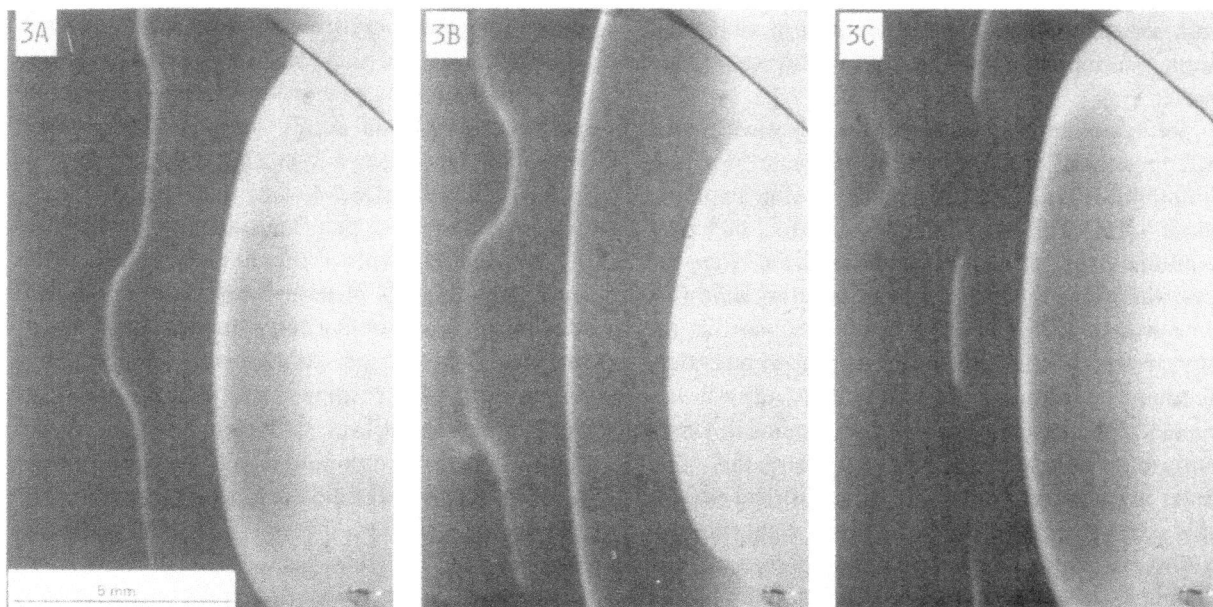

Fig. 3. (A) Top view of the ferroin-immobilized 2D silica gel 230 min after the start of the experiment. Solution I (left) contains 0.33 M of $NaBrO_3$ and malonic acid, solution II (right) contains 0.33 M of $NaBrO_3$ and H_2SO_4. (B), (C) Same system 100 and 160 s later than in (A), respectively.

160 s after fig. 3A the first wave front had disappeared at the reductive end of the gel, and the second wave front was broken into three fragments (fig. 3C) – an appearance of a new structure which evolves *perpendicularly* to the direction of macroscopic concentration gradients.

This fragmentation seems to be caused by the suppression of wave propagation in the domain of the retarding part of the advancing wave front, and this effect is probably pronounced under conditions of low excitability [10]. The retarding part of the distorted first wave interferes with the corresponding succeeding part of the second wave in its propagation, resulting in the fragmentation of the wave front into three pieces. Subsequently, the third wave turns into two fragments. Actually this process proceeds for 2 to 3 h with changes of the fragmentation pattern to some extent.

3.2. Helicoidal chemical waves

There is a theoretical prediction [3, 4] that a *twisted chemical scroll* wave is unstable under isotropic conditions but can be stable under certain anisotropic conditions, i.e., mild gradients of some parameters along the axis of a scroll-type wave.

We observed such a twisted scroll wave in the BZ reaction in a micro-cylindrical, ferroin-immobilized silica gel formed in a thin capillary glass tube of 5 mm length. Initially, the tube contained only the catalyst-immobilized silica gel and was then immersed in the reacting solution. The two ends of the tube were open, so this gel system was in principle under center-symmetrical boundary conditions. Soon after the gel was immersed into the reacting solution, chemical waves started to propagate repetitively from the open ends towards the center of the cylindrical gel. At this early stage, the waves followed each other in a discrete manner and the dominant shape of the chemical waves was planar. Occasionally there appeared a scroll wave at the open end, and with increasing distance from the boundary it turned into a moving helix of a fine pitch; in this case the

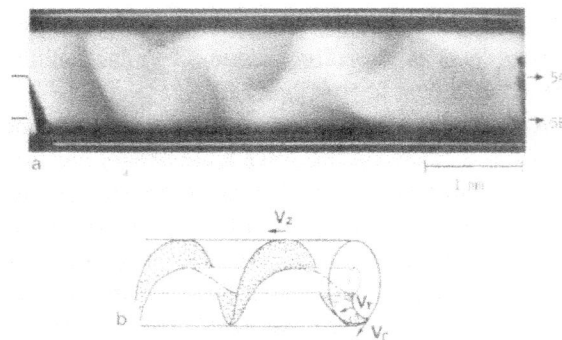

Fig. 4. (a) Side-view of a chemical helix observed 75 min after the gel in a capillary glass tube was immersed into the reacting solution containing 0.33 M of malonic acid, $NaBrO_3$ and H_2SO_4. The length of the gel is 5 mm and the diameter is 1.1 mm. The chemical helix propagates towards the left. The arrows indicate the positions of cuts for the construction of the space–time representations in figs. 5a and 5b. (b) A schematic 3D structure of the chemical helix. Here V_z is the velocity of the helicoidal wave along the cylindrical axis, V_c is the apparent rotational speed along the circumference, and V_r is the velocity normal to the wave front in a circular section.

wave front is no more discrete but continuous. In both cases the wave fronts met at the center of the gel, and usually one of the pacemakers at the open ends conquered the other in the end. During later stages of this transient process, i.e., 1 to 2 h after the gel was in contact with the reacting solution, a twisted scroll wave – in the present cylindrical geometry a *helicoidal chemical wave* or simply a *chemical helix* – was observed along the whole gel column. Such an example is shown in fig. 4a, together with a schematic drawing of the corresponding 3D structure (fig. 4b). The reproducibility for observing this chemical helix covering over a whole gel column was not low (22% based on 57 independent experiments).

Usually a scroll wave existed in a distance up to 0.6 mm from the open edge (see the right end of fig. 4a). Apparently, this acts as a pacemaker for the helical structure. The mild concentration gradients along the cylindrical axis (further discussed below), which decreases the rotation speed of a scroll wave with increasing distance from the open end, cause a winding of the chemical scroll wave into a twisted, helicoidal structure. Close to

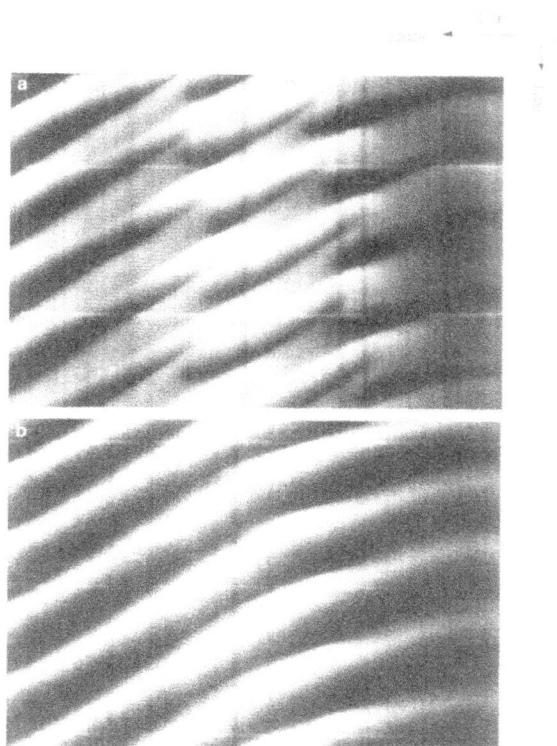

Fig. 5. (a) A space–time representation of the chemical helix of fig. 4a along the cylindrical axis of the capillary tube, constructed from 250 strips of subsequent video images taken at 0.36 s intervals (total time period of 90 s). Each strip consists of 2 × 460 pixels. (b) A space–time representation of the same chemical helix constructed along a cut close to the glass wall (0.12 mm away from the wall) for the same time period of 90 s.

the opposite end of the gel (the left end of fig. 4a), the helicoidal structure is re-wound and the chemical helix disappears. At this side, however, there appeared not a scroll wave but a train of planar discrete waves. The center-symmetry, initially given by the boundary conditions, was broken spontaneously by the system itself. This metamorphosis in a tube is analogous to a transition from helicoidal to banded structures observed in the Liesegang phenomenon [11].

In fig. 5 we show a space–time representation of the moving helix of fig. 4a along different cuts in the axial direction of the system ((a) along the center line; (b) along the line 0.12 mm inside

from the wall of the glass capillary. The positions of these cuts are indicated in fig. 4a). The reductive and oxidative phases of the chemical waves appear as the continuous black and white domains, respectively. In this 2D representation, a scroll wave existing at the right edge of the system gives a pattern parallel to the spatial axis, and a moving helicoidal wave (or planar waves also) results in a pattern diagonal to the space and the time axes.

Fig. 5a suggests that the structure of the central part of the chemical helix is quite complex. If it were a simple helix, then the redox phases should appear as continuous tilted lines from one end of the gel to the other, like those in fig. 5b. But the spatio-temporal pattern seen in fig. 5a contains more complex features – the redox pattern along the center axis of the chemical helix is also twisted on the space–time plane. The elucidation of the detailed three-dimensional structure of this chemical helix remains as a challenge for experimental and theoretical works.

From these space–time representations some basic features of the chemical helix can be analyzed. The pitch and the rotational period of the helix are 1.23 mm and 21.5 s, respectively. The velocity of the helix along the axis (V_z, see fig. 4b) is therefore 5.7×10^{-2} mm s^{-1}. As the diameter of the gel is 1.1 mm, the apparent rotational speed along the circumference (V_c in fig. 4b) obtained as 1.6×10^{-1} mm s^{-1}. The velocity normal to the wave front in a circular section (V_r in fig. 4b) cannot yet be clearly determined.

In our experiments the chemical helix disappeared after more than 2 h and a train of planar waves with wavelength of about 1–2 mm appeared in the gel. The value of V_z in these planar waves was 8.4×10^{-2} mm s^{-1} and was 1.5 times higher than that of the chemical helix. This later morphological change of the chemical wave from helical to planar is correlated with the disappearance of the anisotropy extending over the system. According to calculations based on the simple diffusion processes of bromate and malonic acid, we estimate that the concentration of the reactants at the center of the gel column should reach 95% of the homogeneous distribution 1 h after the diffusion process starts, and 99.8% after 2 h (assuming the diffusion coefficients to be 2×10^{-5} cm^2 s^{-1}). This means that there are very mild concentration gradients of these reactants along the cylindrical axis of the gel (less than 6.6×10^{-3} M mm^{-1}) when the chemical helix extends over the whole excitable gel column.

In addition to these mild anisotropies of concentrations, there seem to be two other factors which are responsible for producing the stable chemical helix (which exists for more than 1 h) with a considerably high reproducibility: a high probability of spiral initiation at the open boundaries of the gel and the stable existence of scroll waves under rotationally symmetrical conditions. Firstly, without any artificial treatment, the spiral center appears at the open boundaries with probability much higher than that of the chemical helix (22%). In general, the initiation of the spiral center is strongly affected by microinhomogeneities, concentration fluctuations and other factors, and there is no clear method of controlling its initiation in gels as well as in aqueous 2D solutions. Under the current experimental circumstances it is very difficult to produce smooth open surfaces of the gels. Although having little influence on wave propagation (it is known that the BZ wave propagates across narrow gap between discrete gels on which the catalyst is immobilized, and produces target or spiral patterns [6, 7, 12–14]), these microinhomogeneity and roughness on the open surfaces of the current gel system may be responsible for initiating a center of spiral wave with a high probability, which is a necessary condition for the relatively high reproducibility of the chemical helix. Secondly, the rotational symmetry of the cylindrical system is another candidate for stabilizing the helicoidal structure. A spiral wave once **initiated** at the open surface of the gel develops inside the tube along the cylindrical axis to turn into a scroll **wave, which becomes a pacemaker of the chemical** helix. The core **of the** scroll wave does not

easily drift away from the central axis of the system because of its geometrical symmetry. This stably existing scroll wave seems to be a key structure for producing the stable chemical helix.

Acknowledgements

Fruitful discussions with Professor B. Hess, Drs. L. Kuhnert, P. Borckmans, G. Dewel and D. Walgraef and the technical assistance by Dr. H. Hashimoto and B. Neumann are gratefully acknowledged. One of the authors (T.Y.) thanks NEDO (The New Energy and Industrial Technology Development Organization, Japan) for the financial support of staying at the Max-Planck-Institut, Dortmund, Germany. This work was supported by the Stiftung Volkswagenwerk, Hannover, Germany.

References

[1] G. Dewel, D. Walgraef and P. Borckmans, J. Chim. Phys. 84 (1987) 1335.
[2] D. Walgraef and C. Schiller, Physica D 27 (1987) 423.
[3] A.V. Panfilov, A.N. Rudenko and A.M. Pertsov, in: Self-Organization – Autowaves and Structures far from Equilibrium, ed. V.I. Krinsky (Springer, Berlin, 1984) p. 103.
[4] C. Henze, E. Lugosi and A.T. Winfree, Can. J. Phys., in press.
[5] Z. Noszticzius, W. Horsthemke, W.D. McCormick and H.L. Swinney, Nature 329 (1987) 619.
[6] T. Yamaguchi, L. Kuhnert, Zs. Nagy-Ungvarai, S.C. Müller and B. Hess, J. Phys. Chem., submitted for publication.
[7] H. Linde, H. Brandtstadter, G. Wessler and J. Bielecki, to be published.
[8] K.I. Agladze, V.I. Krinsky, A.V. Panfilov, H. Linde and L. Kuhnert, Physica D 39 (1989) 38.
[9] P. Foerster, S.C. Müller and B. Hess, Science 241 (1988) 685.
[10] Zs. Nagy-Ungvarai, S.C. Müller and B. Hess, in: Spatial Inhomogeneities and Transient Behaviour in Chemical Kinetics, eds. P. Gray, G. Nicolis, F. Baras, P. Borckmans and S.K. Scott (Manchester Univ. Press, Manchester, 1990) p. 644.
[11] S.C. Müller, S. Kai and J. Ross, Science 216 (1982) 635.
[12] J. Maselko and K. Showalter, Nature 339 (1989) 609.
[13] J. Maselko, J.S. Reckley and K. Showalter, J. Phys. Chem. 93 (1989) 2774.
[14] O.A. Mornev, H. Engel, J. Enderlein, H. Linde and V.I. Krinsky, in: Springer Series in Senergetics, to be published.

Physica D 49 (1991) 47–51
North-Holland

Pattern formation in a two-dimensional reaction–diffusion system with a transversal chemical gradient

Anatol M. Zhabotinsky[1] Stefan C. Müller and Benno Hess

Max-Planck-Institut für Ernährungsphysiologie, Rheinlanddamm 201, W-4600 Dortmund 1, Germany

In a thin layer of chromatographic silica gel saturated by Belousov–Zhabotinsky reagent, in which a transversal gradient of oxygen is set up, chemical waves appear first only at the bottom of the layer. Later, waves are also induced at the top of the layer. The top waves propagate almost independently, but weakly interact with the bottom waves. As a result, fronts with unusual sawtooth- or staircase-like geometry are formed.

1. Introduction

Most of the work on wave propagation in chemically active media was performed in homogeneous aqueous solutions [1–3]. These are sensitive to hydrodynamic disturbances [4], but since several years gel systems are available that do not only avoid these disturbances, but also create new possibilities and types of patterns [5, 6]. For instance, specific chemical compounds can be immobilized, and in this context wave spreading has been reported recently in a layer of chromatographic cation exchanger [7, 8]. The propagation of chemical waves in the homogeneous Belousov–Zhabotinsky (BZ) reaction is usually investigated in macroscopically uniform, thin layers with a thickness below 1 mm. In these cases the reaction–diffusion system behaves as a quasi-two-dimensional one. Concentric and spiral-shaped waves travel in such systems which cancel each other under collision.

While studying the propagation of chemical waves in a thin layer of chromatographic silica gel on a glass support saturated by BZ solution, a completely unexpected wave pattern was observed, which consists of two sets of mutually transversal waves. In this system a gradient of O_2

concentration is established vertically to the layer such that waves can propagate the same way as in the homogeneous system only within the lower part of the gel layer along the support. In the upper part the threshold of excitation is strongly elevated and standard wave spreading is not possible. In some parts the bottom waves propagate not only forward but also upward due to local nonuniformities of the gel layer. Thus, new wave fronts arise at the top of the layer which are approximately transversal to the initial bottom waves. Quite remarkably, the two wave systems propagate more or less independently from each other, but due to weak interaction the top waves show sawtooth- or staircase-like front geometry.

At first glance, this complex mode of wave propagation appears to be inconsistent with the basic features of reaction–diffusion waves. Here, we present a first experimental characterization of the phenomenon by summarizing some of the salient features of unusual front geometries evolving in this layered excitable system. These data form the basis of future theoretical analysis which should be possible by introducing a vertical gradient of excitability into existing reaction–diffusion models. The analysis of the reported phenomenon is a necessary step for an understanding of the behavior of excitable systems in three-dimensional space.

---

[1]National Scientific Centre of Hematology, Moscow, USSR.

2. Experimental

Stock solutions of $NaBrO_3$, $CH_2(COOH)_2$, H_2SO_4, $Fe(phen)_3$ and $Fe[batho(SO_3)_2]_3$ were prepared with triply distilled water and analytical grade chemicals. Bathophenantroline disulfonic acid was from Serva, silica gel G (particle size 10–40 μm) from Merck.

The bromate–malonic–ferroin (or bathoferroin) system was prepared as a thin layer of homogeneous solution being excitable but non-self-oscillating while exposed to the open air. The basic set of initial concentrations was: $NaBrO_3$ – 0.3 M, $CH_2(COOH)_2$ – 0.03 M, $FE(phen)_3$ (or $Fe[batho(SO_3)_2]_3$) – 0.002 M, H_2SO_4 – 0.1 M. After preparation of a homogeneous solution of the BZ reaction 7.6 g of silica gel was added to 15 ml of the above solution and the system was intensively stirred for 7–10 min. After stirring the uniform swollen gel was placed as a thin layer in a petri dish or between plane glass plates.

The wave propagation was observed in petri dishes at 25.0 ± 1.0°C. Photographs were taken at timed intervals with green light transmitted through the gel layer, that is, both patterns in the bottom and top sheets were projected on the same picture. Further data were obtained by recording on video tapes and with a digital imaging system [9].

3. Results

The system with the basic set of initial concentrations exposed to open air is excitable. Pacemakers appear spontaneously at the edges of the gel layer or waves can be initiated by a silver wire previously immersed in 5 M H_2SO_4. At the beginning the wave propagation is observed only in the bottom part of the gel layer, where standard patterns of target and spiral waves appear. The only difference to patterns in the homogeneous case is a rather frequent appearance of single spirals. 10–15 min after the start of the reaction in the gel some wave fronts begin to propagate in the vertical direction in separate small regions

until they emerge at the top of the layer. Upon reaching the top they propagate forward along with the initial bottom fronts. At the same time the oxidized state lasts much longer in the upper part of the layer. As a result, elongated stripes of the oxidized state appear (fig. 1) which lead to formation of new wave fronts propagating transversally to the initial bottom waves. The new autowaves forming in the top part of the layer propagate almost independently of the wave system at the bottom, as shown in fig. 2 in an early (fig. 2a) and in a late stage of an experiment (fig. 2b).

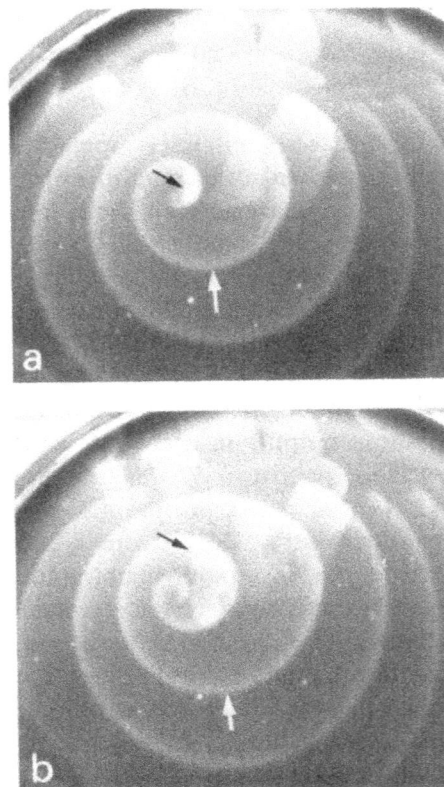

Fig. 1. Simultaneous observation of a spiral wave on the bottom of a thin layer of BZ solution in chromatographic silica gel (white arrow) and of local spreading of wave pieces along the top of the layer (black arrow). Interval between snapshots (a) and (b): 100 s. Diameter of petri dish = 70 mm. Initial concentrations before addition of silica gel: 0.3 M $NaBrO_3$, 0.03 M $CH_2(COOH)_2$, 0.002 M Fe(bathophen)$_3$, 0.1 M H_2SO_4 (see section 2).

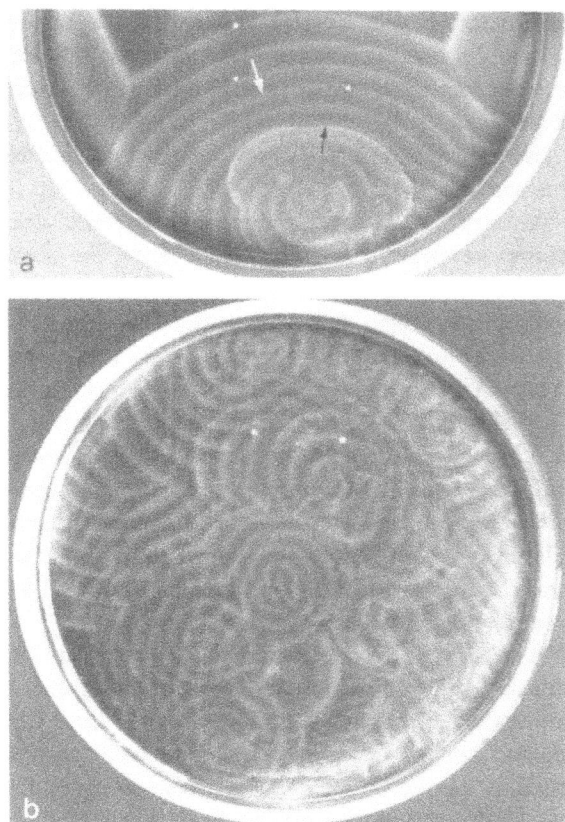

Fig. 2. Independent propagation of top and bottom waves with conditions as in fig. 1, about 10 min (a) and 40 min (b) after the start of the experiment. In picture (a) the white arrow points to the bottom wave, the black arrow to the top wave.

Transversal propagation of the top waves tends to prevail if the initial waves have a sufficiently short wavelength. After 50–60 min transverse fronts crossing 4–10 bottom waves have evolved (fig. 3a). During the following 30–40 min periodic changes take place in the character of the wave propagation. Sometimes a mutual acceleration is seen in the places where oxidized wave fronts overlap. This leads to a transient appearance of a sawtooth shape of the wave front (fig. 3b) which later is smoothed out again (fig. 3c).

In some cases a distinct staircase-like geometry of transversal top fronts forms. Its basic structure is maintained for 10 min or longer (fig. 4a). For the given initial concentrations, the propagation velocity of waves at the bottom of this gel layer is about two times smaller than in pure aqueous solution. In most of the samples so far examined, top and bottom waves propagate with equal velocity in transversal direction, but there are cases where the transversal wave fronts at the top remain stationary for an extended period of time. An example is indicated in the video image in fig. 4b. The front indicated by the arrow remains stationary for at least 4 min.

The patterns presented above were observed in a rather wide range of the initial reagent concentrations and using relatively large particle size of the silica gel. The thickness of the gel layer plays

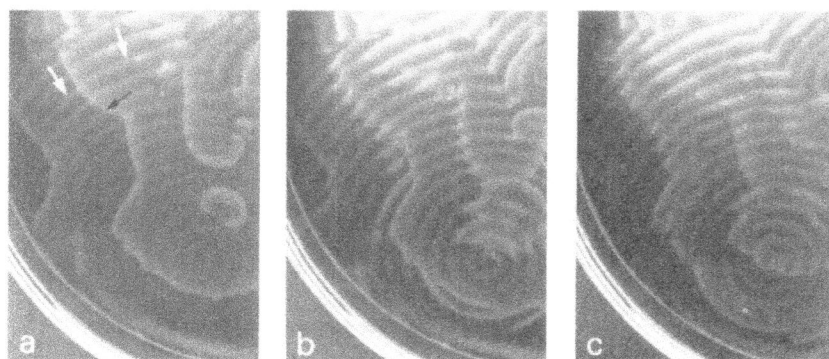

Fig. 3. Temporal transition from smooth (a) to sawtooth-shaped (b) and back to smooth (c) transversal wave fronts at the top of the layer. Conditions as in fig. 1. Interval between successive photographs: 40 s. In picture (a) the white arrows point to a front of the wave system at the bottom, the black arrow to the smooth front at the top of the layer.

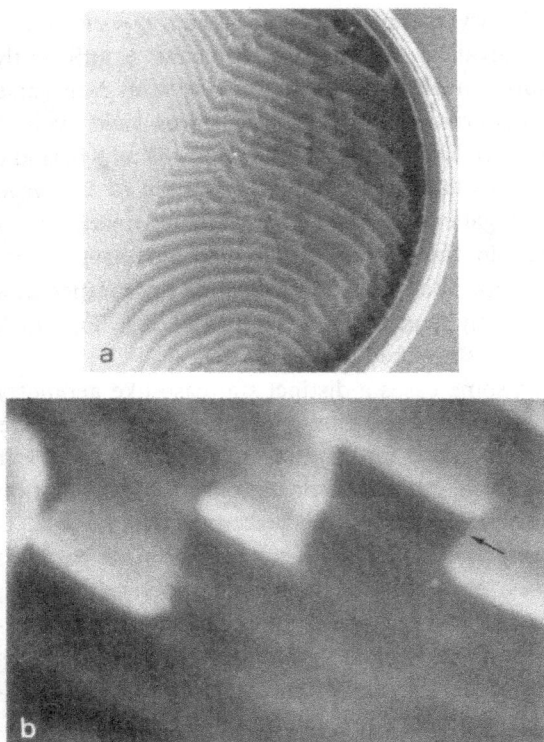

Fig. 4. Staircase shape of top wave fronts observed visually (a) and recorded in a 3.7×2.4 mm^2 section with a video camera at 490 nm (b). Conditions as in fig. 1. The black arrow in (b) points to a stationary front.

an important role. At a thickness equal to 0.5 mm the wave propagation is uniform along the vertical direction like in a homogeneous solution. A 1 mm layer offers optimum conditions for observation of the phenomena reported here. Layers with further increased thickness are inconvenient for observation. The oxygen gradient is of crucial importance for all the phenomena: they disappear when the gel layer is placed between two flat glass plates or exposed to pure N$_2$ in a petri dish. In these cases the wave patterns are the same as in a homogeneous solution.

4. Discussion

The mechanism of the transverse wave front formation is still unclear. It is possible to suppose

partial scrolling of the wave front in the upper part of the layer according to the simplest scheme of a trigger wave propagation. However, the shape of oxidized "tails" at the layer top, as seen in fig. 1, cannot be completely explained on the basis of this scheme.

In this quasi-two-dimensional system the smooth vertical chemical gradient along the third axis causes the separation of the layer into two weakly interacting sheets. A system division into qualitatively distinct parts has been previously observed under the influence of smooth longitudinal gradients in quasi-one-dimensional systems [6, 10, 11]. In our case the transversal chemical gradient leads to the appearance of new shapes of wave fronts and to a new type of spatiotemporal self-organization. It is noteworthy that the weak coupling between the sheets gives rise to wave fronts with remarkably sharp edges.

For the initial mathematical analysis of the reported phenomenon the simplest two-component model of the BZ reaction based on the FKN mechanism will be used [2]. The vertical oxygen gradient responsible for the gradient of excitability in our experiments can be easily introduced in the model as a gradient of the stoichometric factor which connects the bromide production with the ferriin consumption.

While in the experiments reported here the gradient of excitability across the layer arises from the gradients of O$_2$ concentration, similar effects could be also obtained due to gradients of other physical nature, for instance due to a temperature gradient. As an interesting application, the analysis of chemical waves in an excitable chromatographic gel should be useful for modelling the wave propagation in micro-heterogeneous biological systems, like heart tissues [12].

References

[1] A.N. Zaikin and A.M. Zhabotinsky, Concentration wave propagation in two-dimensional liquid-phase self-oscillating system, Nature 225 (1970) 535.

[2] R.J. Field and M. Burger, eds., Oscillations and Traveling Waves in Chemical Systems (Wiley, New York, 1985).

[3] P. Foerster, S.C. Müller and B. Hess, Critical size and curvature of wave formation in an excitable chemical medium, Proc. Natl. Acad. Sci. USA 86 (1989) 6831.

[4] H. Miike, S.C. Müller and B. Hess, Oscillatory deformation of chemical waves induced by surface flow, Phys. Rev. Lett. 31 (1989) 2109.

[5] Z. Noszticzius, W. Horsthemke, W.D. McCormick, H.L. Swinney and W.Y. Tam, Sustained chemical waves in a Turing–Nicolis–Prigogine ring reactor, Nature 329 (1987) 619.

[6] Q. Ouyang, J. Boissonade, J.C. Roux and P. De Kepper, Sustained reaction–diffusion structures in an open reactor, Phys. Lett. A 134 (1989) 282.

[7] J. Maselko, Y.S. Reckley and K. Showalter, Regular and irregular spatial patterns in an immobilized–catalyst Belousov–Zhabotinskii reaction, J. Phys. Chem. 93 (1989) 2774.

[8] J. Maselko and K. Showalter, Chemical waves on spherical surfaces, Nature 339 (1989) 609.

[9] S.C. Müller, Th. Plesser and B. Hess, Two-dimensional spectrophotometry of spiral wave propagation in the Belousov–Zhabotinskii reaction. I. Experiments and digital data representation, Physica D 24 (1987) 71.

[10] G. Nicolis and I. Prigogine, Self-Organization in Nonequilibrium Systems (Wiley, New York, 1977) p. 490.

[11] W.Y. Tam, J.A. Vastano, H.L. Swinney and W. Horsthemke, Regular and chaotic chemical spatiotemporal patterns, Phys. Rev. Lett. 61 (1988) 2163.

[12] A.T. Winfree, When Time Breaks Down (Princeton Univ. Press, Princeton, 1987).

Physica D 49 (1991) 52–60
North-Holland

Chapter 2. Spiral, ring and scroll patterns: Theory and simulations

Target and spiral waves in oscillatory media in the presence of obstacles

A. Babloyantz and J.A. Sepulchre[1]

Université Libre de Bruxelles, CP 231, Campus Plaine, Boulevard du Triomphe, 1050 Bruxelles, Belgium

Propagation of target and spiral waves in the presence of walls and windows in a two-dimensional reaction–diffusion model is considered. The time evolution of the system is such that for a range of parameter values a supercritical Hopf bifurcation leads to bulk oscillations. It is shown that in a finite system, for sufficiently small passages, no target waves are triggered. The passage of target waves through a window induces in the next compartment spiral or target waves. In this case a new bulk frequency appears and quasi-periodic motion is observed. In presence of two windows, propagation through a large opening can inhibit the onset of waves from smaller windows.

1. Introduction

Spatio-temporal self-organized phenomena such as target or spiral waves have been observed in many experimental systems. They appear for example in reaction–diffusion systems, electrical activity of heart muscles and brain waves [1–3]. The mechanisms of their onset has also been the focus of extensive theoretical investigations. Several papers of this special issue of Physica D and enclosed references address the spatio-temporal events in chemical media. Winfree has shown the importance of spiral waves in cardiac arrhythmias. Recently, Destexhe and Babloyantz showed that target and spiral activity may also appear in model systems which describe a thalamo-cortical tissue subject to nondiffusive interactions [4].

In this paper we concentrate solely on the investigation of target and spiral waves in oscillatory media in the presence of obstacles. The reaction kinetics are described by Ginzburg–Landau type of equations and evolve in a two-dimensional oscillating medium. A partition and windows are introduced in the system. Target waves are produced in one compartment. We

show that in a finite system, there is a critical window length for which target waves propagate in the next compartment with very large wavelengths, compared to the size of the vessel. For appropriate window size and position, spiral waves appear behind the partition. In the presence of two windows of different size, the waves emerging from one window can inhibit propagation from the other window.

In section 2 we introduce the chemical reaction model and the conditions for the onset of target waves. In the third section a partition with one and two windows is considered.

2. The model

2.1. Amplitude equation

It is assumed that the reaction kinetics are nonlinear and exhibit auto- or cross-catalytic steps. Moreover for a range of parameter values a supercritical Hopf bifurcation may appear and drive the system into an oscillatory mode. Throughout this paper we shall be interested in these oscillatory media which are subject to zero-flux boundary conditions. The general equation describing

---

[1]Fellow from the Institut pour l'Encouragement de la Recherche Scientifique dans l'Industrie et l'Agriculture.

this system in a homogeneous medium is

$$\frac{\partial X}{\partial t} = F(X) + \mathbf{D}\nabla^2 X, \tag{1}$$

where X is a vector representing the concentration variables and $\mathbf{D}$ is the diffusion matrix, which is taken to be a multiple of the identity matrix.

It is assumed that the time evolution of the system is such that for a range of parameter values, a supercritical Hopf bifurcation leading to bulk oscillations with frequency Γ takes place. Near the instability point, such a system may be reduced to a Ginzburg–Landau type equation which characterizes the behaviour of the slowly varying complex amplitude $A(r,T)$ defined as

$$X(r,t) = X_0' + \varepsilon\eta A(r,T)\,e^{i\gamma t} + \text{c.c.} + \mathscr{o}(\varepsilon^2) \tag{2}$$

with the slow time scale $T = \varepsilon^2 t$ and $0 < \varepsilon \ll 1$. In this expression X_0 describes the concentration values of the reference state and c.c. means complex conjugate; η is the critical mode, i.e. the eigenvector corresponding to the eigenvalue $i\gamma$ of the linear part of the operator $F(X)$ in eq. (1), at the bifurcation point. Substituting expression (2) in eq. (1), and keeping only terms proportional to ε, we find in the postcritical regime [5, 6]

$$\frac{\partial A}{\partial T} = \mu A - (g_1 + ig_2)|A|^2 A + D'\nabla^2 A, \tag{3}$$

where $\mu > 0$, g_1 and g_2 are real numbers which depend on the parameters of $F(X)$. Under the scaling of time $\tau = \mu T$ and amplitude $\Psi = (g_1/\mu)^{1/2}A$, the amplitude equation derived from eq. (1) reads

$$\frac{\partial \Psi}{\partial \tau} = \Psi - (1 + i\beta)|\Psi|^2\Psi + D\nabla^2\Psi. \tag{4}$$

Here $\Psi(r,\tau)$ is a complex field associated with the concentration variables at a given point in the

oscillating bulk. $\beta = (g_2/g_1)$ is a nonlinear dispersion coefficient which is taken to be the unity in all our simulations.

At the lowest order in ε, the bulk frequency of the concentration variables is, with these notations,

$$\Gamma = \gamma - \beta\mu\varepsilon^2$$

In an oscillating bulk, target waves may be generated around pacemaker centres by several mechanisms. Here we consider the case where target waves appear as a result of spatial inhomogeneities due to e.g. dust particles or impurities acting locally as pacemaker centres. The local inhomogeneities change slightly the frequency or the amplitude of the concentration oscillations in a small region of the space [7]. This situation can be described by adding a space-dependent term to eq. (1):

$$\frac{\partial X}{\partial t} = F(X) + \mathbf{D}\nabla^2 X + g(X,r), \tag{5}$$

where $g(X,r)$ is assumed to be nonzero only in a small region of space defining the pacemaker. In this paper a particular case is considered, for which $g(X,r)$ is chosen such that the amplitude equation derived from eq. (5) becomes simply

$$\frac{\partial \Psi}{\partial \tau} = [1 + i\Delta\omega(r)]\Psi - (1 + i\beta)|\Psi|^2\Psi + D\nabla^2\Psi, \tag{6}$$

where $\Delta\omega(r)$ takes a nonzero constant value in a small region of the system. If $\Delta\omega$ is sufficiently small, this local frequency shift induces a radial phase distribution around the pacemaker centre, giving rise to concentric waves with frequency $\Gamma' = \Gamma + \Delta\Gamma$. We define the scaled frequency shift $\Delta\Omega$ by $\Delta\Omega = (\Delta\Gamma/\mu)\varepsilon^2$. Far from the pacemaker, the wave number k of these target waves is given by $k^2 = \Delta\Omega/\beta D$. Such phenomenon was analyzed by Hagan [8] in the framework of the phase description of eq. (5). This description is a more simplified approximation to eq. (5) than the amplitude equation considered here, but it offers the advantage of leading to analytical results. More recently eq. (6) was discussed by Sakagushi for the case of a one-dimensional system. He showed that if $\Delta\omega$ becomes too large, the system may become desynchronised and the oscillation amplitude may even vanish [9, 10].

2.2. Simulation of target waves

To perform numerical simulations of the partial differential eq. (6), we divide the vessel into a network of $N \times N$ cells. In this network, the diffusion operator $D\nabla^2$ as well as the zero-flux boundary conditions may be approximated by finite differences [11]. It leads to a large system of coupled ordinary differential equations which can be integrated numerically by classical methods. We used a Runge–Kutta scheme, with controlled time step [11].

In numerical simulations of eq. (6), we have found that it is more convenient to follow the evolution of the complex variable $W(r, \tau) = \Psi(r, \tau)e^{i\beta\tau}$, which is just the variable Ψ viewed in a rotating frame in the complex plane. In this rotating frame, the bulk oscillation appears as a steady state: $W = $ constant.

In the network, the pacemaker is represented by a small cluster of cells for which $\Delta\omega \neq 0$ in eq. (6). Two shapes of pacemaker are considered: a square shape build on four adjacent cells and a cross shape formed by a five-cells cluster. Qualitatively similar results were obtained with both shapes. Fig. 1 shows target waves generated by eq. (6) with $N = 79$ and a cross shaped pacemaker at point P.

Let us study the variation of the frequency shift $\Delta\Omega$ when the intensity of the local frequency shift $\Delta\omega$ is varied. Our simulations reported in fig. 2 show that $\Delta\Omega$ is a highly nonlinear function of $\Delta\omega$. The frequency shift is also a function of the diffusion coefficient D, the nonlinear dispersion β and the diameter of the perturbed region. It is seen from this relationship that $\Delta\Omega$ exhibits

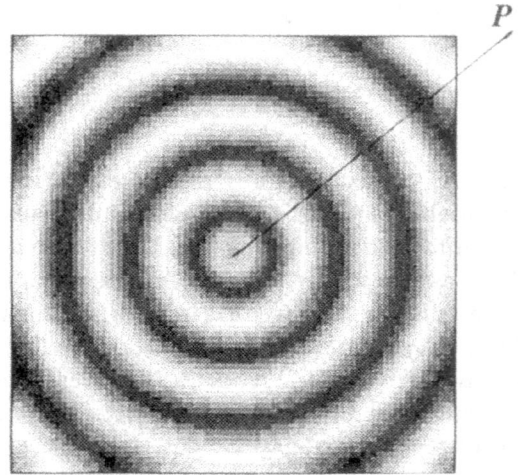

Fig. 1. Target waves in an oscillatory media. A pacemaker region at P with a frequency shift $\Delta\omega = 1$, generates target waves with a frequency shift $\Delta\Omega$. The parameters of eq. (6) are $\beta = 1$, $D = 1.25 \times 10^{-4}L^2$. The discretization of the vessel is 79×79. The width of the wall is $1.25 \times 10^{-2}L$.

a maximum for a given value of $\Delta\omega_r$ and decreases for $\Delta\omega > \Delta\omega_r$.

This result does not depend on the particular discretization scheme. The same behaviour of $\Delta\Omega(\Delta\omega)$ is also achieved with a vessel with cylindrical symmetry and using a discretization for the diffusion operator in cylindrical coordinates. The existence of a maximum in the curve of fig. 2 could be important when there are several pacemaker centres in the medium, as seen in the next paragraph.

We now consider the case in which two simultaneous pacemaker regions are active. The extent of the pacemaking regions are identical to the one used for the establishment of the graph in fig. 2. However, presently two different local frequency shifts $\Delta\omega_1$ and $\Delta\omega_2$ are considered and $\Delta\omega_1 < \Delta\omega_2$. These values correspond respectively to the frequency shifts $\Delta\Omega(\Delta\omega_1)$ and $\Delta\Omega(\Delta\omega_2)$ as seen from fig. 2. Target waves start to propagate from both centres. The waves emerging from the first centre are faster than ones emerging from the second one. However, after a while, one sees that one of the centres continues to emit

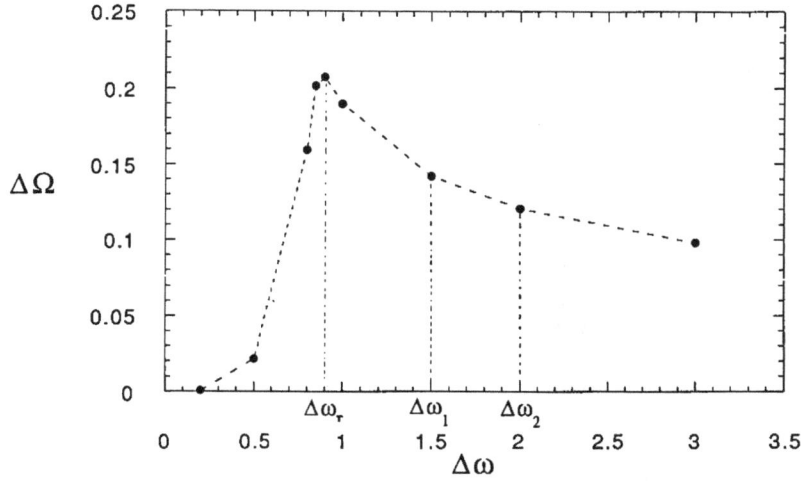

Fig. 2. Variation of the frequency shift $\Delta\Omega$ as a function of local frequency shift $\Delta\omega$. The parameters are as in fig. 1.

target waves whereas the second one becomes inactive. As was predicted previously, the pacemaker centre which wins the competition is the one whose frequency shift $\Delta\Omega$ is the largest [8, 12]. It is important to note that $\Delta\omega_1 < \Delta\omega_2$ and that at first sight one would expect that $\Delta\Omega(\Delta\omega_2)$ should always be the winner if $\Delta\Omega(\Delta\omega)$ was a monotonic increasing function. However, because of the decreasing part of function depicted on fig. 2, the reverse situation can prevail and a frequency shift $\Delta\Omega(\Delta\omega_1)$ can be seen in the system if $\Delta\omega_1$ is closer to $\Delta\omega_r$ than $\Delta\omega_2$. More generally, if several pacemakers act together with different $\Delta\omega$ taken in the domain of the graph of fig. 2, it is expected that after a transient regime only target waves with the closest frequency to $\Delta\Omega(\Delta\omega_r)$ will be observed.

3. Target waves and obstacles

3.1. Single window

In this section we shall investigate the behaviour of the target waves in the presence of various obstacles. As a starting point a partition is introduced in the reaction vessel which divides the system into two compartments communicat-

ing via a single opening. The length of the window changes in our simulations and as we shall see the phenomena observed are critically dependent not only on the window length but also on the position of the window relative to the pacemaker position.

Let us study the case where the window position is in the centre of the wall and the source is aligned perpendicular to the window. In the same manner as before, we produce target waves with frequency shift $\Delta\Omega_I$ in the first compartment.

Let us now follow the events in the second compartment. In the first compartment only one pacemaker centre is considered at point P of fig. 3. Fronts propagating from point P eventually reach the window and enter the space between the two compartments. From this space they generate a new set of target waves which propagate in the compartment II with a frequency shift $\Delta\Omega_{II}$. The frequency of emerging waves in the second compartment is again a function of several parameters such as for example $\Delta\Omega_I$ and the window length l.

For a fixed value of $\Delta\Omega_I$ numerical simulations reported in fig. 4 show the functional relationship between $\Delta\Omega_{II}$ and l. From fig. 4 it is seen that for large l, $\Delta\Omega_{II}$ has a maximum constant value equal to $\Delta\Omega_I$ independent of window size. The

P

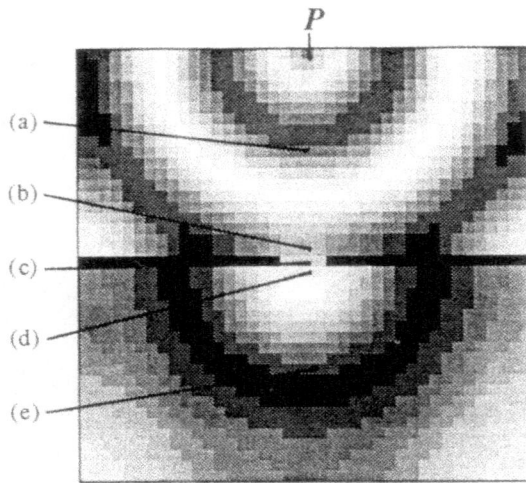

Fig. 3. A square vessel of side L is partitioned by a wall and communicates through a window of size $l = 0.01L$. A pacemaker region at P generates target waves in compartment I, with a frequency shift $\Delta\Omega_I$. New targets emerge from l with a frequency shift $\Delta\Omega_{II}$. The parameters of eq. (6) are $\beta = 1$, $D = 5 \times 10^{-4}L^2$. The discretization of the vessel is 40×40. The width of the wall is $0.025L$.

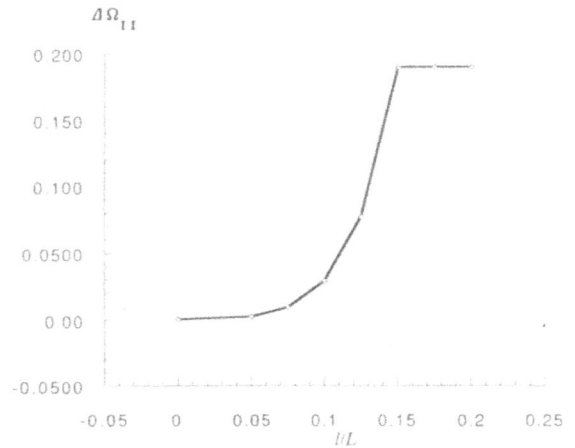

Fig. 4. Frequency shift $\Delta\Omega_{II}$ in compartment II as a function of the window size. The parameters are the same as in fig. 3.

frequency shifts in the two compartments are equal. This is not surprising since, in the presence of a large window, all happens as if we were in the presence of a single medium with somewhat modified configuration. However, as the size of the window l decreases, from fig. 4 it is seen that $\Delta\Omega_{II}$ tends to zero, and the wavelength of the target waves tends to an infinite value. Thus if the length of the vessel is much smaller than the wavelength of the target waves, one sees only a weak concentration gradient. Therefore, in a finite system target patterns are not observed in compartment II when the size l of the window becomes too small. This results from the fact that the wavelength is much larger than the length L of the vessel.

Let us presently monitor the time evolution of events at different locations in the system. The window length in this experiment is $0.1L$. From fig. 4 we see that this corresponds approximately to a value of $\Delta\Omega_{II} \approx \Delta\Omega_I/6.5$. Time evolution of five points is monitored (fig. 5). The first point (a) is in the compartment I and far away from the

partition. The second point (b) is much closer to the window, the third one (c) is in the opening and the fourth point (d) is in the compartment II but still close to the preceding one. A fifth location (e) is taken far away from the window in compartment II. Fig. 5 shows the time behaviour of these points. As expected, point (a) has periodic dynamics with frequency $\Delta\Omega_I$. The point at (b) already shows some distortion effect. The distortion is much more pronounced inside the window and increases further at point (d). At point (e) again an oscillatory behaviour is seen with periodic amplitude modulation.

Fig. 6 shows the power spectra of the signals at points (b), (d) and (e). It is seen that the frequency $\Delta\Omega_I$ of the waves at point (a) appears in the dynamics of all other points. However close to the wall a new bulk frequency $\Delta\Omega_{II}$ appears. As the spectra show, frequencies seem to be a linear combination of $\Delta\Omega_I$ and $\Delta\Omega_{II}$, and the motion seems to be of quasi-periodic type. This fact is confirmed through the stroboscopic plot at frequency $\Delta\Omega_I$ of signal from point (d), which is a closed curve.

We also monitored the amplitude and phase at points (a), (d) and (e). As expected no amplitude change is seen at point (a). On the contrary, at point (d) there is a strong variation of amplitude

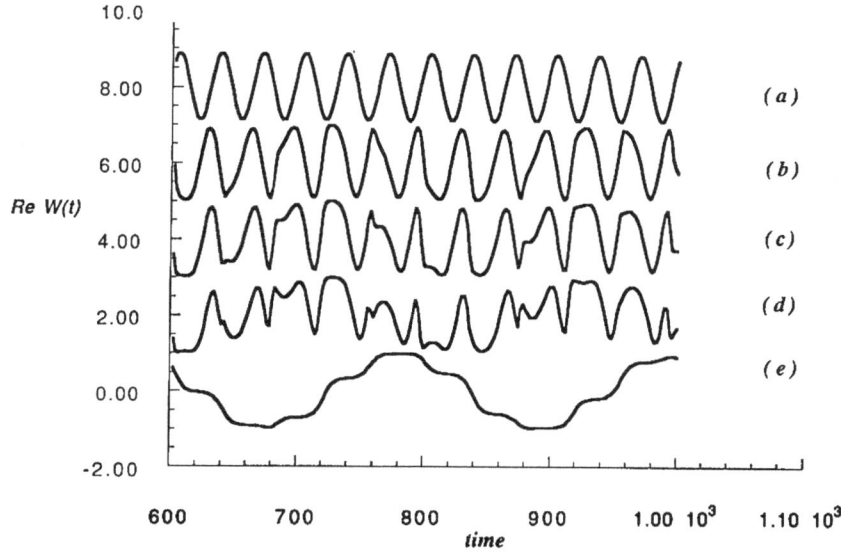

Fig. 5. Time variation of Re $W(t)$ at various points in the two compartments of fig. 3.

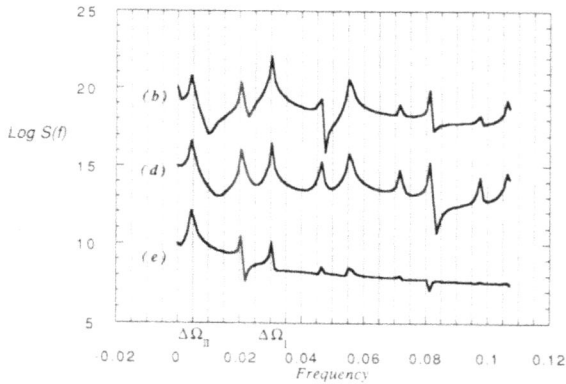

Fig. 6. Power spectra of signals shown in fig. 5. Close to and beyond the window a new bulk frequency appears and the dynamics is quasi-periodic. From (b) to (e) the contribution of the frequency $\Delta\Omega_I$ decreases and the contribution of $\Delta\Omega_{II}$ increases.

in a quasi-periodic fashion. At point (e) again the amplitude is not constant; however, now we see a slight periodic modulation of the amplitude. Again, as expected inside the first compartment, far from the barriers and the window, the phase varies in time linearly proportional to $\Delta\Omega_I$. Inside the window this variation is periodic, however with a mean value which is again linear and exhibits the same slope as at point (a). A similar type of variation is seen for the phase at point (e), but with a mean slope equal to $\Delta\Omega_{II}$.

In compartment I we are not in the presence of plane wave propagation. From our simulations it turns out that the relative position of the pacemaker and of the window may be of importance to the type of propagation seen in compartment II. To see this we simulate the reaction in a medium where the pacemaker is not aligned perpendicular to the partition in the center of that partition. Fig. 7 shows the variation of $\Delta\Omega_{II}$ as a function of l in this case. Contrary to fig. 4 in the present case we observe a plateau region for moderate values of l. For values $l < l_a$ and $l > l_b$ the behaviour of the system is similar to the case discussed in fig. 3. However, for window size $l_a < l < l_b$ one observes the onset of spiral wave activity as shown in fig. 8.

3.2. Two windows

Let us now consider the same vessel and the same chemical reactions as described above. We introduce two openings l_1 and l_2 in the system. Again a single local frequency shift $\Delta\omega$ is created

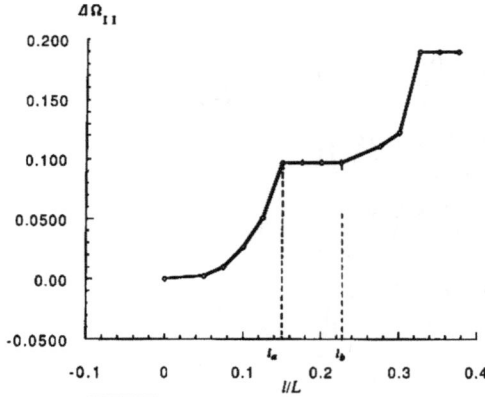

Fig. 7. Frequency shift $\Delta\Omega_{II}$ in compartment II as a function of the window size. The parameters are as in fig. 1 with $N = 80$. The distance between the window and the left boundary is $0.125L$ as in fig. 8. Spiral waves appear if the window size is between l_a and l_b.

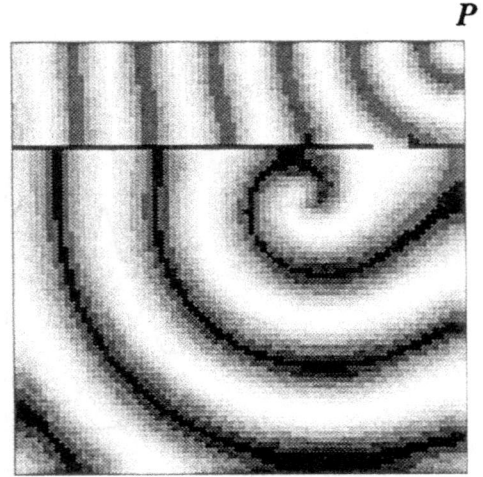

Fig. 8. For the specific configuration considered, in compartment II spiral waves are triggered. $l = 0.075L$ and other parameters are as in fig. 1 with $N = 80$.

at point P. Target waves start to propagate in compartment I.

Target waves reach successively the two openings l_1 and l_2. For $D = 0.8$ and $\Delta\omega = 3$, the two windows act as new pacemaker regions and generate in turn target waves in the second compartment with $\Delta\Omega_{II}(l_1) \approx \Delta\Omega_{II}(l_2) \approx \Delta\Omega_I$. After a while, the target waves propagating from the two centres collide and cusp like structures are formed (fig. 9a).

In the next experiment, we increase slightly the extent of the pacemaker region, increase the diffusion coefficient D and decrease $\Delta\omega$. A higher value of $\Delta\Omega_I$, as compared with the preceding case, is obtained in the first compartment for the propagating target waves. In this experiment, when the wavefront reaches windows l_1 and l_2, new fronts are again generated. However, very soon the waves emerging from window l_2 take over the entire system and inhibit all propagation from l_1 (fig. 9b).

A tentative explanation to this phenomenon could be given as follows. In this case one observes that $\Delta\Omega_{II}(l_1) < \Delta\Omega_{II}(l_2) < \Delta\Omega_I$. We saw in a preceding paragraph that if two pacemakers emit simultaneously, then the fastest waves inhibit the slower propagation and take over the

entire system. As $\Delta\Omega_{II}(l_1) < \Delta\Omega_{II}(l_2)$ it is reasonable to think that a similar explanation prevails here. The frequency of the waves emerging from l_2 is fastest and, therefore, such waves take over the slowly evolving target waves which emerge from l_1.

In section 3.1 we showed that for a specific configuration of source P and a single window length l, spiral waves may appear in the second compartment. Let us again consider two windows l_1 and l_2. The window l_1 and P have the same configuration as in fig. 8. When the fronts starting from compartment I reach l_1 again spiral type activity is generated in the second compartment. However when a new front reaches l_2 cusp-like fronts are seen (fig. 10a). In time the tip of the spiral moves toward the opposite boundary and finally disappears, giving rise to the ordinary target waves as shown in fig. 10b.

4. Discussion

The formalism introduced here is general as eq. (6) describes a normal form and therefore may be used for modeling of all oscillatory media

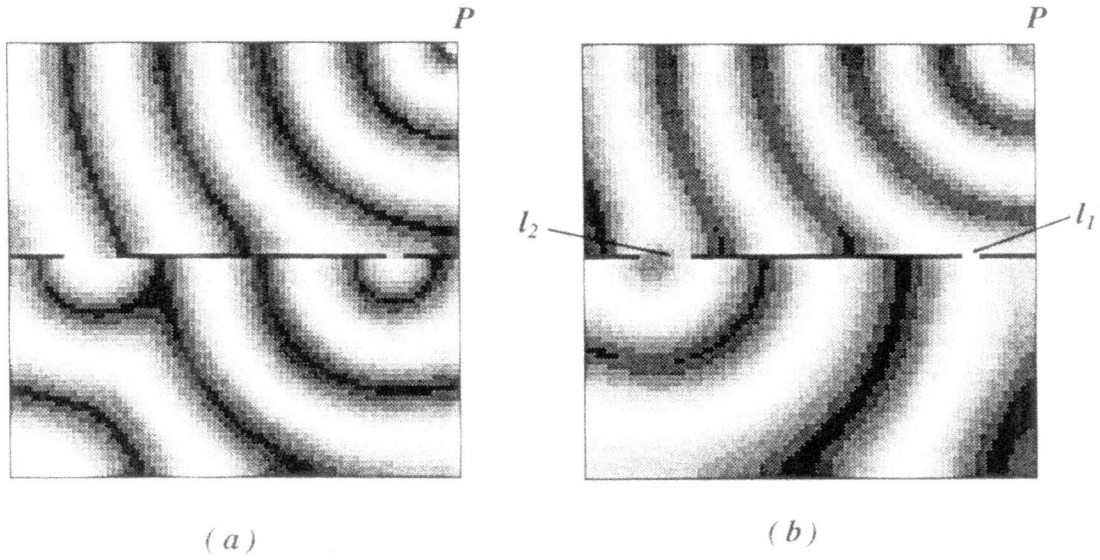

Fig. 9. Wave propagation in the presence of two windows. Parameters are: $\beta = 1$, window width = $0.0125L$, window lengths $l_1 = 0.0375L$, $l_2 = 0.1125L$ and a grid of 80×80. (a) Waves propagate into compartment II from both windows and a cusp-like structure is formed. $D = 1.25 \times 10^{-4}L^2$, $\Delta\omega = 3$. (b) Waves propagate from the largest window and inhibit the propagation from the small window. $D = 2.22 \times 10^{-4}L^2$, $\Delta\omega = 1$.

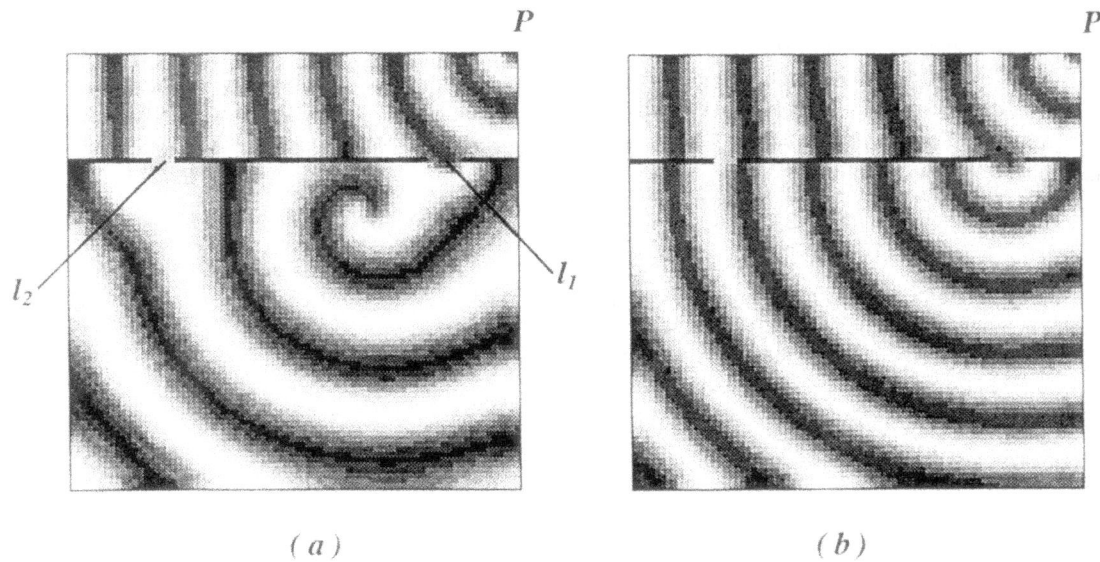

Fig. 10. A two-window partition produces spiral waves which in time give way to target waves. All happens as if again the propagation from the small window is inhibited. Parameters are as in fig. 8.

whose oscillations come from a supercritical Hopf bifurcation. The numerical simulations reported in this paper show that target waves propagating in an oscillatory medium may exhibit unexpected behaviour in the presence of obstacles[#1]. Several of these properties are due to the fact that the waves evolve in a finite medium which is the case in most experimental situations.

The fact that waves emerging from large windows inhibit the ones propagating from smaller openings must be contrasted to the property of the sound waves or electromagnetic waves which would be propagating from both apertures producing interference effects. No such interference phenomena are seen with target waves.

In a previous paper we have discussed the front propagation into a uniform state unstable with respect to a supercritical Hopf bifurcation [13]. It was shown that due to the existence of a phase gradient, target waves were generated behind the front. According to the sign of β in eq. (4), the waves and the front travel in the same or in the opposite direction. In the latter case on sees incoming target waves traveling toward the pacemaker region.

Let us note that in unstable media, contrary to the oscillating systems described in this paper, there is no critical window length for target wave propagation. As the medium is unstable, the smallest perturbation may trigger wave trains in compartment II. This fact was verified by several simulations.

The numerical simulations of reaction–diffusion systems requires the discretization of the space into small cells which are locally connected to their first neighbours. A coarse discretization gave qualitatively the same spatiotemporal patterns as the one described in the preceding paragraphs. Thus a network of oscillators coupled to their first neighbours and forming a square lattice, exhibits identical properties to that of continuous systems considered here. Such networks can be used as analogical representation for solving specific problems encountered in Artificial Intelligence [14].

References

[1] J.M. Davidenko, P. Kent and J. Jalife, Physica D 49 (1991) 182–197, these Proceedings.
[2] A.T. Winfree, Physica D 49 (1991) 125–140, these Proceedings.
[3] M. Shibata and J. Bures, J. Neurophysiol. 38 (1985) 158.
[4] A. Destexhe and A. Babloyantz, J. Neural. Comput., in press.
[5] P.H. Richter, I. Procaccia and J. Ross, in: Advances in Chemical Physics, Vol. 43, eds. I. Prigogine and S.A. Rice (Wiley, New York, 1980) p. 217.
[6] Y. Kuramoto and T. Tzuzuki, Prog. Theor. Phys. 55 (1976) 356.
[7] A.T. Winfree, Theoret. Chem. 4 (1978) 1.
[8] P.S. Hagan, Adv. Appl. Math. 2 (1981) 400.
[9] H. Sakaguchi, Prog. Theor. Phy. 80 (1988) 743.
[10] H. Sakaguchi, Prog. Theor. Phys. 83 (1990) 169.
[11] M. Kubicek and M. Marek, Computational Methods in Bifurcation Theory and Dissipative Structures (Springer, Berlin, 1983).
[12] Y. Kuramoto and T. Yamada, Prog. Theor. Phys. 56 (1976) 724.
[13] A.M. Pertsov, E.A. Ermakova and E.E. Shnol, Physica D 44 (1990) 178.
[14] J.A. Sepulchre, G. Dewel and A. Babloyantz, Phys. Lett. A 147 (1990) 380.
[15] L. Steels, in: Proceedings of the European Conference on Artificial Intelligence 88, ed. Y. Kadratoff (Pitman, London, 1989).

[#1] The work reported in this paper was first presented at a conference in Leeds in July 1989. Related results on the diffraction of waves by obstacles have also been reported by Pertsov et al. [13].

Physica D 49 (1991) 61-70
North-Holland

A model for fast computer simulation
of waves in excitable media

Dwight Barkley

Applied and Computational Mathematics Department, Princeton University, Princeton, NJ 08544, USA

Starting from a two-variable system of reaction–diffusion equations, an algorithm is devised for efficient simulation of waves in excitable media. The spatio-temporal resolution of the simulation can be varied continuously. For fine resolutions the algorithm provides accurate solution of the underlying reaction–diffusion equations. For coarse resolutions, the algorithm provides qualitative simulations at small computational cost.

1. Introduction

Wave propagation in excitable media provides an important and beautiful example of spatio-temporal self-organization. Spiral waves in unstirred Belousov–Zhabotinsky reagent and impulse propagation along nerve axons are two well-known examples of this phenomenon. Numerous other examples can be found throughout the literature[#1]. Given the widespread interest in these waves, numerical techniques for fast computer simulations of excitable media are of considerable importance. This is particularly true if numerical studies are to be made of large two- and three-dimensional excitable systems.

The fundamental difficulty in simulating excitable media is the separation of spatio-temporal scales in such systems. The time scale on which variables change as the system becomes locally excited is typically several orders of magnitude faster than the time scale on which interesting behavior occurs in the extended medium. Similarly, spatial gradients at fronts where excitation occurs are vastly larger than gradients elsewhere in the medium. If a numerical scheme is to accurately simulate a given excitable system, then it

must resolve the fast dynamics of excitation, and unfortunately this requires a fine spatial mesh and a time step which is very small in comparison with the time scale of ultimate interest.

Cellular-automaton models have been proposed as method of circumventing the scale-separation problem [4–10]. In effect, these models reduce the fast dynamics of excitation to a single, discontinuous, jump in some state variable. Recently, very impressive high-speed simulations of waves in excitable media have been obtained using the cellular-automaton approach [8–10]. Despite the success of automaton models, they have some disadvantages: (1) they are governed by somewhat ad hoc rules, and (2) because they do not resolve the fast dynamics, they cannot obtain arbitrarily fine spatio-temporal resolution. This second point means that it is difficult to assess the validity of results from automaton simulations without resorting to simulations by other methods.

Here we propose a simple numerical model which offers most of the advantages of cellular automata, but in addition has the following virtues: (1) it is based directly on a system of reaction–diffusion equations and (2) the spatio-temporal resolution can be adjusted continuously. For coarse resolutions, the fast dynamics is not accurately resolved by the model, but the impor-

---

[#1]These Proceedings give the most recent review of the field. Refs. [1–3] are other reviews of interest.

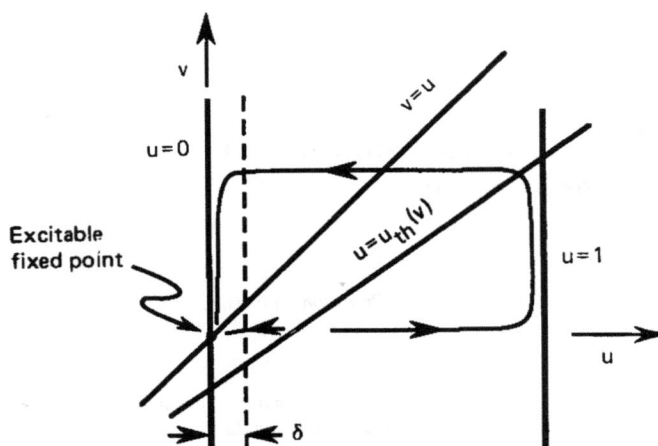

Fig. 1. Illustration of the local dynamics. The axes are the variables u and v. Shown are the system nullclines: the v nullcline $g(u,v) = 0$ is the line $v = u$, and the u nullcline $f(u,v) = 0$ consists of three lines: $u = 0$, $u = 1$, and $u = u_{th}(v) = (v + b)/a$. An excitable fixed point sits at the origin where the u and v nullclines intersect. u_{th} is the excitability threshold for the fixed point. Initial conditions near the fixed point and to the left of the threshold, decay directly to the fixed point. Initial conditions to the right of the threshold undergo a large excursion before returning to the fixed point. δ denotes a small "boundary layer" flanking the left branch of the u nullcline. If the system is outside the boundary layer then is excited, otherwise it is recovering.

tant characteristics of excitability are maintained[#2]. In the limit of a fine resolution, the numerical scheme provides accurate solutions to the underlying reaction–diffusion equations. Thus low-resolution simulations can be used to gain insight and explore parameter space at small computational expense, while high-resolution simulations are available with the same compute code and can be used to assess the validity of low-resolution results.

2. The model

Our starting point is a two-variable system of reaction–diffusion equations modeling the dynamics of an excitable medium:

$$\frac{\partial u}{\partial t} = f(u,v) + \nabla^2 u, \quad \frac{\partial v}{\partial t} = g(u,v), \quad (1)$$

where the functions $f(u,v)$ and $g(u,v)$ express

[#2] For coarse spatio-temporal discretizations, it is perhaps appropriate to consider our model as a coupled-map lattice, see e.g. ref. [11].

the local kinetics of the two variables u and v. By choice of length scales, the diffusion coefficient for the u-variable is scaled to unity. For simplicity, we assume that v does not diffuse and discuss the applicability of this approximation at the conclusion. At present the spatial domain is arbitrary; later we shall specialize to a two-dimensional square domain with Neumann boundary conditions.

We model the local kinetics with equations

$$f(u,v) = \epsilon^{-1} u(1-u)\left[u - u_{th}(v)\right],$$

$$g(u,v) = u - v, \quad (2)$$

where $u_{th}(v) = (v + b)/a$, and a, b, and ϵ are parameters, with ϵ generally small. The local dynamics, that is the dynamics in the absence of diffusion, is illustrated in fig. 1 with a typical nullcline picture of phase space. The system has a stable but excitable fixed point at the origin where the nullclines, $f(u,v) = 0$ and $g(u,v) = 0$, intersect. The variables u and v are known as the excitation and recovery variables, respectively. Two-variable models of this kind are ubiquitous in the study of excitable systems [2, 3][#3]; these particular local kinetics have been used recently in the study of the periodic–quasiperiodic transition for spiral waves [12]. Only three parameters appear in the local kinetics and this is of considerable advantage when it comes to exploring parameter space and classifying the model dynamics. We return to this point at the conclusion.

Any numerical scheme for solving (1) should recognize and take advantage of the two distinct dynamical states of the system: excitation and recovery. To be precise, consider a small "boundary layer", of size δ, flanking the left branch of the u nullcline as shown in fig. 1. For our purposes here, we call a given point in the spatial domain excited if (u,v) lies outside the boundary layer, i.e. if $u > \delta$. A point in the spatial domain is recovering if it is within the boundary layer. Regions of excitation and recovery for two cases are shown in plates Ic and Ie discussed below[#4].

The following features of excitation and recovery should be considered in the design of efficient numerical schemes. (i) Within the excited region, the time scale of u can be very fast due to the small parameter ϵ. In particular, transitions between the vertical branches of the u-nullcline are very quick. Thus, either small time steps must be taken to accurately resolve the fast dynamics, or if large (less accurate) time steps are to be used, then care must be taken to prevent numerical instabilities and to preserve the qualitative character of the fast jumps in u. (ii) Within the recovery region the dynamics is very simple. The excitation variable u is essentially zero, so that the local dynamics effectively reduces to exponen-

[#3]It should be noted that problems can arise in our model if the system gets close to the "corners" where the diagonal segment of the u-null-cline intersects the vertical segments. There are no such difficulties for the spiral waves presented here.

[#4]For some purposes, one might want to called a point excited if it lies to the right of the excitability threshold u_{th}, and recovering if it lies to the left of the threshold. Also, in some cases the term "boundary layer" is used for regions of space where u makes fast jumps.

tial decay of the recovery variable v. Moreover, in the interior of recovery regions, diffusion is negligible because the spatial profile of u is basically flat. With these points in mind, we design an algorithm for solving (1).

2.1. Local dynamics

For the moment we ignore the diffusion term and consider a scheme for time stepping the local dynamics. With the approximation that $u = 0$ within the boundary layer, we obtain the following simple algorithm for the local dynamics:

if $u^n < \delta$

$$u^{n+1} = 0,$$
$$v^{n+1} = (1 - \Delta t)v^n,$$

else

$$u_{th} = (v^n + b)/a,$$
$$v^{n+1} = v^n + \Delta t(u^n - v^n),$$
$$u^{n+1} = F(u^n, u_{th}),$$

where u^n and v^n are the value of variables u and v at the nth time step (at some point in the spatial domain) and Δt is the time step. The function F gives the time stepping of u outside the boundary layer and will be discussed momentarily. Inside the boundary layer, u is simply set to zero.

The variable v is stepped both inside and outside the boundary layer by the explicit-Euler method, with the condition that u is set to zero within the boundary layer. The time scale of v is so slow in comparison with the time scale of u, that even with time steps very large relative to the u time scale, explicit-Euler stepping of v is both stable and accurate.

The function F for stepping the u-dynamics may be either of explicit or implicit form. With explicit-Euler time stepping, we have found that $\Delta t \approx \epsilon$ is the maximum Δt for obtaining reasonably accurate results in the fast regions. This Δt

is also near the stability limit for an explicit-Euler step. In order to take very large, less accurate, time steps a semi-implicit form for F can be used. This is obtained from:

$$u^{n+1} = u^n + (\Delta t/\epsilon)u^{n+1}(1-u^n)(u^n - u_{th})$$

if $u^n \leq u_{th}$,

$$= u^n + (\Delta t/\epsilon)u^n(1-u^{n+1})(u^n - u_{th})$$

if $u^n > u_{th}$,

where u at the future time is used in those factors on the right-hand side which undergo largest change as the system approaches the stable branches of the u-nullcline. This semi-implicit form keeps the time stepping of the kinetics from overshooting the stable branches of the u-nullcline even if a large time step is taken in the fast region. Solving the above expressions for u^{n+1}, we obtain

$$F(u^n, u_{th}) = \frac{u^n}{1 - (\Delta t/\epsilon)(1-u^n)(u^n - u_{th})}$$

if $u^n \leq u_{th}$,

$$= \frac{u^n + (\Delta t/\epsilon)u^n(u^n - u_{th})}{1 + (\Delta t/\epsilon)u^n(u^n - u_{th})}$$

if $u^n > u_{th}$. $\qquad(3)$

For small $\Delta t/\epsilon$ the denominators in the above expression can be expanded to recover the explicit-Euler form for F:

$$F(u^n, u_{th}) = u^n + (\Delta t/\epsilon)u^n(1-u^n)(u^n - u_{th})$$
$$+ \mathcal{O}(\Delta t^2).$$

For large $\Delta t/\epsilon$, F goes over to the limit

$$F(u^n, u_{th}) = 0 \quad \text{if } u^n < u_{th},$$
$$= u_{th} \quad \text{if } u^n = u_{th},$$
$$= 1 \quad \text{if } u^n > u_{th},$$

while remaining continuous for any finite Δt. This large-Δt (or small-ϵ) limit models the fast excitation dynamics in the crudest way possible, and is reminiscent of cellular automata [9] in which the u-variable takes on just two values, 0 and 1. Thus, (3) is a representation for u-dynamics which in the small-$\Delta t/\epsilon$ limit gives an accurate time step, while in the large-$\Delta t/\epsilon$ limit preserves the important characteristics of excitability.

2.2. Diffusion

Having presented a scheme for stepping the local dynamics, we turn to the efficient treatment of the diffusion term. With the approximation that $u = 0$ within the boundary layer, the u-field is flat in the interior of recovery regions, and hence the Laplacian is zero there and need not be evaluated. To avoid unnecessary computation, the Laplacian can be evaluated "actively" rather than "passively". By this we mean the following. Consider the five-point finite-difference Laplacian formula

$$h^2 \nabla^2 u_{ij} = u_{i+1,j} + u_{i-1,j} + u_{i,j+1} + u_{i,j-1} - 4u_{ij},$$

where u_{ij} with the value of u at grid point (i, j) and h is the grid spacing. (We now restrict attention to regular square lattices.) Passive evaluation of the Laplacian is obtained by looping grid indices and evaluating directly the above formula at each point in the spatial domain, that is,

for each i, j

$$\text{lap}_{ij} \leftarrow u_{i+1,j} + u_{i-1,j} + u_{i,j+1} + u_{i,j-1} - 4u_{ij}.$$

The factor of h^2 **is absorbed into the diffusion** coefficient.

Alternatively, active evaluation is obtained by considering the contribution that each point makes to the Laplacian of nearby points, that is,

for each i, j

$$\text{lap}_{ij} \leftarrow 0,$$

for each i, j

$$\text{lap}_{ij} \leftarrow \text{lap}_{ij} - 4u_{ij},$$
$$\text{lap}_{i+1,j} \leftarrow \text{lap}_{i+1,j} + u_{ij},$$
$$\text{lap}_{i-1,j} \leftarrow \text{lap}_{i-1,j} + u_{ij},$$
$$\text{lap}_{i,j+1} \leftarrow \text{lap}_{i,j+1} + u_{ij},$$
$$\text{lap}_{i,j-1} \leftarrow \text{lap}_{i,j-1} + u_{ij}.$$

Clearly the two methods of evaluating the Laplacian give the same result. What makes active evaluation of the Laplacian desirable is that it can be incorporated into the algorithm for the local dynamics in such a way that unnecessary computation is avoided at points which make zero contribution to the Laplacian of the u-field. Specifically, the following algorithm updates a single grid point in the spatial domain and computes its contribution to the Laplacian of neighboring points for use at the next time step:

if $u_{ij} < \delta$

$$u_{ij} \leftarrow D \, \text{lap}_{kij},$$
$$v_{ij} \leftarrow (1 - \Delta t)v_{ij}$$

else

$$u_{\text{th}} \leftarrow (v_{ij} + b)/a,$$
$$v_{ij} \leftarrow v_{ij} + \Delta t \, (u_{ij} - v_{ij}),$$
$$u_{ij} \leftarrow F(u^n, u_{\text{th}}) + D \, \text{lap}_{kij},$$
$$\text{lap}_{k'ij} \leftarrow \text{lap}_{k'ij} - 4u_{ij},$$
$$\text{lap}_{k',i+1,j} \leftarrow \text{lap}_{k',i+1,j} + u_{ij},$$
$$\text{lap}_{k',i-1,j} \leftarrow \text{lap}_{k',i-1,j} + u_{ij},$$
$$\text{lap}_{k',i,j+1} \leftarrow \text{lap}_{k',i,j+1} + u_{ij},$$
$$\text{lap}_{k',i,j-1} \leftarrow \text{lap}_{k',i,j-1} + u_{ij},$$
$$\text{lap}_{kij} \leftarrow 0 \tag{4}$$

where $D = \Delta t / h^2$. The Laplacian $\text{lap}_{k,i,j}$ has three subscripts, the first of which takes on just two values (zero and one, say). The values of k and k' are interchanged at every time step. In effect, there are two Laplacian fields which are used in alternation: the first (unprimed) is used to update the u-field at the current step and is then set to zero for use at the next time step. The second (primed) Laplacian is computed for use at the next time step. A complete subroutine for taking one time step is given in the appendix. In the limit $\Delta t, h, \delta \to 0$, and with appropriate boundary conditions, we expect the numerical solution obtained from (4) to converge to the solution of PDE (1) [13].

3. Results

We have simulated spiral waves using the subroutine given in the appendix with the implicit form for the function F. The spatial domain is a square grid of area L^2 containing N^2 grid points. Hence, the grid spacing is $h = L/(N - 1)$. No-flux boundary conditions are imposed on the domain boundary. There are seven parameters for the problem: the four "physical parameters", a, b, ϵ, and L, and three "numerical parameters", Δt, δ, and N. As a practical matter, we have found it convenient to fix the relationship between spatial and temporal discretizations. For all results reported here $\Delta t = L^2/5(N - 1)^2$.

Plate I shows some representative results from our model. Plates Ia and Ib show a single spiral wave at two different resolutions. The parameter values for (a) are: $a = 0.3$, $b = 0.01$, $1/\epsilon = 200$, $L = 40$, $N = 81$, and $\delta = 10^{-4}$. With these parameters, $h = 0.5$ and $\Delta t / \epsilon = 10$, and this is the coarsest resolution found to produce meaningful results. For coarser resolutions grid effects become dominant and there is significant slowing in the wavespeed along the grid diagonals. Even at the resolution of Ia significant grid effects are

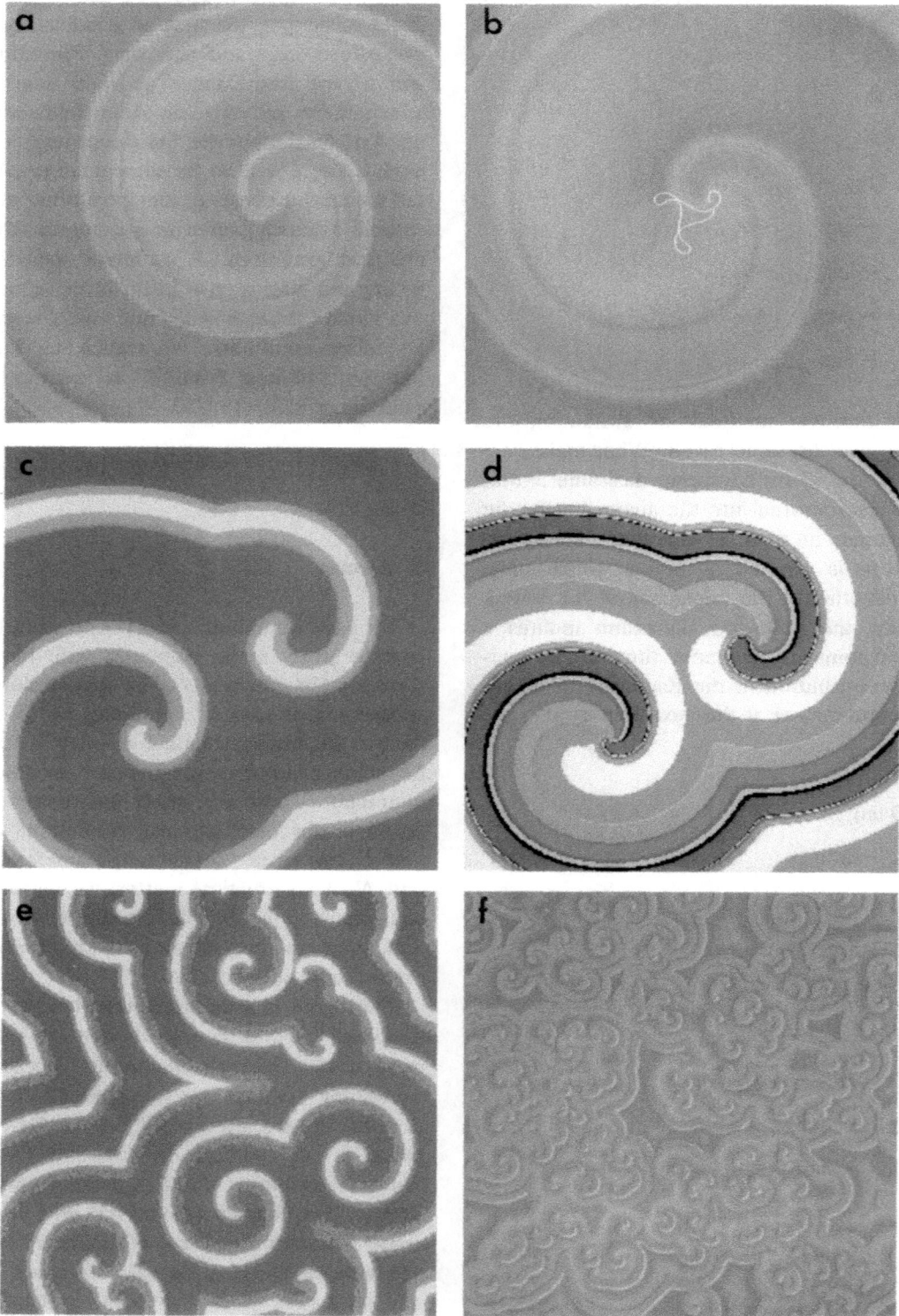

sometimes observed[#5]. Note that the time step is an order of magnitude larger than that which is possible with explicit time stepping of the reaction kinetics.

Simulations at the resolution of plate Ia are extremely fast: the execution time required for one wave rotation is about 2 s using single precision arithmetic on a Silicon Graphics 4D/200 series workstation with a Mips R3000 processor. This execution time is comparable to that recently reported for cellular-automata simulations [9]. An exact comparison of computational speeds is not possible, however, because of differences between cellular-automaton approach and that used here. While we do not expect our simulations to achieve quite the speed of well-optimized cellular automata (due primarily to the number of floating-point operations required to compute the function F), we do expect our model to be competitive on machines with good floating-point hardware.

The advantage of our model is that the spatio-temporal resolution of the simulation can be increased in a well controlled manner. For example, the spiral tip in plate Ia exhibits complex motions (meandering [14]), but at the resolution of the figure it is not possible to obtain quantitative results on this motion. Plate Ib shows the same situation as Ia except that the number of grid points had been increased to $N = 121$ and the box size has been decreased to $L = 25$ so that the motion of the spiral tip is easily discernible. The white curve shows the path of the spiral tip

over the two wave rotations leading to the state shown. The spiral tip is defined as the intersection of the two contours $u = 1/2$ and $f(u = 1/2, v) = 0$, where f is given in (2). Increasing the resolution further produces only small quantitative changes in the wave dynamics[#6].

Shown in plates Ic and Id are contours for the excitation and recovery variables for a pair of co-rotating spiral waves. The parameter values are the same as in Ia except that $N = 241$. This gives $h = 0.167$ and $\Delta t / \epsilon \simeq 1$, corresponding to a well resolved simulation. The point we wish to make here is that even at fine resolutions our algorithm is very efficient. Throughout the blue region in Ic the system is within the boundary layer illustrated in fig. 1. Hence throughout this region (more than 70% of the domain), our algorithm requires just one conditional evaluation and two floating-point multiplications per grid point per time step. The execution time on the aforementioned Silicon Graphics workstation for one wave rotation is only 80 s at this resolution.

Plates Ie and If show representative simulations in large boxes containing many spiral waves. The parameter values for Ie are the same as for Ic except that $a = 0.4$, $1/\epsilon = 150$, $L = 100$ and $\delta = 10^{-3}$. The waves shown were obtained by breaking the waves of Ic many times. (We have found that wave breaking is easily accomplished by transposing the right and left halves of the domain.) The parameter values for If are the same as for Ic except that $a = 0.5$, $L = 200$, $N =$

[#5]Grid anisotropy can be reduced by using a nine-point Laplacian formula and operator splitting as discussed in ref. [12].

[#6]The path of the spiral tip provides a very stringent diagnostic of numerical solutions. For completely converged, anisotropic results the resolution must be increased beyond that of plate Ib.

◄ Plate I. Representative results from model simulations. Parameter values are given in the text. (a) and (b) show the v-field for a single spiral wave at two different resolutions. Colors range from red at the minimum value of v to blue at the maximum value of v. The white curve in (b) is the path of the spiral tip during two wave rotations. (c) and (d) show the u- and v-fields, respectively, for a pair of co-rotating spiral waves. Those points in (c) which are within the boundary layer of fig. 1 are shown in blue, points outside the boundary layer, but to the left of the threshold are green, points to the right of the threshold are yellow. Colors in (d) range from white at the minimum value of v to read for the maximum value of v. (e) and (f) show two different simulations in large domains. In (e) the u-field is shown with the same color map as (c). In (f) the v-field is shown with the same color map as (a).

401, and $\delta = 10^{-3}$. The initial condition for If was a random (u, v) field. Both Ie and If are coarse-resolution simulations and are quite fast: the execution time per wave rotation is about 25 s in the case of Ie and about 55 s in the case of If. These low-resolution simulations illustrate how results can be obtained for large domains at relatively high speed using our model. This, in turn, will make the future investigation of such large domains practical.

4. Discussion and conclusion

A few comments on the applicability of our model are in order. The model, specifically the local dynamics (2), has been chosen both for simplicity and for efficient numerical implementation. While the model is not derived from any particular excitable system, it is based on widely recognized characteristics of excitability and the model dynamics should be generally representative of excitable media. There are only three model parameters, a, b, and ϵ, by which the properties of the medium can be adjusted, thus making a complete classification of parameter space a possibility. By the same token, certain details of any particular system might elude our model, for there undoubtedly is not enough freedom with only three parameters to accurately match the detailed characteristics of any specific medium. For this it will be necessary to change the form of the local kinetics.

The numerical parameters δ, Δt, and h enter the model parameter space only in so far as results might vary somewhat with these parameters if they are not sufficiently small for results to be converged to the solution of the underlying PDE. When using low spatio-temporal resolution to explore parameter space, one hopes that the same structure (bifurcation points, etc.) exists at low resolutions as at high resolutions and that only quantitative variation arises as the resolution is varied. This must be verified in practice and this is easily addressed using the proposed model.

It should be kept in mind that in PDE (1) only the excitation variable diffuses. This is appropriate for simulating systems such as neuro-muscular tissue [1–3, 15] and catalytic surfaces [16, 17] in which the recovery variable is immobile. However, caution should be observed in using (1) to simulate systems, such as chemical reactions, in which all species diffuse with approximately the same diffusion coefficient. Our motivation for leaving v-diffusion out of the model is that the simulations are considerably faster with only the excitation variable diffusing. The hope is that under many circumstances neglecting diffusion of the recovery variable does not result in much quantitative error when simulating chemical systems such as the Belousov–Zhabotinsky reaction. In any case, it is a simple matter to add v-diffusion to the simulations[#7].

In conclusion, we have presented an efficient algorithm for simulating waves in excitable media and have shown a variety of results from such simulations. We have focused on the ability of our model to provide both high-speed qualitative results and accurate simulations to the underlying system of reaction–diffusion equations. The numerical scheme is based on a model which is simple and yet contains the essential features of a broad class of excitable systems. The algorithm itself is easily implemented and requires less than 50 lines of computer code for the time-stepping subroutine. Thus, the model presented here offers much in terms of simplicity, speed, and wide applicability.

Acknowledgements

I wish to thank M. Kness, L.S. Tuckerman, and G. Zanetti for helpful discussions, and A. Konstantinov for assistance with color photography. This work has been supported by DARPA grant number N00014-86-K-0759.

[#7]We note that the character of the transition from simple to compound rotation can be affected critically by diffusion of v, and probably should be included when modeling this transition in chemical systems.

Appendix

The following is a complete subroutine, in the language C, for taking one time step of the model.

```c
float u[N+1][N+1], v[N+1][N+1], lap[2][N+2][N+2] ;

void step()
{
int i, j ;
float u_th ;

/* interchange k and kprm */
ktmp = kprm ;
kprm = k ;
k = ktmp ;

/* main loop */
for( i=1; i<=N; i++ )
  for( j=1; j<=N; j++ )
    {
    if ( u[i][j] < DELTA )
      {
        u[i][j] = D * lap[k][i][j] ;
        v[i][j] = one_m_dt * v[i][j] ;
      }
    else
      {
        u_th = one_o_a * v[i][j] + b_o_a ;
        v[i][j] = v[i][j] + dt * ( u[i][j] - v[i][j] ) ;
        u[i][j] = F( u[i][j], u_th ) + D * lap[k][i][j] ;
        lap[kprm][i][j] = lap[kprm][i][j] - 4.* u[i][j] ;
        lap[kprm][i+1][j] = lap[kprm][i+1][j] + u[i][j] ;
        lap[kprm][i-1][j] = lap[kprm][i-1][j] + u[i][j] ;
        lap[kprm][i][j+1] = lap[kprm][i][j+1] + u[i][j] ;
        lap[kprm][i][j-1] = lap[kprm][i][j-1] + u[i][j] ;
      }
    lap[k][i][j] = 0. ;
    }

/* impose no-flux boundary conditions */
for( i=1; i<=N; i++ )
  {
  lap[kprm][i][1] = lap[kprm][i][1] + u[i][2] ;
  lap[kprm][1][i] = lap[kprm][1][i] + u[2][i] ;
  lap[kprm][i][N] = lap[kprm][i][N] + u[i][N-1] ;
  lap[kprm][N][i] = lap[kprm][N][i] + u[N-1][i] ;
  }
}
```

where the parameters have the following meanings: $dt = \Delta t$, $one\_m\_dt = 1 - \Delta t$, $one\_o\_a = 1/a$, $b\_o\_a = b/a$.

Note that while the grid indices run from 1 to N, the second two subscripts on the variable lap run from 0 to N + 1. This over-dimensioning of the Laplacian is necessary in active Laplacian evaluation to keep the subscripts from going out of bounds at the domain boundaries. For simplicity, the variables u and v are also over-dimensioned (because C uses zero-based indexing).

The function F is only used symbolically in the preceding subroutine. We suggest in-line coding of the function in the following way:

```
#if EXPLICIT   /* explicit form for F  */

    u[i][j] = u[i][j] + dt_o_eps * u[i][j] * (1.0-u[i][j]) * (u[i][j] - u_th)
              + D * lap[k][i][j];

#else          /* implicit form for F  */

    if( u[i][j] < u_th )
      u[i][j] = u[i][j] / (1.- dt_o_eps * (1.0-u[i][j]) * (u[i][j] - u_th) )
              + D * lap[k][i][j];
    else
      {
        temp = dt_o_eps * u[i][j] * (u[i][j] - u_th) ;
        u[i][j] = (u[i][j] + temp) / (1. + temp ) + D * lap[k][i][j];
      }

#endif
```

where $dt\_o\_eps = \Delta t/\epsilon$. The explicit form should be used for $dt\_o\_eps \leq 1$ because it is more efficient than the implicit form. The implicit form should be used for larger time steps.

References

[1] A.T. Winfree, When Time Breaks Down (Princeton Univ. Press, Princeton, 1987).

[2] V.S. Zykov, Modelling of Wave Processes in Excitable Media (Manchester Univ. Press, Manchester, 1988).

[3] J.J. Tyson and J.P. Keener, Physica D 32 (1988) 327.

[4] N. Wiener and A. Rosenblueth, Arch. Inst. Cardiol. Mex. 16 (1946) 205.

[5] V.I. Krinsky, Biophysics (USSR) 11 (1966) 776.

[6] J.M. Greenberg and S.P. Hastings, SIAM J. Appl. Math. 34 (1978) 515.

[7] V.S. Zykov and A.S. Mikhailov, Sov. Phys. Dokl. 31 (1986) 51.

[8] M. Gerhardt and H. Schuster, Physica D 36 (1989) 209.

[9] M. Gerhardt, H. Schuster and J.J. Tyson, Science 247 (1990) 1563; Physica D 46 (1990) 392, 416.

[10] M. Markus and B. Hess, Nature 347 (1990) 56.

[11] M.-N. Chee, R. Kapral and S.G. Whittington, J. Chem. Phys. 92 (1990) 7315, and references therein;
D. Barkley, in: Nonlinear Structures in Physical Systems, eds. L. Lam and H.C. Morris (Springer, New York, 1990) p. 192.

[12] D. Barkley, M. Kness and L.S. Tuckerman, Phys. Rev. A 42 (1990) 2489.

[13] G.D. Smith, Numerical Solution of Partial Differential Equations: Finite Difference Methods (Clarendon Press, Oxford, 1985).

[14] A.T. Winfree, Science 181 (1973) 937.

[15] A.T. Winfree, J. Theor. Biol. 138 (1989) 353.

[16] J.R. Brown, G.A. D'Netto and R.A. Schmitz, in: Temporal Order, eds. L. Rensing and N.I. Jaeger (Springer, Berlin, 1985) p. 86.

[17] J. Maselko and K. Showalter, Nature 339 (1989) 609; Physica D 49 (1991) 21–32, these Proceedings.

Physica D 49 (1991) 71–74
North-Holland

Kinematics of spiral waves on nonuniformly curved surfaces

V.A. Davydov

Department of Physics, Moscow Institute of Radio Engineering, Electronics and Automation, 117454 Moscow, USSR

and

V.S. Zykov

Institute of Control Sciences, Moscow, USSR

The evolution of spiral waves in a two-dimensional excitable medium with nonuniform curvature is considered. Using an analytical approach based on approximate kinematical relations, we find that spiral waves drift at a rate proportional to the gradients of the Gaussian curvature of the surface. The theoretical predictions agree well with numerical simulations of the full reaction–diffusion equations.

1. Introduction

Excitable media are usually described by a system of nonlinear partial differential equations of the reaction–diffusion type:

$$u = f_1(u, v) + D_u \Delta u,$$
$$v = f_2(u, v) + D_v \Delta v. \qquad (1)$$

For chemical systems u and v have the meaning of concentrations of certain reagents; D_u and D_v are the diffusion coefficients; $f_1(u, v)$ and $f_2(u, v)$ are nonlinear functions.

One of the most important elementary autowave patterns in two-dimensional excitable media is a rotating spiral wave [1–4]. Spiral waves are created from the breaks of an autowave front. In the stationary regime such a break (or a free end) rotates with constant angular velocity along a circle, which gives the boundary of the core of a spiral wave. Inside the core the medium is in the rest state.

The evolution of spiral waves on a plane has been studied in detail. In this paper we will describe some new effects related to spiral wave circulation on nonuniformly curved surfaces. Particularly we show that spiral waves should drift on a nonuniformly curved surface, and we determine both the speed and the direction of such a drift.

2. Spiral waves on a sphere

The kinematical approach [5–10] is a highly effective method for investigating the evolution of spiral waves. In the framework of this approach it is assumed that an autowave is completely defined by the oriented curve of its front. The front is specified by $K = K(l, t)$, which gives the curvature in terms of the length of arc l (measured from the free end) and the time t. Each point of the front moves in the normal direction with velocity V. In general this velocity depends both on the curvature of the front and on the time period T of the wave train. Below we restrict our consideration to situations when T is larger than the refractoriness of the excitable medium. In this case the velocity V depends only on the curvature of the front and, for small curvatures, the linear dependence $V(K) = V_0 - DK$ is valid. At the free end the front might sprout or contract in the tangential direction with the velocity C. This value is not a constant (as it was supposed in ref. [11]) but depends on the curvature K_0 of the front curve at the free end: $(C = \gamma(K_{br} - K_0)$. Hence in the framework of the kinematical approach any excitable medium is characterized by a small number of phenomenological parameters: the velocity of the flat front V_0, the coefficient D in the dependence $V(K)$, and the parameters γ and K_{br} in the dependence $C(K_0)$.

An equation for $K(l, t)$ can be easily derived [5, 7] from the above assumptions. In particular, evolution of the autowave front on a sphere obeys the following equation [6]:

$$\frac{\partial K}{\partial t} + \left(\int_0^l KV(K)\,dl + C \right) \frac{\partial K}{\partial l}$$

$$+ K^2 V(K) + \frac{\partial^2 V(K)}{\partial l^2} = -K_G V. \tag{2}$$

Here K_G is the Gaussian curvature of the sphere; it can be expressed in terms of the sphere radius as $K_G = 1/R_0^2$.

The stationary solution of eq. (2) on a spherical surface with curvature $K_G = K_0$ describes a spiral wave rotating around some fixed core (see fig. 1). The angular velocity of the spiral wave on a sphere is higher than that of one on a plane [10]:

$$\omega = \omega_0 \left(1 + \frac{1}{2\xi^2} \frac{V}{DK_{br}} \frac{K_0}{K_{br}^2} \right), \tag{3}$$

where ω_0 is the angular velocity of the spiral wave on the plane and $\xi = 0.685$.

3. Spiral waves on a nonuniformly curved surface

Suppose for a moment that the radius of the sphere oscillates with time. This would lead to oscillation of the angular velocity of the spiral wave. As a result the center of the spiral wave core would drift on the spherical surface as it does in the case of the spiral wave resonance on a plane [7, 8].

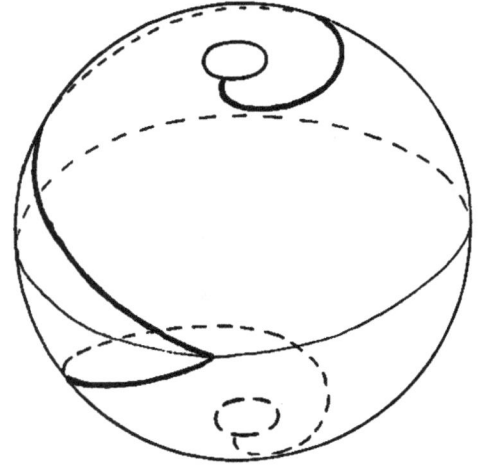

Fig. 1. Steadily rotating spiral wave on a spherical surface.

Now let us investigate evolution of a spiral wave on a static spherical surface which is slightly deformed, so that its Gaussian curvature is a certain function of the polar angle θ. Then, the spiral wave would drift on this surface. Indeed, in this case the free end point passes regions with different values of the Gaussian curvature. Hence, the momentary angular velocity would oscillate with time and this would lead to drift of the center of a spiral wave core as well as in the case of an oscillating sphere.

The description of spiral wave drift is greatly simplified if the quasistationary approximation is satisfied, i.e., if $\gamma/D \ll (V_0/DK_{br})^{1/2}$. This condition is not very restrictive and holds in a wide interval of parameters of excitable media. Within the quasistationary approximation the motion of the free end point of the spiral wave is completely determined by a system of four ordinary differential equations (see refs. [7–9]). This system specifies the position of the free end, the orientation of the front, and its curvature at this point. The analysis of various quasistationary effects was described in detail in refs. [7–10].

The following equations were derived [10] to describe the drift of the spiral waves on a nonuni-

formly deformed spherical surface:

$$\dot{\theta}_0 = 0,$$

$$\dot{\phi}_0 = \frac{V_0 r_0}{4\sin\theta_0} \frac{dK_G}{d\theta}\bigg|_{\theta=\theta_0} \left(1 - \frac{V_0}{\xi^2 D K_{br}} \frac{K_0}{K_{br}^2}\right), \quad (4)$$

where θ_0 and ϕ_0 are the polar and azimuthal angles that determine the position of the spiral wave center on the spherical surface and $r_0 = V_0/\omega$ is the core radius. Eqs. (4) were derived under the assumptions that $\theta_0 - \pi/2 \ll 1$ and $r_0 \ll (K_0)^{-1/2}$.

Examination of eqs. (4) allows one to formulate [10] the following laws of spiral wave drift on nonuniformly curved surfaces. First, the drift velocity is proportional to the magnitude of the gradient of the Gaussian curvature. Second, the motion takes place in the direction which is orthogonal to the gradient. Note that the term $V_0 K_0/\xi^2 D K_{br}^3$ in eqs. (4) is small with respect to unity, since it is of order $r_0^2 K_0$. Therefore, the sign of the angular velocity $\dot{\phi}$ is determined by

the sign of $dK_G/d\theta$. For example, on a surface of a stretched ellipsoid a spiral wave would drift with velocity $\dot{\phi} < 0$ if it rotates in the counter clockwise direction in the northern hemisphere.

Eqs. (4) show that the drift of spiral waves does not occur on nonuniformly curved surfaces if their Gaussian curvature is constant. A cone and a cylinder with an arbitrary form of their base are two examples of such surfaces. Thus, a necessary condition for drift is nonuniformity of the Gaussian curvature of the surface.

The effect of the spiral wave drift on nonuniformly curved surfaces was also investigated by us in computer simulations using model (1) with

$$f_1(u,v) = f(u) - v,$$

$$f(u) = -k_1 u, \qquad u < \sigma,$$

$$= k_f(u - p), \quad \sigma \le u \le 1 - \sigma,$$

$$= k_r(1 - u), \quad 1 - \sigma < u,$$

$$f_2(u,v) = \varepsilon_k(k_G u - v),$$

$$\varepsilon_k = \varepsilon, \qquad k_G u - v > 0,$$

$$= \varepsilon k_\varepsilon, \quad k_G u - v < 0,$$

where $k_f = 1.7$, $k_G = 2$, $p = 0.1$, $\varepsilon = 0.15$, $k_\varepsilon = 6$, $\sigma = 0.01$, and $D_u = D_v = 1$.

Computations for the case of stretched ellipsoid confirm the predictions of the theory. The drift occurs only along a parallel of the ellipsoid in the direction, determined by eqs. (4) (see fig. 2). Moreover, an increase of drift velocity is observed if the gradient of the Gaussian curvature increases.

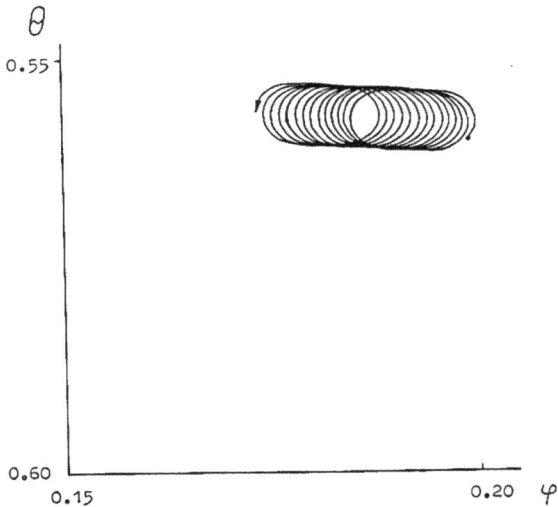

Fig. 2. Trajectory of the free end point of a spiral wave on the surface of the stretched ellipsoid with semiaxes $a = 2$ and $b = 2.9$.

4. Discussion

There now exists a good opportunity to observe the effect of the spiral wave drift. This is provided by the modifications of the Belousov–Zhabotinsky reaction using an immobilized catalyst. In this case autowave propagation takes place not in the bulk of the solution but only within a thin layer near the bottom surface. In the experiment [12] the spiral waves were observed on a spherical surface. However, the parameters of the excitable medium in this case apparently fell in the region of nonstationary circulation, and one could observe "meandering" [4, 13, 14] of the free end rather than circular motion. In addition, the radius of the spherical surface was very small with respect to the spiral wave core.

For observation of the spiral wave drift it is sufficient to put the immobilized catalyst on a nonuniformly curved surface. The estimate of the drift velocity shows [10] that after ten rotations of a spiral wave on a stretched ellipsoid with the semiaxes $a = 2$ mm and $b = 3$ mm the total displacement will be approximately equal to the radius of the spiral wave core.

With a curved surface of a given shape, it is possible to control efficiently spiral waves, to direct their motion, and to induce annihilation of spiral waves with opposite rotation directions. It is very important that in order to observe this, there is no need for heterogeneity of an excitable medium.

References

[1] G. Nicolis and I. Prigogine, Self-Organization in Non-Equilibrium Systems (Wiley, New York, 1977).

[2] L.S. Polak and A.S. Mikhailov, Self-Organization in Non-Equilibrium Physico-Chemical Systems (Nauka, Moscow, 1983).

[3] V.A. Vasiljev, Yu.M. Romanovsky, D.S. Chernavsky and V.G. Yakhno, Autowave Processes in Kinetic Systems (VEB Deutsche Verlag der Wissenschaft, Berlin, 1987).

[4] A.T. Winfree, When Time Breaks Down (Princeton Univ. Press, Princeton, 1987).

[5] V.S. Zykov, Simulation of Wave Processes in Excitable Media (Manchester Univ. Press, Manchester, 1988).

[6] P.K. Brazhnik, V.A. Davydov and A.S. Mikhailov, Teor. Mat. Fiz. 74 (1988) 440–447.

[7] V.A. Davydov, A.S. Mikhailov and V.S. Zykov, in: Nonlinear Waves in Active Media, ed. J. Engelbrecht (Springer, Berlin, 1989) pp. 38–51.

[8] V.A. Davydov, V.S. Zykov, A.S. Mikhailov and P.K. Brazhnik, Izv. VUZ Radiofizika 31 (1988) 574–582.

[9] P.V. Brazhnik, V.A. Davydov, V.S. Zykov and A.S. Mikailov, Zh. Eksp. Teor. Fiz. 93 (1987) 1725–1736.

[10] A.Yu. Abramychev, V.A. Davydov and V.S. Zykov, Zh. Eksp. Theor. Fiz. 97 (1990) 1188–1197.

[11] E. Meron and P. Pelce, Phys. Rev. Lett. 60 (1988) 1880–1883.

[12] J. Maselko and K. Showalter, Nature 339 (1989) 609–611.

[13] V.S. Zykov, Biofizika 31 (1986) 862–865.

[14] V.S. Zykov and O.L. Morozova, Nonlinear Biology 1 (1990), in press.

Physica D 49 (1991) 75–81
North-Holland

Stability of vortex rotation in an excitable cellular medium

V.G. Fast and I.R. Efimov

Institute of Biological Physics, The USSR Academy of Sciences, 142292, Pushchino, Moscow Region, USSR

Stability of a vortex was studied in a modified cellular automaton model. The model includes dependence of wave velocity on excitation interval and front curvature. Two kinds of vortex instability are observed, as found also in models based on partial differential equations: meandering of the vortex and vortex drift in a heterogeneous medium.

1. Introduction

Vortices of excitation are commonly observed in various excitable media. Perhaps the most important example is the heart, where the vortices form the basis of some dangerous arrhythmias [1]. One of the important questions is the stability of vortices in the heart. Vortices of this type exist in tissue without an unexcitable obstacle; therefore, they can be unstable in the sense that they may change their localization from one cycle to another. Actually, as experiments on the heart tissue show, sometimes vortices are stable and sometimes they are unstable.

In general, two mechanisms of vortex instability in various two-dimensional excitable media are known: meandering of the wave tip and drift of the vortex in inhomogeneous media.

Stability of vortices was most extensively studied in mathematical models of an excitable medium based on partial differential equations (PDE) of the reaction–diffusion type. Particularly, in the FitzHugh–Nagumo model [2, 3] and Oregonator model [4] transition from the stable circular to meandering (cycloidal) rotation of the vortex has been studied. A similar phenomenon was also observed experimentally in the Belousov–Zhabotinsky reaction [4, 5]. Drift of the vortices was studied numerically in the FitzHugh–Nagumo model [6–8].

With respect to the heart muscle it would be desirable to study the case when the period of the vortex is close to the absolute refractory period (the smallest possible interval between two waves). In a PDE model this case corresponds to very stiff equations; hence simulations require large amounts of computer time.

For this case models based on cellular automata are more suitable. Such models represent a system of elements that can take one of the discrete values representing the quiescent, excited and refractory states. Transitions from one state to another are determined by simple automaton rules. An element becomes excited if it is in quiescent state and has one of its nearest neighbors in excited state. An excited element becomes refractory and a refractory element becomes quiescent in the next time step. This model simulates some but not all essential features of excitation in heart tissue; for example, the dependence of wave velocity on excitation interval inherent to myocardium [9, 10] are not exhibited by this model.

In the present work we used a modified cellular automata model [11] which simulates the dependence of wave velocity on excitation interval and front curvature by including nonlocal excitation connections between elements of the media [12–15] and relative refractory period. It results in the appearance of two kinds of vortex instability: meandering of the wave tip and drift of vortices due to inhomogeneity of the refractory period.

2. Model

The model consists of a 100 by 100 array of elements. Each element is attributed an integer value A representing either excitable, absolute refractory, relative refractory or quiescent states:

$$A_{k,l} = 0 \qquad\qquad \text{quiescient state,}$$
$$= 1 \qquad\qquad \text{excited state,}$$
$$= 2, \ldots, R_a \qquad \text{absolute refractory state,}$$
$$= R_a + 1, \ldots, R \quad \text{relative refractory state,}$$

where (k, l) is number of the given element, $R\ (= R_a + R_r)$ is the duration of the refractory period, R_a and R_r are respectively the durations of the absolute refractory and relative refractory periods.

The time evolution of the cell is given by the following excitable medium automaton dynamics:

$$A_{k,l}^{t+1} = A_{k,l}^t + 1, \quad \text{if } A_{k,l}^t \in [1, R_a] \quad \text{or} \quad \left(A_{k,l}^t \in (R_a, R) \quad \text{and} \quad S \le \text{Th}(A) \right),$$

$$= 1, \qquad\quad \text{if } \left(A_{k,l}^t = 0 \text{ or } A_{k,l}^t \in (R_a, R] \right) \quad \text{and} \quad S > \text{Th}(A),$$

$$= 0, \qquad\quad \text{if } \left(A_{k,l}^t = 0 \text{ or } A_{k,l}^t = R \right) \quad \text{and} \quad S \le \text{Th}(A),$$

where $A_{k,l}^t$ and $A_{k,l}^{t+1}$ are states of the (k, l) cell at moments t and $t + 1$; S is an excitation sum (see below); and $\text{Th}(A)$ is the excitation threshold, which depends on A.

These rules state that after having been excited, the cell evolves through a series of absolute refractory states (duration R_a) and through a series of relative refractory states (duration R_r) before returning to the quiescent state. A cell can be excited only if it is in the quiescent state or in one of the relative refractory states and the excitation strength defined by the sum S exceeds a threshold Th. S is given by

$$S = \sum W(k,l) E(k,l), \quad k = 1, \ldots, K, \quad l = 1, \ldots, K,$$

where $E(k,l)$ describes the states of the cells in the $K \times K$ neighborhood:

$$E(k,l) = 1, \quad \text{if } A_{k,l} = 1,$$
$$= 0, \quad \text{if } A_{k,l} \neq 1,$$

and $W(k,l)$ describes the weights of interaction between a given cell and cells in its neighborhood. In natural excitable media the strength of interaction depends on the distance between the elements: the influence of more distant elements is less than proximal. Therefore, weights in an array W decrease with the distance from the central element. For simplicity we assume a linear dependence with rotational symmetry (isotropic medium). The following weight array was used in our simulations:

$$
W(k,l) = \begin{pmatrix}
0 & 0 & 0 & 0 & 1 & 1 & 1 & 0 & 0 & 0 & 0 \\
0 & 0 & 1 & 2 & 3 & 3 & 3 & 2 & 1 & 0 & 0 \\
0 & 1 & 3 & 4 & 5 & 5 & 5 & 4 & 3 & 1 & 0 \\
0 & 2 & 4 & 6 & 7 & 8 & 7 & 6 & 4 & 2 & 0 \\
1 & 3 & 5 & 7 & 9 & 10 & 9 & 7 & 5 & 3 & 1 \\
1 & 3 & 5 & 8 & 10 & 0 & 10 & 8 & 5 & 3 & 1 \\
1 & 3 & 5 & 7 & 9 & 10 & 9 & 7 & 5 & 3 & 1 \\
0 & 2 & 4 & 6 & 7 & 8 & 7 & 6 & 4 & 2 & 0 \\
0 & 1 & 3 & 4 & 5 & 5 & 5 & 4 & 3 & 1 & 0 \\
0 & 0 & 1 & 2 & 3 & 3 & 3 & 2 & 1 & 0 & 0 \\
0 & 0 & 0 & 0 & 1 & 1 & 1 & 0 & 0 & 0 & 0
\end{pmatrix}.
$$

The dimension K of W was chosen to ensure the biggest range of velocities in the dispersion curve on the one hand and on the other hand reasonable time of computer simulations.

The threshold of excitation $\mathrm{Th}(A)$ was a descending linear function of $A \in (R_a, R_a + R_r]$. It changed from a value Th_a at the beginning of the relative refractory period to a value $\mathrm{Th}_q < \mathrm{Th}_a$ at the quiescent state. The maximal value for Th_a was 76. This is the biggest threshold at which propagation of waves (velocity 2 cells/time unit) is still possible.

Boundary conditions, similar to impermeable boundary conditions in PDE models, were imposed at the edge of the medium. For this, at each time step cells in the layer (width $(K-1)/2$) adjacent to the border were mirrorly reflected to the additional layer of cells beyond the border.

For the initiation of vortices, a planar wave was excited at one of the borders of the medium. When the wave propagated to the center of the medium, half of the medium was set up in the rest state so that a wave break was created. The wave break then gave rise to a vortex.

The wave tip is defined as an excited point which has three nearest neighbors, one of which is in the excited state, a second is in the absolute refractory state, and a third is in a relative refractory state (or in the excited, absolute refractory and quiescent states).

Simulations were performed on a PC, an IBM-compatible AT.

3. Results

Fig. 1 demonstrates the dependence of wave velocity on excitation interval measured by the double-pulse method. To obtain this relation two planar waves were initiated with interval I, and the velocity V of the second wave was measured at the first moment. The low excitability in the relative refractory state ($10 < I < 15$) provides the lower velocity of the wave. The maximal and minimal velocities of the wave are 5 and 2 cells/time unit, a factor of 2.5 difference, which is close to that measured in heart muscle [9, 10].

Fig. 2a shows the dependence of the vortex period on the excitation threshold in the medium with the smallest possible refractory period ($R_a = 3$, $R_r = 0$). It can be seen that over the range of thresholds $\mathrm{Th}_q \geq 32$ the rotation period is greater than the refractory period of the medium. In this case the excitation front of the vortex is not affected by refractory tail; see fig. 2b. In this range the vortex period depends on excitability (excitation threshold Th_a) only and does not depend on refractoriness. This property is explained by effects of front curvature on velocity, which is well known for PDE models [16] and also observed in cellular automata [11, 12, 14].

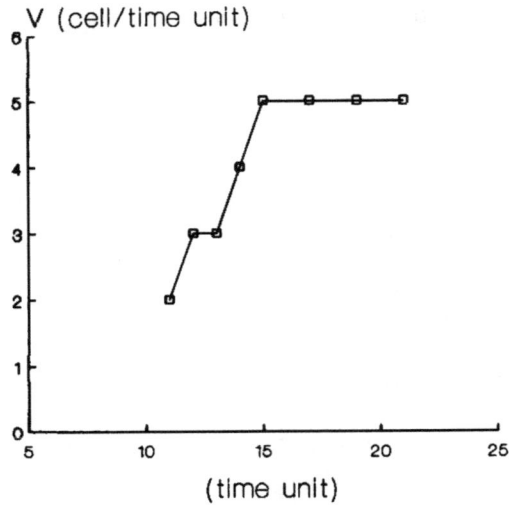

Fig. 1. The dependence of velocity (V) of the planar waves on excitation interval (I). Parameters of the model are $R_a = 10$, $R_r = 5$, $Th_q = 2$, $Th_a = 76$.

We examined the dependence of the vortex stability on parameters of the model (excitation thresholds Th_q and Th_a and refractory periods R_a and R_r) and found that rotation may be either stable or meandering, depending on the parameters.

Four different types of vortices were observed: circular or nearly circular, cycloidal, rotation of "Z"-type, and linear rotation; see fig. 3. The "Z"-type rotation was first observed in our previous work [11]. In this case, the wave tip path represents a combination of two parts – circular and linear ones. The linear part of the core is formed when wave front moves along the absolute refractory tail. The circular part of the core is formed when the wave performs the turn. The path of turning differs from circular because the wave front moves through the media in the relative refractory state in which excitability and

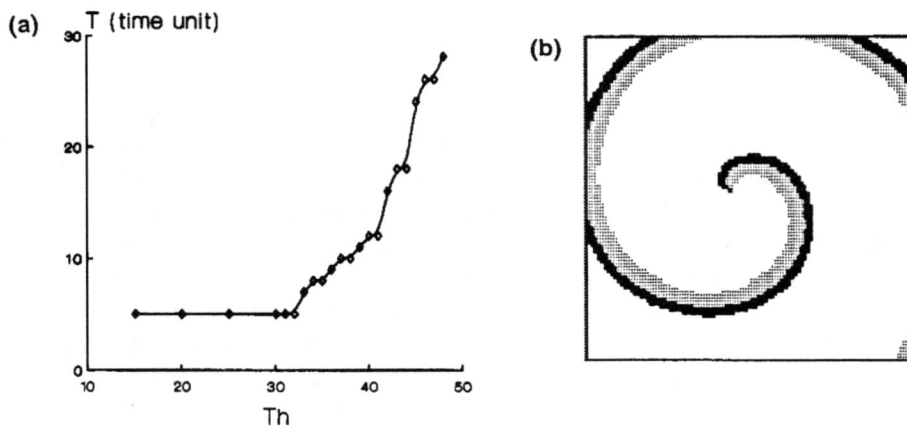

Fig. 2. (a) The dependence of the vortex period (T) on excitation threshold (Th). (b) Shape of the vortex in the medium with smallest possible refractory period. Excited, absolute refractory, and relative refractory states of the medium are indicated by black, grey and white, respectively. Parameters of the model are: $R_a = 3$, $R_r = 0$. Excitation threshold in (b): $Th_q = 44$.

T/R

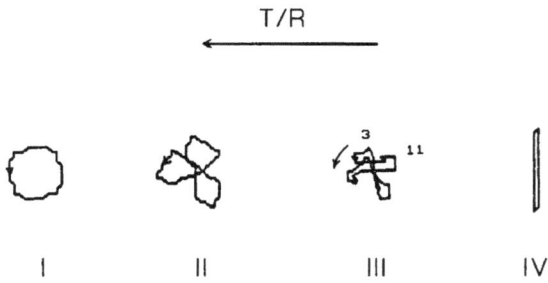

Fig. 3. Types of vortices: (I) circular rotation (period of rotation 28); (II) cycloidal rotation (rotation time 25–26); (III) rotation of the "Z"-type (rotation time 20–21); (IV) linear rotation (period of rotation 16). Parameters are: $R_a = 15$, $R_r = 10$; (I) $Th_q = 48$, $Th_a = 76$; (II) $Th_q = 44$, $Th_a = 72$; (III) $Th_q = 22$, $Th_a = 50$; (IV) $Th_q = 2$, $Th_a = 30$.

hence turning radius are not constant. Subsequently, as rotation proceeds in the case of unstable rotation the linear part of tip path is deformed (not shown).

Circular and linear rotation are stable, while cycloidal motion meanders. Rotation of the Z type can be either stable or meandering, depending on model parameters. Meandering was absent when the duration of the relative refractory period was equal zero (not shown). In this case the wave velocity does not depend on excitation interval.

Transitions from one type of rotation to another were accompanied by the decreasing of the ratio T/R, where T is the period of vortex rotation and R is the refractory period. When rotation is circular, T is greater than the full refractory period ($T > R_a + R_r$). When cycloidal rotation is observed, T falls into the end of relative refractory period ($T \cong R_a + R_r$). For Z-type rotation, T becomes closer to the absolute refractory period ($R_a + R_r > T > R_a$). Finally, when rotation along a line is observed, $T \cong R_a$.

Fig. 4 shows the dynamics of the vortices of the different types shown in fig. 3 for a medium with a step inhomogeneity of refractory period. It can be seen that the inhomogeneity results in a drift of the

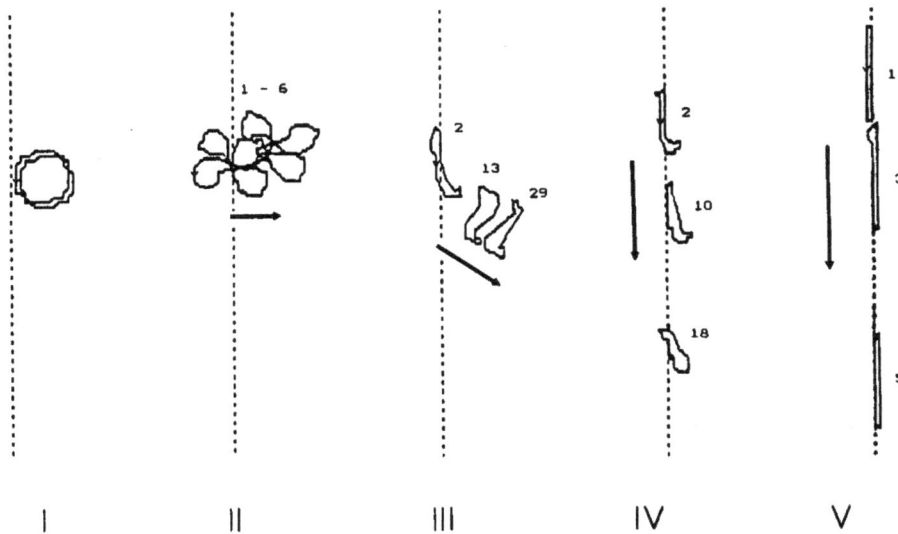

Fig. 4. Dynamics of vortices of different types in a medium with a step inhomogeneity of refractory period. $R_a = 15$, $R_r = 10$ in the left part of the medium and $R_a = 17$, $R_q = 10$ in the right part of the medium. Other parameters are the same as in fig. 3.

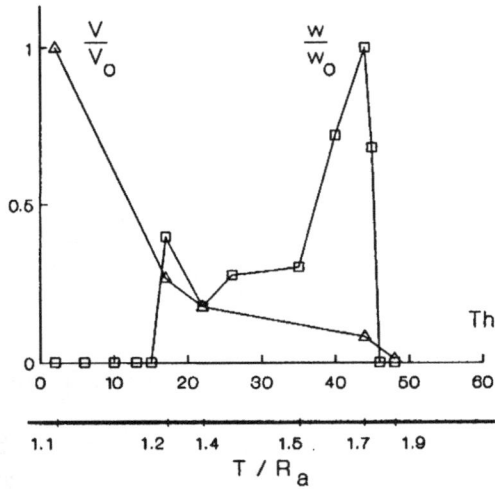

Fig. 5. Dependence of drift velocity V and the velocity of meandering w on excitation threshold Th. V_0 and w_0 are the maximal values: $V_0 = 0.9$ cells/time unit and $w_0 = 2.5°$/time unit. At the bottom the ratio between vortex period T and absolute refractory period R_a is shown.

vortices. Generally the drift has two components: longitudinal and transversal to the gradient of refractory period. The longitudinal drift is directed to the region with longer refractory period. A marked longitudinal drift takes place when vortex rotation is cycloidal or of the Z type (cases II and III). Due to this drift the vortex is shifted entirely to the region with longer refractoriness and drift ceases.

Transversal drift is more noticeable when vortices are of the Z type or linear (cases IV and V). As a consequence of the transversal drift, the vortex reaches the edge of the medium and disappears.

Fig. 5 compares two mechanisms of vortex instability: meandering and drift. V is a drift velocity due to inhomogeneity in the refractory period included both components of the drift: longitudinal and transversal. Because the longitudinal component disappears with time, we measured V at the beginning of vortex rotation when both components were present. V was measured as the distance through which the vortex shifted during the first five cycles of rotation, divided by the time. w is an average angular velocity of the vortex core rotation (due to meandering) in homogeneous medium. It was measured as the angle of turn of the vortex core as a whole through several cycles. It can be seen that at high excitabilities ($\text{Th}_q < 17$), when rotation is linear or of Z type, meandering is absent while drift is strong.

4. Discussion

To apply the results obtained in our model to heart tissue let us consider the dimensionless parameter which can be measured both in the model and in the heart tissue, namely, the ratio between vortex period and absolutely refractory period T/R_a. For the heart muscle this ratio is believed to be usually equal or slightly more than 1 [17]. In the model this corresponds to the case of high excitability when vortices have linear cores or cores of the Z type (see figs. 3 and 5). For these vortices there is a range of parameters ($\text{Th}_q < 17$, $T/R_a < 1.2$) for which meandering of the wave tip is absent. It suggests that in myocardium the instability of the vortex due to meandering can be negligible and in homogeneous myocardium rotation can be rigid. This can explain the stability of vortex rotation in the experiments of

Frazier et al. [17] and Davidenko et al. [18], in which vortices were induced in relatively homogeneous myocardium.

On the other hand, if the vortices be induced due to inhomogeneity of the refractory period in the myocardium, the model predicts their instability as a result of the drift along the border of inhomogeneity. This drift has been experimentally observed in heart muscle [19].

Acknowledgements

We thank Dr. V.I. Krinsky and anonymous referees for their comments.

References

[1] R. Lazzara and B.J. Scherlag, Generation of arrhythmias in myocardial ischemia and infarction, Am. J. Card. 61 (1988) 20A–26A.

[2] V.S. Zykov, Cycloidal circulation of spiral waves in excitable medium, Biofizika 31 (1986) 862–865.

[3] E. Lugosi, Analysis of meandering in Zykov kinetics, Physica D 40 (1989) 331–337.

[4] W. Jahnke, W.E. Skaggs and A.T. Winfree, Chemical vortex dynamics in the Belousov–Zhabotinskii reaction and in the two variable Oregonator model, J. Phys. Chem. 93 (1989) 740–749.

[5] G. Skinner and H.L. Swinney, Periodic to quasiperiodic transition of chemical spiral rotation, Physica D 48 (1991), press.

[6] A.M. Pertsov and E.A. Ermakova, Mechanism of the drift of a spiral wave in an inhomogeneous medium, Biofizika 33 (1988) 338–342.

[7] G.R. Ivanitsky, V.I. Krinsky, A.V. Panfilov and M.A. Tsyganov, Two regimes of the drift of the reverberators in nonhomogeneous active media, Biofizika 34 (1989) 297–299.

[8] A.V. Panfilov and B. Vasiev, Physica D 49 (1991) 107–113, these Proceedings.

[9] M.S. Spach, J.M. Kootsey and J.D. Sloan, Active modulation of electrical coupling between cardiac cells of the dog: A mechanism for transient and steady-state variations in conduction velocity, Circ. Res. 51 (1982) 347–362.

[10] J.L.R.M. Smeets, M.A. Allessie, W.J.E.P. Lammers, F.I.M. Bonke and J. Hollen, The wavelength of the cardiac impulse and reentrant arrhythmias in isolated rabbit atrium, Circ. Res. 58 (1986) 96–108.

[11] V.G. Fast, I.R. Efimov and V.I. Krinsky, Transition from circular to linear rotation of a vortex in an excitable cellular medium, Phys. Lett. A 151 (1990) 157.

[12] V.S. Zykov and A.S. Mikhailov, Rotating spiral waves in simple model of excitable medium, Dokl. AN SSSR 286 (1986) 341–344.

[13] M. Gerhardt, H. Schuster and J.J. Tyson, A cellular automaton model of excitable media including curvature and dispersion, Science 247 (1990) 1563–1566.

[14] M. Gerhardt, H. Schuster and J.J. Tyson, A cellular automaton model of excitable media II. Curvature, dispersion, rotating waves and meandering waves, Physica D 46 (1990) 392–415.

[15] M. Markus and B. Hess, in: Nonlinear Wave Processes in Excitable Media, Nato ASI Series eds. A.V. Holden, M. Markus and H.G. Othmer (Plenum Press, New York), in press.

[16] J.J. Tyson and J.P. Keener, Singular perturbation theory of traveling waves in excitable media (a review), Physica D 32 (1988) 327–361.

[17] D.F. Frazier, P.D. Wolf, J.M. Wharton, A.S.L. Tang, W.M. Smith and R.E. Ideker, Mechanism for electrical initiation of reentry in normal canine myocardiu, J. Clin. Invest. 83 (1989) 1039.

[18] J.M. Davidenko, P. Kent and J. Jalife, Spiral waves in normal isolated ventricular muscle, Physica D 49 (1991) 182–197, these Proceedings.

[19] V.G. Fast and A.M. Pertsov, Drift of vortex in myocardium, Biofizika 35 (1990) 478–482.

Physica D 49 (1991) 82–89
North-Holland

The eikonal equation:
stability of reaction–diffusion waves on a sphere

Jagannathan Gomatam and Derek A. Hodson

Department of Mathematics, Glasgow College, Cowcaddens Road, Glasgow, G4 0BA, Scotland, UK

The eikonal approach to excitable reaction–diffusion systems provides an analytical hold on the phenomenon of wave propagation on curved surfaces and the evolution of toroidal scroll structures in the neighbourhood of organising filaments. The extent to which the eikonal equation contains information on the stability of these wave solutions is important from a theoretical and experimental point of view. We initiated this study in 1987 by demonstrating the stability of an expanding spherical wave. Recent experiments on wave propagation on a sphere encouraged us further to investigate stability of waves on surfaces. We analyse in this paper the geometrical stability of symmetric counter-rotating spiral waves propagating on the unit sphere; we demonstrate that these solutions are stable under small perturbations normal to the wave front, lying on the unit sphere.

1. Introduction

The eikonal approximation to excitable reaction–diffusion equations captures the essential geometrical properties of the kinematics of wave propagation in two- and three-dimensional media [1–3]. The most striking aspect of this theme is the relationship between the normal velocity of the wave surface and its mean curvature [4–8]. The predominance of the spiral and toroidal wave forms in excitable chemical [9–12] and biological [12, 13] systems highlights the importance of mathematical formulations that directly address the geometry of wave motion in three-dimensional space. This is provided by the eikonal approach to reaction–diffusion equations.

The main objective of this paper is to analyse the problem of stability of a class of solutions to the eikonal equation, displayed in section 2 in eq. (8).

In principle, the variational equations are easily formulated:

$$\delta N + \varepsilon \, \delta K = 0, \tag{1}$$

where δN and δK are small perturbations of the normal velocity N and the (twice) mean curvature K of the wave surface. The stability of an expanding spherical wave was discussed in ref. [2] and this already indicates the need for careful geometrical interpretation of results in the light of symmetries enjoyed by the unperturbed solution.

We discuss here the stability of the symmetric counter-rotating spiral waves on the unit sphere obtained in ref. [1]. Chemical waves that spiral from the north pole to the south pole have been observed in recent experiments [14] in which the catalyst ferroin is localised on a spherical surface immersed in the Belousov–Zhabotinsky (BZ) reagent. We believe that it is important to analyse and classify the diversity and stability of wave forms on spherical and other surfaces in the hope that experiments similar to the one reported in ref. [14] might provide a further test of validity of the eikonal approach to higher-dimensional problems.

Section 2 contains a summary of results leading to three-dimensional eikonal equations; the representation of this equation on the unit sphere, the rotating spiral solutions and the variational equation are presented in section 3. The stability of counter-rotating spiral waves and the associated numerical results are discussed in section 4. A summary of results and future outlook are presented in section 5.

2. The eikonal equation for the three-dimensional reaction–diffusion systems

We provide a brief summary of the eikonal representation for the class of reaction–diffusion (RD) systems:

$$\varepsilon \frac{\partial u}{\partial t} = \varepsilon^2 \Delta u + A(u,w), \qquad (2a)$$

$$\frac{\partial w}{\partial t} = B(u,w), \qquad (2b)$$

$$0 < \varepsilon \ll 1, \qquad (2c)$$

where A and B model excitable kinetics. The rate of evolution of u is assumed to be very large compared to w. Generally u is referred to as the fast variable, and w as the recovery variable.

The physical meaning of the choice of the nondiffusing variable w and the attendant analytical tractability of the eikonal equation, i.e. the reduction of the stiff higher-dimensional partial differential equations to numerically manageable ordinary differential equations, is well expounded in the literature [6, 7]. Therefore the system of equations (2) provide a reasonable starting point for the derivation of the eikonal equation in three dimensions [1, 2].

Following refs. [1, 2], we introduce the moving coordinate system

$$r = r(\mu, \eta, \lambda, t), \qquad (3a)$$
$$t = t, \qquad (3b)$$

where the position vector r is a function of curvilinear coordinates represented by a triple orthonormal set of surfaces μ, η and λ:

$$r_\mu \cdot r_\lambda = 0, \quad r_\lambda \cdot r_\eta = 0, \quad r_\eta \cdot r_\mu = 0. \qquad (4a)$$

Here the abbreviation $r_\mu \equiv \partial r / \partial \mu$ is employed. Without loss of generality we let

$$|r_\mu|^2 = 1. \qquad (4b)$$

The scale factors are defined by

$$h_2^2 \equiv r_\eta^2, \quad h_3^2 \equiv r_\lambda^2. \qquad (4c)$$

The use of stretched coordinates $\mu = \varepsilon \xi$ and the restriction $u(r,t) = V(\xi)$ lead to the approximate representation

$$V_{\xi\xi} + (N + \varepsilon K)V_\xi + f(v, w_0) = 0 \qquad (5)$$

to order ε for the fast variable u in eq. (2a). A detailed discussion of the eikonal equation on a plane can be found in refs. [6, 7].

Here the normal velocity N is defined by

$$N = \frac{\dot{r} \cdot (r_\eta \times r_\lambda)}{h_2 h_3}, \quad \dot{r} \equiv \frac{\partial r}{\partial t} \qquad (6)$$

and K, twice the mean curvature of the wave surface $\mu = 0$, by

$$K = -\frac{r_{\eta\eta} \cdot (r_\eta \times r_\lambda)}{h_2^3 h_3} - \frac{r_{\lambda\lambda} \cdot (r_\eta \times r_\lambda)}{h_2 h_3^3} \qquad (7)$$

Comparing eq. (5) with the equation for a one-dimensional travelling front moving with velocity c, leads to the eikonal equation

$$N + \varepsilon K = c. \qquad (8)$$

A few remarks on what had been achieved by a study of (8) and why we embark on a further analysis of the same equations here are in order. A review of (8) in the context of spiral wave propagation on a plane [6, 7] and the motion of the singular filament in space is contained in ref. [15]. The eikonal equation allows the existence of spiral waves on a sphere [1], the analogue of double spirals seen in the Belousov–Zhabotinsky (BZ) reagent [9] in a petri dish. Other wave forms, such as rotating waves on a torus, expanding spherical waves are discussed in refs. [1, 2]. Using physically plausible approximations it is possible to demonstrate that twisted toroidal waves [2, 3] and helical waves are solutions of eq. (8). The fact that both symmetric and asymmetric spiral waves have been observed [14, 16] on spheres coated with the BZ reagent make the study of wave propagation and its stability on curved two-dimensional surfaces worthwhile. A study of the stability of solutions of (8) was initiated in ref. [2], where the stability of an expanding spherical wave was demonstrated. In the next section we formulate the stability problem for the rotating spiral waves on a sphere.

3. Rotating spiral waves on the unit sphere: the variational equations

The aim of this section is to obtain a representation of the eikonal equation (8) on a sphere of radius λ and to formulate the variational equation for the stability problem.

Let

$$r(\mu, \eta, \lambda, t) = \lambda e(\mu, \eta, t), \tag{9}$$

where

$$|e| = 1 \tag{10}$$

and $\{\mu, \eta\}$ provide an orthonormal grid on the sphere. The following relations follow immediately from eq. (9):

$$r_\lambda = e, \quad r_{\lambda\lambda} = 0, \tag{11a}$$

$$r_\eta = \lambda e_\eta, \quad r_{\eta\eta} = \lambda e_{\eta\eta}, \tag{11b}$$

$$\dot{r} = \lambda \dot{e}, \tag{11c}$$

$$r_\mu - \frac{e_\eta \times e}{|e_\eta|} \tag{11d}$$

Substituting (11a)–(11d) in (8) and setting $\lambda = 1$ (the unit sphere), we arrive at

$$\dot{e} \cdot \frac{e_\eta \times e}{|e_\eta|} - \varepsilon \frac{e_{\eta\eta} \cdot (e_\eta \times e)}{|e_\eta|^3} = c. \tag{12}$$

Eq. (12) can now be used to study wave propagation on a solid unit sphere with an excitable surface.

We stipulate the rotational motion

$$\dot{e} = \omega(\hat{k} \times e), \tag{13}$$

where ω is the angular velocity of the rotating wave and $\hat{k}$ is the unit vector pointing in the positive z-direction; this reduces eq. (12) to

$$\omega \frac{\hat{k} \cdot e_\eta}{|e_\eta|} - \varepsilon \frac{e_{\eta\eta} \cdot e_\eta \times e}{|e_\eta|^3} = c. \tag{14}$$

With no loss of generality, we can choose η as the arc length of the curve $\mu = 0$; therefore

$$|e_\eta| = 1 \quad \text{and} \quad e_{\eta\eta} \cdot e = -1. \tag{15}$$

Eqs. (10), (14) and (15) imply the vector differential equations

$$e_{\eta\eta} = -e - fn, \tag{16}$$

where

$$f \equiv \frac{c - \omega \hat{k} \cdot e_\eta}{\varepsilon} \tag{17}$$

$$n \equiv e_\eta \times e. \tag{18}$$

It is easily seen from (13) that

$$e(\eta, t) = R(\omega t) e(\eta, 0),$$

where $e(\eta, t) \equiv$ column vector $(x(\eta, t), y(\eta, t), z(\eta, t))$ and the rotation operator $R(\omega t)$ has the matrix representation

$$R(\omega t) = \begin{pmatrix} \cos \omega t & -\sin \omega t & 0 \\ 0 & 0 & 1 \end{pmatrix}.$$

This implies that eq. (14) is autonomous; therefore it is enough to solve (16) (see ref. [1]) for $t = 0$, with

$$e(0,0) = \hat{k} = -e(l,0), \tag{19a}$$

$$e_\eta(0,0) = \hat{i}, \tag{19b}$$

where l is the total length of the solution curve; boundary conditions (19) imply an eigenvalue problem for ω/ε and c/ε. The full time-depen-

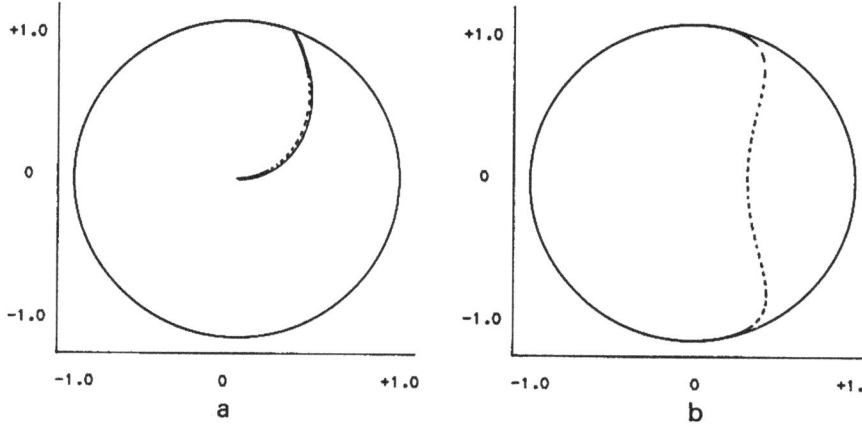

Fig. 1. Rotating spirals on the unit sphere at $t = 0$, $c/\varepsilon = 3.0$, $w/\varepsilon = -4.13$; dotted curve is on the rear. (a) North polar view. (b) Side view.

dent solution could be reconstructed using eq. (13) (see figs. 1 and 2).

The variational equation for stability of solutions of (8) are easily derived, but the actual analysis would have to exploit the geometrical properties of the unperturbed wave surface.

We start with the perturbed solution $F(\eta, t)$ on the unit sphere

$$\dot{F} \cdot \frac{F_\eta \times F}{|F_\eta|} - \varepsilon \frac{F_{\eta\eta} \cdot (F_\eta \times F)}{|F|^3} = c, \qquad (20)$$

$$|F(\mu, \eta, t)| = 1. \qquad (21)$$

Assume that

$$F = e + p, \quad |p|^2 \ll 1, \qquad (22)$$

where e is the unperturbed solution of (16) and p is a small perturbation. Eq. (22) implies that

$$F \cdot p = 0 + |p|^2. \qquad (23)$$

The variational equation for p is

$$\omega\left[\hat{k} \cdot p_\eta - (\hat{k} \cdot e)(e \cdot p_\eta) + 2(\hat{k} \cdot e_\eta)(e_\eta \cdot p_\eta)\right]$$
$$+ \dot{p} \cdot n - \varepsilon(n \cdot p - fe_\eta \cdot p_\eta + p_{\eta\eta} \cdot n)$$
$$= 3ce_\eta \cdot p_\eta. \qquad (24)$$

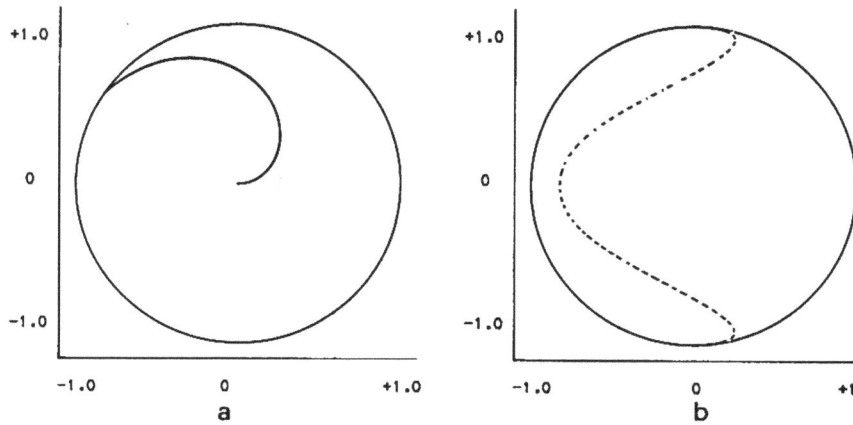

Fig. 2. Rotating spirals on the unit sphere at $t = 0$, $c/\varepsilon = 5.0$, $w/\varepsilon = -8.995$; dotted curve is on the rear. (a) North polar view. (b) Side view.

We expand $p(\eta,t)$ in terms of the orthonormal set $\{e, e_\eta, n\}$ and exploit the relation $e \cdot p = 0$, i.e.,

$$p = \beta(\eta,t)\, e_\eta + \gamma(\eta,t)\, n$$
$$= \beta e_\eta + \gamma n, \qquad (25)$$

suppressing explicitly the dependence on η and t. It can be shown that

$$p_\eta = -\beta e + (\beta_\eta + \gamma f)e_\eta + (\gamma_\eta - \beta f)n, \qquad (26)$$

$$p_{\eta\eta} = -(\beta_\eta + \gamma f)e$$
$$+ \left[(-\beta_\eta + \gamma f)_\eta - \beta f^2 + \gamma_\eta f\right]e_\eta$$
$$+ \left[(\gamma_\eta - \beta f)_\eta - f(\beta_\eta + \gamma f)\right]n, \qquad (27)$$

$$\dot{p} = \omega\left[\beta(k \cdot n) - \gamma(\hat{k} \cdot e_\eta)\right]e + (\beta + \omega\gamma\hat{k} \cdot e)e_\eta$$
$$+ \left[\dot{\gamma} - \omega\beta(\hat{k} \cdot e)\right]n. \qquad (28)$$

Substituting eqs. (26)–(28) in (24), we arrive at

$$\beta\left(\varepsilon f_\eta - \omega f\hat{k} \cdot n - \omega\hat{k} \cdot e\right)$$
$$+ 3\beta_\eta\left[\omega(\hat{k} \cdot e_\eta) - c + \varepsilon f\right]$$
$$+ \dot{\gamma} - \varepsilon\gamma_{\eta\eta} + \omega\hat{k} \cdot n\gamma_\eta + (-\varepsilon - \varepsilon f^2)\gamma = 0. \qquad (29)$$

A simple calculation shows that the coefficients of β and β_η vanish identically, leaving the variational equation for $\gamma(\eta,t)$:

$$\dot{\gamma} - \varepsilon\gamma_{\eta\eta} + \omega(\hat{k} \cdot n)\gamma_\eta + (-\varepsilon - \varepsilon f^2)\gamma = 0. \qquad (30)$$

Note that the rotating solutions $e(\eta,t)$ imply that f and $\hat{k} \cdot n$ are independent of time. This property simplifies the stability analysis to a considerable extent. Let

$$\gamma(\eta,t) = \Gamma(\eta)\exp(-\sigma t). \qquad (31)$$

A further substitution

$$\Gamma(\eta) = G(\eta)\exp\left(\frac{\omega}{2\varepsilon}\int(\hat{k} \cdot n)\,d\eta\right) \qquad (32)$$

reduces (30) to

$$G_{\eta\eta} + \left(\frac{\sigma}{\varepsilon} + 1 + f^2 + \frac{\omega}{2\varepsilon}(\hat{k} \cdot e_\eta)f\right.$$
$$\left. - \frac{\omega^2}{4\varepsilon^2}(\hat{k} \cdot n)^2\right)G = 0. \qquad (33)$$

We first investigate stability for the class of perturbations

$$\beta(\eta,t) \equiv 0, \qquad (34)$$

i.e.

$$p(\eta,t) \equiv \gamma(\eta,t)\, n. \qquad (35)$$

These are perturbations normal to the solution curve on the unit sphere.

The boundary conditions on G are that

$$G(0) = G(l) = 0, \qquad (36)$$

where l is the total arc length of the solution curve from the north pole to the south pole. Therefore (33) and (36) constitute a linear eigenvalue problem for the stability parameter σ introduced in (31).

4. Stability of rotating waves for binormal perturbation: numerical results

Eqs. (33) and (36) can be solved numerically for the solution set $(c/\varepsilon, \omega/\varepsilon, e)$ to obtain the spectrum of eigenvalues σ. A positive set of σ's would indicate stability of the rotating solutions e of (16)–(19). We have examined two sets of $(c/\varepsilon, \omega/\varepsilon, e)$ and found a selection of eigenvalues σ, all of which are positive. With $\sigma = 0$, we were unable to find nontrivial solutions that satisfied the boundary conditions (36). Therefore we conclude that when $\sigma = 0$, only the trivial solution $G(\eta) \equiv 0$ is allowed. The reason for this is that the perturbations p are defined relative to the rotating solutions e through $F = e + p$, and the rotational invariance of the unperturbed solu-

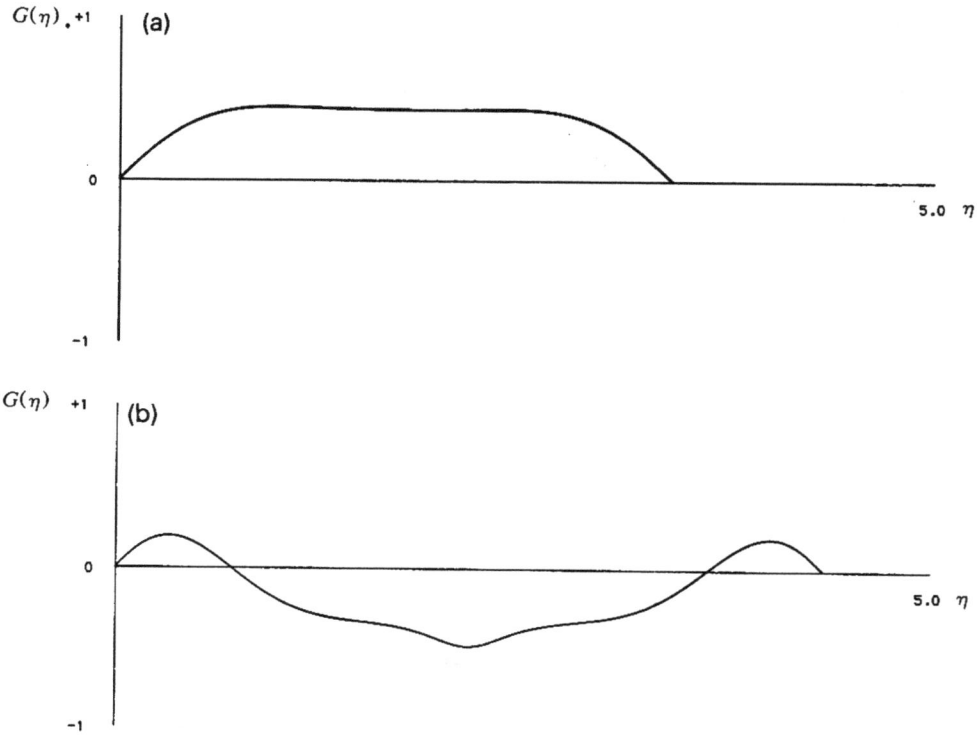

Fig. 3. The graph of $G(\eta)$ versus η for the lowest value of σ (a) with $c/\varepsilon = 3.000$, $w/\varepsilon = -4.13$, $\sigma/\varepsilon = 2.5$; (b) with $c/\varepsilon = 5.000$, $w/\varepsilon = -8.995$, $\sigma/\varepsilon = 8.67$.

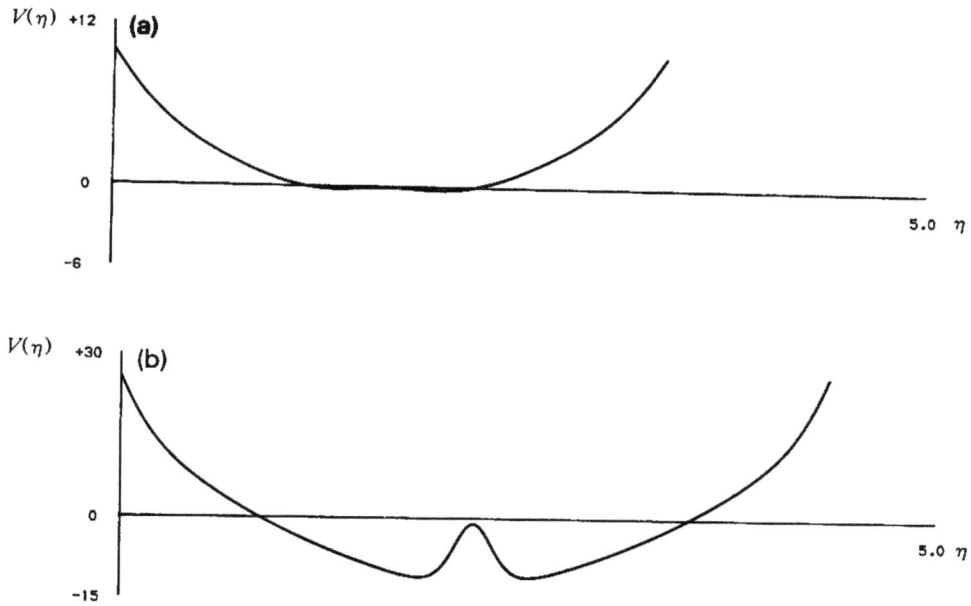

Fig. 4. The graph of $V(\eta)$ versus η (a) with $c/\varepsilon = 3.000$, $\omega/\varepsilon = -4.13$; (b) with $c/\varepsilon = 5.000$, $\omega/\varepsilon = -8.995$.

tions e is encapsulated in the freedom available in the specification of $e_\eta(0, t)$. (In figs. 1 and 2 the choice $e(0, t) = \hat{i}$ is made for computational convenience.) We also found that for selected negative values of σ, $G(\eta)$ become unbounded and hence do not satisfy the boundary conditions (36). The oscillatory character of $G(\eta)$ for positive eigenvalues σ is shown in figs. 3a and 3b. The "potential" curves

$$V(\eta) \equiv 1 + f^2 + \frac{\omega}{2\varepsilon}(\hat{k} \cdot e_\eta)f - \frac{\omega^2}{4\varepsilon^2}(\hat{k} \cdot n)^2$$

are displayed in figs. 4a and 4b.

A few comments on the choice $\beta(\eta, t) = 0$ are in order. The variational equations do not determine tangential perturbations $\beta(\eta, t)e_\eta$. However, note that, with

$$p = \beta(\eta, t) e_\eta$$

and

$$|F_\eta|^2 = 1,$$

we have

$$\beta_\eta(\eta, t) = 0$$

or

β = constant + a function of time.

Now the boundary conditions on β, viz.

$$\beta(0, t) = 0, \quad \beta(l, t) = 0,$$

imply that

$$\beta(\eta, t) = 0. \tag{37}$$

5. Conclusions and outlook

An important property of the eikonal equation is that it encapsulates the essential geometry of the spiral and simple scroll waves in three dimen-

sions. It is therefore of interest to analyse the stability of these structures, as described by (8). Perturbations are defined and classified as direct disturbances on the visually observed patterns. We initiate this program for surface waves in this paper by demonstrating the linear stability of the counter-rotating spiral waves under small perturbations normal to the solution curve on the unit sphere. The fact that such waves can be studied experimentally [14, 16] indicates the need for further theoretical analysis of surface waves.

A fruitful endeavour will be a study of generalisations of the eikonal approach that encompass in some measure the rich variety of solutions discovered recently [17], and a stability analysis of these solutions along the lines indicated in this paper.

Acknowledgements

One of the authors (J.G) would like to thank Dr. P. Grindrod for useful discussions on stability of solutions of the eikonal equation and to referees for critical comments and for providing clarification for the choice (34) made in section 3. He is grateful to the British Council for a travel grant.

References

[1] P. Grindrod and J. Gomatam, J. Math. Biol. 25 (1987) 597.
[2] J. Gomatam and P. Grindrod, J. Math. Biol. 25 (1987) 611.
[3] J. Gomatam and P. Grindrod, in: NATO Advanced Research Workshop on Nonlinear Wave Phenomena in Excitable Media, ed. A. Holden (University of Leeds, 1989).
[4] V.S. Zykov, Biophysics 25 (1980) 329.
[5] V.S. Zykov, Biophysics 25 (1980) 906.
[6] J.P. Keener, SIAM J. Appl. Math. 46 (1986) 1039.
[7] J.P. Keener and J.J. Tyson, Physica D 21 (1986) 307.
[8] W. Jahnke, C. Henze and A.T. Winfree, Nature 336 (1988) 662.
[9] A.T. Winfree, The Geometry of Biological Time (Springer, Berlin, 1980) p. 300.

[10] B.J. Welsh, Ph.D. Thesis, Glasgow College of Technology, Scotland (1984).

[11] B.J. Welsh and J. Gomatam, Diversity of Three-Dimensional Chemical Waves, Physica D 43 (1990) 304.

[12] V.I. Krinsky, in: NATO Advanced Research Workshop on Nonlinear Wave Phenomena in Excitable Media, ed. A. Holden (University of Leeds, 1989).

[13] A.T. Winfree, When Time Breaks Down (Princeton Univ. Press, Princeton, 1987) p. 115.

[14] J. Maselko and K. Showalter, Nature 339 (1989) 609–611.

[15] J.J. Tyson and J.P. Keener, Physica D 32 (1988) 327.

[16] K. Showalter, private communication (1989).

[17] A.T. Winfree, SIAM Rev. 32 (1990) 1.

Physica D 49 (1991) 90–97
North-Holland

Spiral selection as a free boundary problem

David A. Kessler[a] and Herbert Levine[b]

[a]*Department of Physics, University of Michigan, Ann Arbor, MI 48109, USA*
[b]*Department of Physics and Institute for Nonlinear Science, University of California at San Diego, La Jolla, CA 92093, USA*

We present a new formulation of the spiral selection problem for the Belousov–Zhabotinsky reaction. In particular, we focus on deriving an exact integro-differential shape equation and discuss the possible behavior of solutions to this equation. We also present some new results on the asymptotic (far from core) structure of spirals in the Fife scaling regime.

1. Introduction

One of the most striking examples of spatial pattern formation in non-equilibrium systems is the rotating spiral seen in diffusion-coupled excitable media[#1]. The most studied example to date is the Belousov–Zhabotinsky (BZ) case [2–5], but similar structures appear in systems as diverse as catalysis on platinum surfaces [6] and aggregation of slime mold amoeba [7]. Remarkably and regrettably there is as yet no first principles calculation of a spiral shape in any reaction–diffusion system.

This work attempts to formulate and treat spiral patterns in a simplified Oregonator model[#2] to be described shortly. Some of the ideas have appeared elsewhere in preliminary form, but here we offer some new results and a new perspective on this problem. The basic ingredient of our method is the rewriting of the spiral pattern problem as a free boundary problem for the thin reaction zone of the fast reaction. This approach was advocated long ago by Fife, but will be reformulated here in a novel manner.

The outline of this paper is as follows. First, we briefly review the Fife regime of the Oregonator model and explain why this scaling is crucial for the spiral shape determination. Next, we derive asymptotic corrections (far from the core) to the spiral shape and discuss the utility of these results. Finally, we derive an integro-differential shape equation in which the rotation frequency appears as an eigenvalue. We present a general argument as to why the spectrum of frequencies *might* be discrete, corresponding to the experimental observation of unique spiral structures at fixed conditions. This selection occurs despite the fact that the asymptotic expansion allows for a continuous frequency. Finally, we summarize our findings and offer some conjectures regarding spiral tip meandering.

2. Fife scaling

Let us briefly review the Oregonator model with the asymptotic wave velocity in the Fife region [9]. What this means is the following: the Oregonator equations have the form

$$\dot{u} = \epsilon \nabla^2 u + \frac{f(u,v)}{\epsilon} \tag{1}$$

$$\dot{v} = \epsilon \nabla^2 v + g(u,v), \tag{2}$$

[#1] For a general introduction to excitable media, see ref. [1].
[#2] For a discussion of the Oregonator model, see ref. [8].

where the equation $f(u,v) = 0$ admits multiple stable solutions $u = u_\pm(v)$ for v in some range. The ratio of reaction rates, ϵ, is small and hence as reviewed by Keener and Tyson [10], one can assume that $u = u_+(v)$ or $u_-(v)$ except for thin reaction zone. Actually, this methodology crucially depends on the *assumption* that the relevant length scales in the spiral pattern are never comparable to the reaction zone width ϵ; this will be important in the discussion below.

In the reaction zone, v can be taken constant, and a domain wall solution which proceeds from u_+ to u_- (or vice versa) constructed. At one particular value of v, the stall concentration v_s, this domain wall is motionless. For $v \sim v_s$, we can linearize the relationship between velocity and concentration; including curvature effects, this leads to

$$(c_n + \epsilon\gamma\kappa)v_0' = (v - v_s) \tag{3}$$

for some constant v_0', c_n the normal velocity and κ the curvature. The constant γ should equal 1 for the model given in eqs. (1), (2) but we leave it as an arbitrary additional parameter, which can be related to a ratio of unequal diffusivities for the two chemical species.

The Fife limit [9, 11] arises from the assumption that we are dealing with a wave with $c \sim \epsilon^{1/3}$. From the above equation, uniformity demands that lengths scale as $\epsilon^{2/3}$, and $v - v_s \sim \epsilon^{1/3}$. In the dynamical equation for v, the function g can be evaluated at the stall value v_s. Combining all of these comments leads to the free boundary problem [9, 11]

$$\nabla^2 \bar{v} \pm a_\pm = \partial\bar{v}/\partial t \tag{4}$$

with $\bar{v}, \bar{v}'$ continuous across zones and where we must satisfy the rescaled version of eq. (3):

$$(\bar{c}_n + \gamma\bar{\kappa})v_0' = \bar{v}. \tag{5}$$

Here $a_\pm = \pm g(u_\pm(v_s), v_s)$.

To find a spiral solution of the above equations, we assume a uniformly rotating field and replace $\nabla^2 - \partial/\partial t$ by

$$\frac{\partial^2}{\partial r^2} + \frac{1}{r}\frac{\partial}{\partial r} + \frac{1}{r^2}\frac{\partial^2}{\partial\phi^2} + \omega\frac{\partial}{\partial\phi}. \tag{6}$$

Also, $c_n = \omega r\hat{n}\cdot\hat{\phi}$. We must then find a curve $\Phi_0(r)$ for the reaction front such that the boundary condition (5) is satisfied by a continuous field $\bar{v}$. There are two methods to do this; we can eliminate $\bar{v}$ to find a shape equation for Φ_0 or we can map the spiral to a simpler region and attempt to solve the resulting partial differential equation. These will be discussed in subsequent sections.

The most important thing to notice about the Fife limit is that ϵ has completely dropped out of the governing equations and boundary conditions. Let us therefore assume that there are indeed spiral solutions with the scaling assumed here $c \sim \epsilon^{1/3}$. Then, the physical scale of the spiral, being proportional to $\epsilon^{2/3}$, can be made arbitrarily large compared to the reaction zone width, which is of size ϵ. This means that the thin reaction zone approximation and the concomitant recasting of the problem as a free surface one is uniformly valid at the spiral core.

The above concept is critical to our entire approach and therefore it is worth discussing what other possibilities exist. One way of thinking about the Fife scaling is to recognize that it determines the core size by requiring the relevance of v diffusion. The other possibility is that the core size is determined not by v diffusion but instead by the finiteness of the reaction zone; this must occur for example in systems without diffusion of the slow species, but could also occur in general if the selected asymptotic velocity is $\mathcal{O}(1)$ instead of $\mathcal{O}(\epsilon^{1/3})$. Then, the singular perturbation theory used to derive the governing equation (4) is not valid, and one must return to the coupled PDE system (1), (2) in the inner core region. Of course, whether or not spirals will exist (for some choice of kinetic parameters) in the Fife regime could be studied computationally, but this has not yet been attempted.

3. Spiral asymptotics

It is intuitively clear that asymptotically far from the core, the spiral pattern appears to be just a plane wave with some particular asymptotic velocity. In this section, we present a methodology for calculating corrections to the asymptotic shape in a systematic expansion about large r. As we shall see, this calculation will place the spiral selection problem in proper focus.

The basic idea is to define new coordinates:

$$\tilde{\phi} = \phi - \Phi_0(r), \qquad \tilde{r} = r. \tag{7}$$

This transforms the governing PDE to the form

$$\left[\left(\frac{\partial}{\partial \tilde{r}} - \frac{\partial \Phi_0}{\partial \tilde{r}} \frac{\partial}{\partial \tilde{\phi}} \right)^2 + \frac{1}{\tilde{r}} \left(\frac{\partial}{\partial \tilde{r}} - \frac{\partial \Phi_0}{\partial \tilde{r}} \frac{\partial}{\partial \tilde{\phi}} \right) + \omega \frac{\partial}{\partial \tilde{\phi}} + \frac{1}{\tilde{r}^2} \frac{\partial^2}{\partial \tilde{\phi}^2} \right] \tilde{v}_{\pm} \pm a_{\pm} = 0. \tag{8}$$

The boundary conditions are now imposed at $\tilde{\phi} = 0$ and $\tilde{\phi} = 2\pi\lambda_+/(\lambda_- + \lambda_+)$ (for the excited "+" region) and, $-2\pi\lambda_-/(\lambda_- + \lambda_+)$ (for the quiescent "−" region). Note that one full turn of the spiral pattern corresponds to a change of $\tilde{\phi}$ by 2π. Note too that λ_+ and λ_-, the widths respectively of the "+" and "−" phases, depend on $\tilde{r}$. Using the relationship

$$c_n = \frac{\omega}{2\pi} [\lambda_+(\tilde{r}) + \lambda_-(\tilde{r})], \tag{9}$$

we can find λ_- given λ_+ and the curve Φ_0; $\lambda_+(\tilde{r})$ itself must be found as part of the solution.

To make the above discussion less abstract, let us make use of this formulation to find the asymptotic shape of the spiral. We assume the series

$$\tilde{v}_{\pm}(\tilde{r}, \tilde{\phi}) = \sum_{n=0}^{\infty} \frac{v_{\pm}^{(n)}(\tilde{\phi})}{\tilde{r}^n} \tag{10}$$

and

$$\Phi_0(\tilde{r}) = -\frac{\omega \tilde{r}}{c} + D_0 \ln \tilde{r} + \sum_{n=1}^{\infty} \frac{D_n}{\tilde{r}^n}. \tag{11}$$

We substitute these expansions into eq. (8), which we then solve order by order in $1/\tilde{r}$. To leading order

$$v_{\pm}^{(0)} = A_{\pm}^{(0)} + B_{\pm}^{(0)} e^{-c^2\tilde{\phi}/\omega} \pm \frac{a_{\pm}\tilde{\phi}}{\omega}. \tag{12}$$

Applying the boundary conditions gives six equations

$$A_{+}^{(0)} + B_{+}^{(0)} = A_{-}^{(0)} + B_{-}^{(0)} = v_0'c,$$

$$B_{+}^{(0)} - B_{-}^{(0)} = -\frac{a_{+} + a_{-}}{c^2},$$

$$\bar{B}_{+}^{(0)} - \bar{B}_{-}^{(0)} = -\frac{a_{+} + a_{-}}{c^2},$$

$$\bar{B}_{+}^{(0)} + A_{+}^{(0)} - \frac{a_{+}}{c}\lambda_{+}^{(0)} = \bar{B}_{-}^{(0)} + A_{-}^{(0)} - \frac{a_{-}\lambda_{-}^{(0)}}{c} = -v_0'c. \tag{13}$$

Here $\bar{B}_{\pm}^{(0)} = \exp(\pm c\lambda_{\pm}^{(0)}) B_{\pm}^{(0)}$. This is exactly the same system as for plane waves [11, 12] and allows us to determine the unknown coefficients as well as $\lambda_{+}^{(0)}$ and the asymptotic velocity c.

Let us now extend this calculation to the first nontrivial correction. The field equation to order $1/\tilde{r}$ is now inhomogeneous; we find

$$-\left[\left(\frac{\partial \Phi_0}{\partial \tilde{r}} \right)^2 \frac{\partial^2}{\partial \tilde{\phi}^2} + \omega \frac{\partial}{\partial \tilde{\phi}} \right] v_{\pm}^{(1)} = -cB_{\pm}^{(0)} e^{-c^2\tilde{\phi}/\omega} \pm \frac{a_{\pm}}{c} - \frac{2c^3}{\omega} D_0 B_{\pm}^{(0)} e^{-c^2\tilde{\phi}/\omega}. \tag{14}$$

This is easily solved by

$$v_{\pm}^{(1)} = A_{\pm}^{(1)} + B_{\pm}^{(1)} e^{-c^2\tilde{\phi}/\omega} \pm \frac{a_{\pm}}{\omega c}\tilde{\phi} - \frac{c\tilde{\phi}}{\omega} e^{-c^2\tilde{\phi}/\omega} B_{\pm}^{(0)} \left(1 + \frac{2c^2}{\omega} D_0 \right). \tag{15}$$

Again, we have to apply the boundary conditions. To do this, we need the result

$$c_n = \frac{\omega r}{\sqrt{1 + r^2(\partial \Phi_0/\partial r)^2}} \simeq c + \frac{D_0 c^2}{\omega r} \tag{16}$$

along with $\kappa = 1/\tilde{r}$ and the assumption $\lambda_{+}(r)/[\lambda_{+}(r) + \lambda_{-}(r)] \sim \lambda_{+}^{(0)}/(\lambda_{+}^{(0)} + \lambda_{-}^{(0)}) + E/\tilde{r}$. Applying the boundary conditions gives us, as always, a set of six linear equations in the six unknowns $A_{\pm}$, $B_{\pm}$, D_0 and E. These can be solved directly.

As a concrete example, we choose the set of parameters derived in ref. [11] from the original Oregonator model; $a_{+} = 11/16$, $a_{-} = 1/16$, $\gamma = 1$ and $v_0' = \sqrt{2}/30$. We then pick an asymptotic velocity, say $c = 0.5$. Solving (13) for the planar interface, we find the results $\lambda_{-}^{(0)} = 6.223$, $\lambda_{+}^{(0)} = 0.565$ and the rotation frequency $\omega = 2\pi c/(\lambda_{+}^{(0)} + \lambda_{-}^{(0)}) = 0.463$. We can then evaluate all the lowest-order coefficients and derive the resultant linear system for the first-order shifts; we find

$$D_0 = 0.2625, \qquad E = 0. \tag{17}$$

Extending this to second order, our result becomes

$$\Phi_0(r) = -\frac{\omega r}{c} + 0.2625 \ln r - \frac{0.194}{r} + \mathcal{O}(1/r^2), \tag{18}$$

$$\frac{\lambda_+(r)}{\lambda_+(r) + \lambda_-(r)} - \frac{\lambda_+^{(0)}}{\lambda_+^{(0)} + \lambda_-^{(0)}}. \tag{19}$$

One surprising result of the above calculation is the vanishing of the $E = 0$ coefficient to all orders. That is, the relative width λ_+/λ_- does not vary from its asymptotic value to any order in $1/\bar{r}$; this is in distinction to the curve itself, which actually has logarithmically large deviations since $D_0 \neq 0$. It is possible to make the ansatz $\lambda_+(r)/\lambda_-(r) \simeq \lambda_+^{(0)}(r)/\lambda_-^{(0)}(r) + \hat{E} e^{-\alpha r}$ and derive an eigenvalue equation for α. Since this is of only tangential importance, we skip this derivation here.

What is this asymptotic calculation useful for? First, it is obviously possible in principle to measure the asymptotic velocity and compare the above analytic predictions with measured (or computed) shapes. Next, we will see in the next section that this offers a highly non-trivial check on the numerical implementation of the shape equation. Most importantly from our perspective is the statement that these results make regarding the uniqueness of the spiral pattern and in particular of the rotation frequency. Clearly, there is no problem extending this procedure to arbitrarily higher order; we always recover the same structure and can determine the unknown coefficients to each order. Hence, there is no selection of frequency ω. That is, given ω we can find all other unknowns and construct the expansions (10), (11); we will argue subsequently that this is misleading and in fact the full equation does select ω albeit by terms that are exponentially small at large distances.

How can we go beyond this asymptotic expansion? At present, the only viable approach appears to be numerical solution of the integro-differential shape equation [11]. The next section discusses the derivation of this equation and the possible form of its solutions.

4. Shape equation

In other types of free surface problems [13], it has proven advantageous to recast the equations in the form of an integro-differential shape equation for the actual pattern. To derive this equation we return to the original variables and define

$$\Psi = v \pm \tfrac{1}{4} a_\pm r^2. \tag{20}$$

The auxiliary field Ψ then obeys the homogeneous equation $\nabla^2 \Psi + \omega \, \partial \Psi / \partial \phi = 0$, which furthermore is the same in both phases. Green's theorem then demands that any discontinuities in either Ψ or $\hat{n} \cdot \nabla \Psi$ (for normal to the curve $\hat{n}$) give rise to sources in an integral representation of the field. Since v is continuous, the discontinuities of Ψ can be determined directly from eq. (20):

$$\text{disc } \Psi = \tfrac{1}{4}(a_+ + a_-)r^2, \qquad \text{disc}(n \cdot \nabla \Psi) = \tfrac{1}{2}(a_+ + a_-)r\hat{n} \cdot \hat{r}.$$

Using the above results, we can immediately write down a solution for Ψ in terms of the Green's function

$$G(r, r', \phi, \phi') = \int_{-\infty}^{t} \frac{dt'}{4\pi t'} \exp\{-[r^2 + r'^2 - 2rr'\cos(\phi - \phi' - \omega(t - t'))]/4t'\}. \tag{21}$$

Using the easily derived expressions

$$\text{disc } G = 0, \qquad \text{disc}(\hat{n}' \cdot \nabla' G) = 1, \qquad \text{disc}[(\hat{n} \cdot \nabla)(\hat{n}' \cdot \nabla' G)] = -c_{\mathrm{n}},$$

we find

$$\Psi(r, \phi) = \int \mathrm{d}s' (\hat{n}' \cdot \nabla' G) \tfrac{1}{4} r'^2 (a_+ + a_-) - \int \mathrm{d}s' G\left[\tfrac{1}{2}(\hat{n}' \cdot \hat{r}') r'(a_+ + a_-) + \tfrac{1}{4} c_{\mathrm{n}} r'^2 (a_+ + a_-)\right],$$

$$(22)$$

which determines the field everywhere *if* we know the actual reaction zone $\Phi_0(r)$.

The shape equation now follows when we evaluate Ψ on the reaction zone itself and use the known value of the field v to find an expression for the value of Ψ. Doing this, we derive the final shape equation

$$\frac{v_0'(c_{\mathrm{n}} + \gamma\kappa) + \tfrac{1}{4} a_+ r^2}{a_+ + a_-} = -\int \mathrm{d}s' (\hat{n}' \cdot \nabla' G) \tfrac{1}{4} r'^2 + \int \mathrm{d}s' G\left[\tfrac{1}{2}(\hat{n}' \cdot \hat{r}') r' + \tfrac{1}{4} c_{\mathrm{n}} r'^2\right], \qquad (23)$$

where the first integral must be evaluated on the "+" side of the interface. By definition, the normal vector $\hat{n}$ always points from "+" to "−". This is then a closed form equation for the curve $\Phi_0(r)$ in which the rotation frequency ω appears as a parameter, both in the Green's function and in the normal velocity c_{n} given by eq. (16).

In principle, this equation can be solved by discretizing the curve and varying the parameters using some type of Newton's iteration to converge to a final solution. This has not yet been accomplished, but we have performed one specific consistency check. That is, if we take our (lowest order) asymptotic expansion for the spiral shape derived in the last section and evaluate the integral on the right-hand side, it differs from the left-hand side by terms that are no bigger than $\mathcal{O}(1/r)$. This is rather nontrivial since the generic size of both sides of the equation is $\mathcal{O}(r^2)$.

In the absence of an exact solution, we can employ a heuristic argument that equations of this type should require fixed values for the rotation frequency ω. We can rewrite the shape equation in the schematic form $c_{\mathrm{n}} + \gamma\kappa = F[\Phi_0(r)]$, where F is a rather complex functional of the curve. We know of course that as $s \to \pm -\infty$, F will approach $\pm c$, the asymptotic velocity and that furthermore F cannot depend on either absolute angle or absolute arclength. Let us change to treating the curve parametrically by giving the radius R_0 and the angle Φ_0 as functions of the arclength. We can then rewrite the above equation as a dynamical system for the variables $R_0(s)$, $\Phi_0(s)$ and $\Theta(s) \equiv \cos^{-1}(\hat{n} \cdot \hat{y})$, where the $\hat{y}$ coordinate axis has been chosen for a reference direction; after some geometry, this yields [14]

$$\frac{\mathrm{d}R_0}{\mathrm{d}s} = -\cos(\Phi_0 + \Theta), \qquad \frac{\mathrm{d}\Phi_0}{\mathrm{d}s} = \frac{\sin(\Phi_0 + \Theta)}{R_0}, \qquad \frac{\mathrm{d}\Theta}{\mathrm{d}s} = \frac{\omega R_0 \cos(\Phi_0 + \Theta) + F}{\gamma}. \qquad (24)$$

Now, we can write down the most general asymptotic spiral solution in terms of four constants $s_\pm$, $\Phi_\pm$,

$$R_0(s) \sim \sqrt{(2c/\omega)(s - s_\pm)}, \quad s \to \pm\infty, \qquad \Phi_0(s) \sim \sqrt{(2\omega/c)(s - s_\pm)} + \Phi_\pm, \quad s \to \pm\infty,$$

$$\Theta(s) \sim -\Phi_0(s) + 3\pi/2, \quad s \to +\infty, \quad \text{and} \quad \Theta(s) \sim -\Phi_0(s) + \pi/2, \quad s \to -\infty. \qquad (25)$$

Of these four unknowns, one corresponds to an overall rotation of the entire system (and is hence irrelevant) and one goes to fix λ_+/λ_-; we of course have already correctly chosen c and hence $\lambda_+ + \lambda_-$. This leaves two unknowns but three matching conditions to generate a smooth trajectory from $s = -\infty$ to $s = +\infty$. Hence there is generally no solution unless we also pick ω appropriately. In other words, the frequency is fixed by requiring the matching (at the core) of solutions with the correct behavior at $\pm\infty$ and is not visible in the asymptotic large distance expansion.

The reasoning in the previous paragraph serves to explain why several different "geometrical" models of spirals [10, 14–16][#3] have all yielded unique rotation frequencies. In a geometrical model, F is replaced by a function of local objects and hence eq. (23) can be directly integrated. From our current perspective, any "reasonable" geometrical model will give selection via this general mechanism. This does not really answer the question of whether this actually happens in a physical nonlocal model where there are conceivable loopholes in the above approach; but this does offer one possible scenario as to how the asymptotic freedom of choosing ω is broken by the core region solution. Eventually, we hope to answer this question more definitely by finding full solutions of eq. (23).

5. Comments

To summarize, we have presented a novel way of looking at the spiral selection problem within the Fife scaling limit of the Oregonator model. We have calculated the asymptotic spiral shape, discussed the importance of the diffusion controlled core and reviewed the derivation of a spiral shape equation. Finally, we have presented a general argument as to why the rotation frequency *might* be selected via matching near the core. An explicit method capable in principle of demonstrating the validity of our proposed scenario was presented, and we are currently attempting to implement this approach.

The last comment we wish to make concerns the tip meandering instability studied in great detail for the BZ system [5, 17, 18]. Our best guess is that this instability is a remnant of the Hopf bifurcation at zero traverse wavevector known to occur [19] for planar waves as $\lambda_+, \lambda_- \to 0$. For the Fife limit, this instability is again tied up with v diffusion and hence will become important on the same general scale as the core. Specifically, as the spiral moves inward from infinity, the width of the wave shrinks, with of course $\lambda_+ \to 0$ as we reach the spiral core. Imagine doing a WKB calculation where at each radius we consider the spiral as a locally planar problem with some normal velocity. It appears likely that the stability boundary can be crossed as we move towards the core leading to a locally unstable region. Globally, this would lead to the prediction of an oscillatory mode localized near the core, which is what one appears to observe.

The above conjecture explains why it is possible to see the meandering instability even with v diffusion. Most explanations [20] of the meandering have dealt with simplified models most appropriate to the case of no v diffusion and core sizes of order ϵ. Obviously, a great deal of additional work will be necessary in order to see if this idea makes any sense. We have included this discussion, just to point to that once one can deal with the spiral as a free boundary problem, techniques which have proven capable of finding stability spectra for other free boundary systems [21] can be applied here and may offer additional insight into spiral pattern dynamics.

[#3]For a review of attempts to do this in the Russian literature, see ref. [15].

References

[1] R.J. Field and M. Burger, eds., Oscillations and Traveling Waves in Chemical Systems (Wiley, New York, 1985).

[2] S.C. Muller, T. Plesser and B. Hess, Science 230 (1985) 661; Physica D 24 (1987) 87.

[3] A.T. Winfree, When Time Breaks Down (Princeton Univ. Press, Princeton, NJ, 1987).

[4] W.Y. Tam, W. Horsthemke, Z. Noszticzius and H. Swinney, J. Chem. Phys. 88 (1888) 3395.

[5] G.S. Skinner and H. Swinney, Physica D 48 (1991) 1–16.

[6] M.P. Cox, G. Ertl and R. Imbihl, Phys. Rev. Lett. 54 (1985) 1725.

[7] P.C. Newell and F.M. Ross, J. Gen. Microbiol. 128 (1982) 2715.

[8] J.J. Tyson, in: Nonlinear Phenomena in Chemical Dynamics, eds. C. Vidal and A. Pacault (Springer, Berlin, 1981).

[9] P. Fife, in: Nonequilibrium Dynamics in Chemical Systems, eds. C. Vidal and A. Pacault (Springer, Berlin, 1984).

[10] J.P. Keener and J.J. Tyson, Physica D 21 (1986) 307; Physica D 32 (1988) 327;
J.P. Keener, SIAM J. Appl. Math. 46 (1986) 1039.

[11] D. Kessler and H. Levine, Physica D 39 (1989) 1.

[12] J.P. Dockery, J.P. Keener and J.J. Tyson, Physica D 30 (1988) 177.

[13] D. Kessler, J. Koplik and H. Levine, Adv. Phys. 37 (1988) 255.

[14] D. Kessler and H. Levine, Europhys. Lett., in press.

[15] J.P. Keener and J.J. Tyson, Physica D 32 (1988) 327.

[16] P. Pelcé and J. Sun, Physica D 48 (1991) 253.

[17] W. Jahnke, W.E. Skaggs and A.T. Winfree, J. Phys. Chem. 93 (1989) 740.

[18] D. Barkeley, M. Kness and L.S. Tuckerman, Phys. Rev. A 42 (1990) 2489.

[19] D. Kessler and H. Levine, Phys. Rev. A 41 (1990) 5148.

[20] A. Mikhailov, unpublished;
E. Meron, preprint (1990).

[21] D. Kessler and H. Levine, Phys. Fluids 30 (1987) 1246.

Physica D 49 (1991) 98–106
North-Holland

The role of curvature and wavefront interactions in spiral-wave dynamics

Ehud Meron

Department of Chemical Physics, Weizmann Institute of Science, Rehovot 76100, Israel

A theory of rotating spiral waves in excitable media is presented that allows the study of dynamical aspects such as nonsteady rotation. An approximate spiral-wave solution is proposed in the form of a superposition of curved solitary wavefronts, parallel to each other. This form is used, by means of singular perturbation theory, to derive an evolution equation for the spiral arm. Numerical solutions of that equation are found that describe one- and two-frequency rotations. The transition to two-frequency dynamics (tip "meandering") is attributed to a destabilizing curvature effect.

1. Introduction

Dynamical aspects of spiral waves in excitable media have been addressed so far within the framework of geometrical or kinematical theories. In these theories the spiral wavefront is conceived as an infinitely thin curve whose velocity is determined by its local curvature [1–5]. Considerable progress has been made using this approach but its range of validity is limited. Most experimental studies of spiral-wave dynamics have been carried out in regimes where refractory-tail effects are important. In these regimes the width of the wavefront cannot be ignored; the wavefront velocity is affected by the tail of the preceding wavefront. Attempts to extend the geometrical theories to include refractory-tail effects have recently been made by introducing a phenomenological dependence of the wavefront and tip velocities on the local recovery phase [4, 6].

This paper describes an attempt to derive the dynamics of spiral waves from the basic reaction–diffusion equations, so that wavefront interactions appear as an integral part of the theory. A principal result to be presented here is an evolution equation for the spiral wavefront and an expression for its normal velocity. Numerical simulations on the spiral evolution equation allow us to conclude that wavefront interactions play an important role in stabilizing the dynamics of spiral waves, while curvature acts to destabilize it. The competition between the two can lead to the onset of two-frequency dynamics, commonly referred to as "meandering" [7] or "compound rotation" [8][#1]. A short account of some of these results has appeared in ref. [10].

We will be concerned here with isotropic and homogeneous excitable media in two space dimensions. Such media are described by reaction–diffusion equations (RDEs) of the form

$$\partial_t U = \mathbf{L} U + N(U) + \mathbf{D} \nabla^2 U, \tag{1}$$

where U represents a set of fields, $\mathbf{L}$ and N are, respectively, the linear and nonlinear parts of the reaction dynamics, and $\mathbf{D}$ is a matrix of transport coefficients (diffusion, conduction, etc.). It is assumed that $N(0) = 0$, so that $U = 0$ is a solution representing the quiescent state of the medium. In section 2 we will consider rectilinear wavefront solutions of (1) and properties thereof that will be needed in the subsequent analysis. An approximate spiral solution will be constructed in section 3 and will be used in sections 4 and 5 to derive an evolution equation for the spiral arm and an

---

[#1]Other terms to describe meander have been used in ref. [6] ("cycloidal circulation") and in ref. [9] ("looping").

expression for its normal velocity. Section 6 presents results of numerical integration of the spiral evolution equation. Section 7 concludes with a discussion.

2. One-dimensional wavefront solutions

Rectilinear, solitary wavefronts that propagate with constant speed c_0 are one-dimensional solutions, $U(x, t) = H(x - c_0 t)$, of the ordinary differential equation

$$\mathbf{L}H + N(H) + \mathbf{D}H'' + c_0 H' = 0, \qquad (2)$$

where the prime denotes differentiation with respect to the argument $r \equiv x - c_0 t$. Specific forms of such solutions can be evaluated for particular models of excitable media. For our purposes, however, it is sufficient to realize that, in general, such solutions are localized in space and tail off exponentially fore and aft. The reason is that $H(r)$ is bi-asymptotic to the rest state ($U = 0$) as $r \to \pm\infty$, and therefore the asymptotic forms of H can be obtained from a linear analysis of (2) around that state. The outcomes of such an analysis are generically exponential solutions.

In determining the asymptotic forms of $H(r)$ only those eigenvectors corresponding to slowest growth from $r = -\infty$ and slowest decay as $r \to \infty$ should be considered. All other eigenvectors will contribute exponentially smaller terms. In other words, we consider only those eigenvalues of the linearized problem around $U = 0$ with smallest positive real part and least negative real part. We denote these real parts by η_L and η_R, respectively. The imaginary parts of the eigenvalues, if they exist, represent oscillations superposed on the exponential growth or decay. The recovery to the rest state, in excitable media, is sometimes accompanied by such oscillations. A possible consequence of oscillatory recovery has been discussed in ref. [10]. Here, we assume purely expo-

nential asymptotic forms:

$$H(r) \sim A_L e^{\eta_L r}, \qquad r \to -\infty,$$

$$H(r) \sim A_R e^{-\eta_R r} \qquad r \to +\infty, \qquad (3)$$

where A_L and A_R are constant vectors.

Consider now a wavetrain of rectilinear wavefronts. The dynamics of these wavefronts are affected by the fields of their neighbors. Since the growth rate of excitation (η_R) in excitable media is different from the rate of recovery to the rest state (η_L), the effect of a preceding wavefront will be different from the effect of a successive one. In fact, the actual relationship between growth and decay rates in excitable media is such that $\eta_R \gg \eta_L$. If the variance in spacings between nearby wavefronts is relatively small, one may neglect the influence of successive wavefronts. A propagating wavefront is then affected only by the tail of the preceding one. If the tail is monotonic the interaction is repulsive; propagation is slowed down by the refractory period imposed on the medium after the passage of the preceding wavefront [11].

3. Constructing an approximate spiral-wave solution

Experimental shapes of rotating spiral waves in the Belousov–Zhabotinsky reaction have successfully been fitted to involutes of circles [12, 13]. We therefore choose to consider real spiral shapes as perturbations about these geometrical forms. A rotating involute of a circle of radius ρ_0 is described by

$$X = \rho \cos[\vartheta(\rho) - \omega t],$$

$$Y = \rho \sin[\vartheta(\rho) - \omega t], \qquad (4)$$

where $\rho = (X^2 + Y^2)^{1/2}$, ω is frequency of rota-

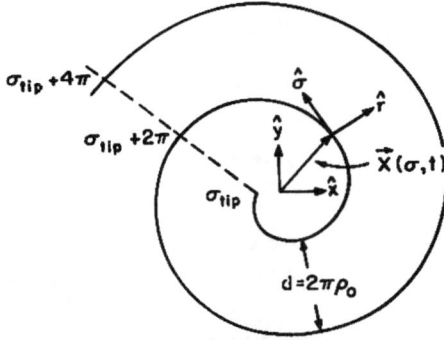

Fig. 1. The involute coordinate system (σ, r). The position vector X is given by eqs. (4).

tion and

$$\vartheta(\rho) = \sigma - \arctan \sigma,$$

$$\sigma = \left[(\rho/\rho_0)^2 - 1 \right]^{1/2}$$

We define now an involute coordinate system (σ, r) by the relation

$$x = X(\sigma, t) + r\hat{r}(\sigma, t), \tag{5}$$

where $X = (X, Y)$ is given by (4)

$$\hat{r} = \frac{\partial_\sigma Y \hat{x} - \partial_\sigma X \hat{y}}{\left[(\partial_\sigma X)^2 + (\partial_\sigma Y)^2 \right]^{1/2}},$$

is a unit vector normal to the involute spiral, and $\hat{x}$ and $\hat{y}$ are unit vectors along the x and y axes, respectively. The range of the normal coordinate r is chosen to be $(-d/2, d/2)$ where $d = 2\pi\rho_0$ is the pitch of the involute spiral. The coordinate system is illustrated in fig. 1.

The real spiral shape will be described here by a displacement function $r = \zeta(\sigma, t)$ or, in the laboratory frame, by

$$x_s(\sigma, t) = X(\sigma, t) + \zeta(\sigma, t)\,\hat{r}(\sigma, t). \tag{6}$$

In choosing to describe the spiral shape in this way we are imposing a constraint on the dynamics of the spiral tip; we allow for displacement in the normal direction, $\hat{r}(\sigma_{\text{tip}})$, but do not introduce an

independent degree of freedom for propagation in the tangential direction, $\hat{\sigma}(\sigma_{\text{tip}})$. This constraint should be taken into account when we come to interpret our results. The benefit of such an approach is the reduction of the two-dimensional problem to a quasi-one-dimensional one in a sense that will become clear in the next section.

We may try to approximate the spiral-wave solution by $H[r - \zeta(\sigma, t)]$ where H solves (2). This form, however, does not take into account the fact that at a given point on the spiral wavefront there is a residual field induced by the passage of the preceding wavefront. To find a better form we split the range of σ into intervals, S_l, representing parallel portions of the involute spiral

$$S_l = \left[\sigma_{\text{tip}} + 2\pi(l-1), \sigma_{\text{tip}} + 2\pi l \right), \quad l = 1, 2, \ldots,$$

and construct solitary wavefronts that are peaked on these intervals

$$H_k(\sigma, r) \equiv H(r + n(\sigma)d - kd), \quad k = 1, 2, \ldots, \tag{7}$$

where

$$n(\sigma) = l \quad \text{if } \sigma \in S_l.$$

Thus, $H_k(\sigma, r) = H(r)$ for $\sigma \in S_k$ and therefore assumes maximal values on the kth segment. In the vicinities of other segments $H_k(\sigma, r)$ is of order $\exp(-m\eta_{\text{R,L}}d)$ where m is a positive integer. We will assume here that the involute pitch, d, is considerably larger than the decay length η_L^{-1}. Thus $H_k(\sigma, r)$ becomes exponentially small for $\sigma \in S_l$, $l \neq k$.

An approximate spiral-wave solution is now written as

$$U_0(\sigma, r, t) = \sum_k H_k(\sigma, r - \zeta_k), \tag{8}$$

where the displacements $\zeta_k \equiv \zeta(\sigma_k, t)$ are evaluated at $\sigma_k = \sigma + 2\pi[k - n(\sigma)]$. To understand better this form consider, for example, the first interval $\sigma \in S_1$. On this interval $n(\sigma) = 1$ and

$$U_0(\sigma, r, t) = H[r - \zeta(\sigma, t)]$$
$$+ H[r - d - \zeta(\sigma + 2\pi, t)] + \dots, \quad \sigma \in S_1.$$

Thus the main contribution comes from the wavefront that is peaked on S_1. In addition, U_0 contains contributions from the tails (or residual fields) of wavefronts ahead of it. The objective of the next section is to determine the displacement function, $r = \zeta(\sigma, t)$. This will be done by requiring that corrections to (8) are small.

4. Evolution equation for the spiral arm

The RDEs (1) in the rotating involute coordinate system become

$$\mathbf{L}U + N(U) + \mathbf{D}\partial_r^2 U + c_0 \partial_r U = P,$$
$$P = -\mathbf{D}K_1 \partial_r U + \partial_t U + \Omega \partial_\sigma U - \mathbf{D}K_1^2 \partial_\sigma^2 U, \quad (9)$$

where

$$K_1(\sigma, r) = \frac{K_0}{1 + rK_0}. \quad \Omega(\sigma, r) = \omega + \rho_0 \mathbf{D}K_1^3,$$

$K_0(\sigma) = (\rho_0 \sigma)^{-1}$ is the involute curvature and we identified $\omega \rho_0$ with the speed, c_0, of a rectilinear solitary wavefront. We consider here spiral waves that can be viewed as perturbations about involute spirals rotating at constant frequencies. The partial derivatives $\partial_t U$, $\partial_\sigma U$ and $\partial_\sigma^2 U$ are therefore small in comparison with $\partial_r U$. If in addition $\mathbf{D}K_0(\sigma_{tip}) \ll c_0$, where D is a numerical estimate of the components of $\mathbf{D}$, the right-hand side of (9) can be considered as a perturbation. Under these assumptions the two-dimensional problem (9) reduces to the one-dimensional problem (2) perturbed by P. We also want to view wavefront interactions as perturbations. Here the small parameter is $\epsilon = \exp(-d\eta_L)$. The effects due to curvature and wavefront interactions are expected to be of the same order of magnitude. We therefore assume $\mathbf{D}K_0(\sigma_{tip}) \sim \epsilon$. In real systems

the curvature of the wavefront near the tip is not small. The smallness of the curvature term, DK, is rather due to D. We will therefore allow $K_0(\sigma_{tip})$ to be as big as $\epsilon^{-1/2}$.

Using the approximate solution (8) we write an exact solution as

$$U(\sigma, r, t) = \sum_k H_k(\sigma, r - \zeta_k) + R(\sigma, r, t), \quad (10)$$

where R is a correction term. We assume the following orders of magnitude

$$R \sim \partial_r R \sim \epsilon, \quad \partial_t R \sim \partial_\sigma R \sim \epsilon^2,$$
$$\zeta \sim \epsilon^{1/2}, \quad \partial_t \zeta \sim \partial_\sigma \zeta \sim \epsilon.$$

Introducing (10) into (9) we get to leading order

$$\mathscr{L}R = J(\{H_k\}) - \sum_k (\partial_t \zeta_k + \mathbf{D}K_1$$
$$+ \Omega \partial_\sigma \zeta_k - \mathbf{D}K_1^2 \partial_\sigma^2 \zeta_k)H_k', \quad (11)$$

where

$$\mathscr{L} = \mathbf{L} + \mathbf{D}\partial_r^2 + c_0 \partial_r + \nabla_H N\left(\sum_k H_k\right), \quad (12)$$

$$J(\{H_k\}) = \sum_k N(H_k) - N\left(\sum_k H_k\right), \quad (13)$$

$H_k = H_k(\sigma, r - \zeta_k)$ and the prime denotes differentiation with respect to the second argument. In deriving (11) we neglected terms of order ϵ^2 and higher, but kept the highest derivative term $\partial_\sigma^2 \zeta$.

Consider now the operator

$$\mathscr{L}_l = \mathbf{L} + \mathbf{D}\partial_r^2 + c_0 \partial_r + \nabla_H N(H_l). \quad (14)$$

It is straightforward to see by differentiating (2) with respect to the argument of H that

$$\mathscr{L}_l H_l' = 0. \quad (15)$$

Since

$$\nabla_H N\left(\sum_k H_k\right) \cdot H_l' = \nabla_H N(H_l) \cdot H_l' + \mathcal{O}(\epsilon),$$

we find that

$$\mathcal{L} H_l' = \mathcal{O}(\epsilon). \tag{16}$$

Eqs. (11) and (16) suggest that the correction term R may contain a "small denominator" component. We will show now that the requirement that R remains of $\mathcal{O}(\epsilon)$ leads to an equation for $\zeta(\sigma, t)$. To this end we define the inner product

$$(F_1, F_2) = \int_{-\infty}^{\infty} dx \, F_1(x) \cdot F_2(x), \tag{17}$$

where the dot denotes the usual scalar product in $\mathbb{R}^n$. For the peaked solitary wavefronts defined by (7), the integral in (17) translates into

$$\sum_{n(\sigma)} \int_{d/2}^{-d/2} dr.$$

The existence of a null vector of $\mathcal{L}_l$ suggests that the null space of the adjoint operator, $\mathcal{L}_l^\dagger$, is nonempty as well. We will assume that a null vector of $\mathcal{L}_l^\dagger$ indeed exists and denote it by $G_l = G(r - \zeta_l + n(\sigma)d - ld)$. Since H_l is localized around the l's wavefront so should G_l. It therefore follows that

$$\mathcal{L}^\dagger G_l = \mathcal{O}(\epsilon). \tag{18}$$

Taking the inner product of (11) with G_l we find that the left-hand side is of $\mathcal{O}(\epsilon^2)$. Hence the right-hand side must also be of $\mathcal{O}(\epsilon^2)$ (if $R \sim \epsilon$). Applying this condition and using the localized nature of H_l' and G_l we get to leading order

$$\partial_t \zeta = -\bar{D} K_1 - \bar{\Omega} \partial_\sigma \zeta + \bar{D} K_1^2 \partial_\sigma^2 \zeta - F_L(\zeta^+ - \zeta + d), \tag{19}$$

where $K_1 = K_1(\sigma, \zeta) = (\rho_0 \sigma + \zeta)^{-1}$, $F_L(-x)$ is the asymptotic tail of the solitary wavefront solution

$H(x)$, $\zeta^+ \equiv \zeta(\sigma + 2\pi, t)$ and[#2]

$$\bar{D} = (G, \mathbf{D} H')/(G, H'),$$
$$\bar{\Omega} = \omega + \rho_0 \bar{D} K_1^3. \tag{20}$$

The wavefront-interaction term, F_L, comes from the integral (G, J) where J is given by (13). The derivation is similar to that presented in refs. [11, 14] and is omitted here. We just note that we used the fact that $\eta_R \gg \eta_L$ to neglect the contribution from the successive wavefront. The specific form of F_L is

$$F_L(x) = a_L e^{-\eta_L x},$$
$$a_L = \frac{(\nabla_H N(H) \cdot A_L, G e^{\eta_L x})}{(G, H')}. \tag{21}$$

5. The normal velocity of the spiral arm

The spiral evolution equation (19) can be expressed in a more compact form in terms of the normal velocity and the curvature of the real spiral, $x_s(\sigma, t)$ (see (6)). The normal velocity is given by

$$v_n = \partial_t x_s \cdot \hat{n},$$
$$n = \frac{\partial_\sigma y_s \hat{x} - \partial_\sigma x_s \hat{y}}{\left[(\partial_\sigma x_s)^2 + (\partial_\sigma y_s)^2\right]^{1/2}}, \tag{22}$$

where $\hat{n}$ is a unit vector normal to the real spiral, and x_s and y_s are the components of x_s. The curvature is given by

$$K(\sigma, t) = \frac{\partial_\sigma x_s \partial_\sigma^2 y_s - \partial_\sigma y_s \partial_\sigma^2 x_s}{\left[(\partial_\sigma x_s)^2 + (\partial_\sigma y_s)^2\right]^{3/2}}. \tag{23}$$

Using (4) and (6) in (22) and (23) we find

$$v_n = \omega \rho_0 + \partial_t \zeta + \omega \partial_\sigma \zeta, \tag{24}$$

[#2] The same expression for $\bar{D}$ has been derived in ref. [3].

and

$$K(\sigma, t) = K_1 + \rho_0 K_1^3 \partial_\sigma \zeta - K_1^2 \partial_\sigma^2 \zeta + \ldots, \qquad (25)$$

where $K_1 = K_1(\sigma, \zeta)$ and the ellipses denote higher-order contributions. Comparing (24) and (25) with (19) and (20) we get

$$v_n = c_0 - F_L(\zeta^+ - \zeta + d) - \bar{D}K(\sigma, t), \qquad (26)$$

where $c_0 = \omega \rho_0$ is the velocity of a rectilinear solitary wavefront. Thus, the normal velocity of the spiral wavefront is affected by the tail F_L of the preceding wavefront (or the recovery phase of the medium ahead of it) and by its local curvature.

Eq. (26) can be written in the more familiar form [15]:

$$v_n = c(\lambda) - \bar{D}K, \qquad (27)$$

where $c(\lambda) = c_0 - F_L(\lambda)$ is the dispersion relation [10], and λ is the pitch of the real spiral wave, x_s, except that here the pitch is not a constant parameter but a function of σ and time and is given by $\lambda = \lambda(\sigma, t) = \zeta(\sigma + 2\pi, t) - \zeta(\sigma, t) + 2\pi c_0/\omega$.

The transport functional $\bar{D}$ is invariant to reaction parameters (i.e. parameters appearing in the reaction part of (1)) when all transport coefficients (diffusion constants) assume the same constant value, D. For in this case $\mathbf{D} = D\mathbf{I}$, where $\mathbf{I}$ is the unit matrix, and according to (20) $\bar{D} = D$. Even in the case of nonequal diffusion constants, $\bar{D}$ has been considered to be a constant number independent of the reaction kinetics [15, 16]. In that case $\bar{D}$ has been associated with the diffusion constant of the fast variable in two-variable type models of excitable media. If we denote quantities pertaining to the fast and slow variables by the indices 1 and 2, respectively, we have according to (20):

$$\bar{D} = \frac{\sum_{i=1}^2 D_i \int_{-\infty}^\infty dx\, G_i H_i'}{\sum_{i=1}^2 \int_{-\infty}^\infty dx\, G_i H_i'}. \qquad (28)$$

Thus, $\bar{D} \approx D_1$ when

$$\int_{-\infty}^\infty dx\, G_1 H_1' \gg \int_{-\infty}^\infty dx\, G_2 H_2', \qquad (29)$$

assuming D_2 is not much larger than D_1. When (29) does not hold, $\bar{D}$ depends on the spatial profile of the spiral wavefront, and therefore on the reaction parameters (see discussion in section 7).

6. Numerical solutions

To solve the spiral evolution equation (19) we have to specify the parameters that appear in this equation as well as boundary conditions. The former include the rotation frequency, ω, the velocity of a rectilinear solitary wavefront, $c_0 = \omega \rho_0$, the transport functional $\bar{D}$, given by (20), and the wavefront-interaction parameters η_L and α_L that appear in (21). The parameters c_0 and η_L can be evaluated by solving the one-dimensional problem (2) for a particular model. The evaluation of $\bar{D}$ and a_L calls in addition for a solution of the one-dimensional adjoint problem $\mathscr{L}_1^\dagger G_1 = 0$. Alternatively, one can evaluate the dispersion relation $c = c(\lambda)$ (either numerically or experimentally), extract from this relation the velocity c_0, and identify $c_0 - c(\lambda)$ with the wavefront interaction term $F_L(\lambda)$. If, in addition, one can assume that all transport coefficients (diffusion constants) assume the same constant value, or when (29) is satisfied, then $\bar{D}$ becomes a known constant. In both cases we are left with one undetermined parameter, ω.

The solutions of (19) are expected to be sensitive to the boundary condition imposed at the spiral tip. If we are interested in steadily rotating spiral solutions, for which $\partial_t \zeta(\sigma, t) = 0$, a proper boundary condition would be

$$\zeta(\sigma_{\text{tip}}, t) = \left(\rho_{\text{tip}}^2 - \rho_0^2 \right)^{1/2} - \sigma_{\text{tip}} \rho_0,$$

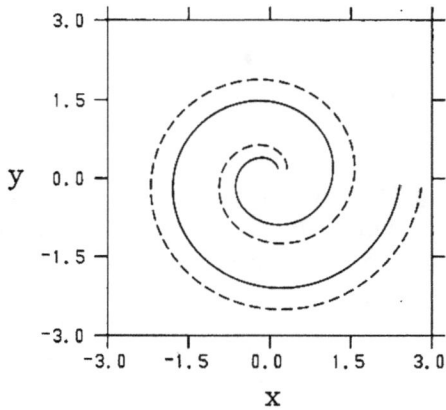

Fig. 2. Asymptotic form of a steadily rotating spiral wave obtained by numerical integration of (19) (solid curve), and the involute spiral (dashed curve). The distance between the two spiral curves is given by $\zeta = \zeta(\sigma)$. Parameter values used: $\rho_0 = 0.2$, $\sigma_{min} = 1.5$, $\omega = 0.9$, $\tilde{D} = 0.001$, $a_L = 1.0$ and $\eta_L = 3.0$.

where

$$\rho_{tip} \equiv |x_s(\sigma_{tip}, t)| = \text{constant}$$

is the radius of the core around which the tip rotates. Our interest here, however, is in dynamical aspects of spiral waves, and in addressing these we want to avoid restrictive boundary conditions. We therefore use in the numerical integration of (19) one-sided derivatives to evaluate boundary terms. We integrated (19) for a system consisting of two wavefronts (i.e. $\sigma \in S_1 \cup S_2$), using an implicit finite-difference scheme. We describe now a few results and discuss their physical meanings.

In a wide range of parameters we find stable, stationary solutions of (19), corresponding to steadily rotating spiral waves. A typical form of such a spiral wave is shown in fig. 2. The convergence to steady rotation is shown in fig. 3a, where the distance of the spiral tip from the center of rotation, $\rho_{tip} \equiv |x_s(\sigma_{tip}, t)|$, is plotted as a function of time. As the parameter $\tilde{D}$ is increased beyond a critical value, $\tilde{D}_c$, steady rotation becomes unstable and two-frequency rotation (compound rotation) sets in. Fig. 3b displays ρ_{tip} as a function of time for $\tilde{D} > \tilde{D}_c$. The relaxation time

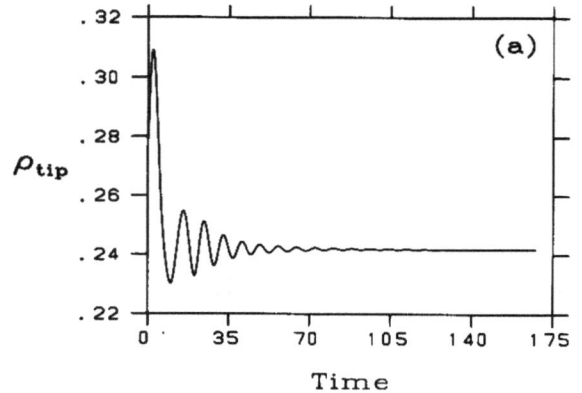

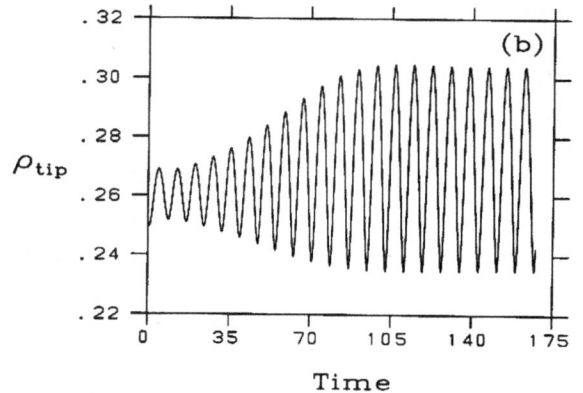

Fig. 3. The distance of the tip from the center of rotation, ρ_{tip}, as a function of time, for values of $\tilde{D}$ corresponding to (a) steady and (b) compound rotations. Parameter values as in fig. 2 except that $\tilde{D} = 0.001$ in (a) and $\tilde{D} = 0.003$ in (b).

to stable rotation becomes very long as $\tilde{D} \to \tilde{D}_c$, suggesting a supercritical Hopf bifurcation. A detailed study of this bifurcation will appear elsewhere.

The transition to compound rotation seems to originate from a competition between curvature and wavefront-interaction effects. Away from the spiral tip curvature acts to smooth out local or short-wavelength perturbations [5]. Near the tip, however, curvature becomes a destabilizing factor: upon straightening a small segment that contains the tip, curvature is reduced, normal velocity is enhanced, and further straightening is favored. Fig. 4 illustrates this instability. Wave-

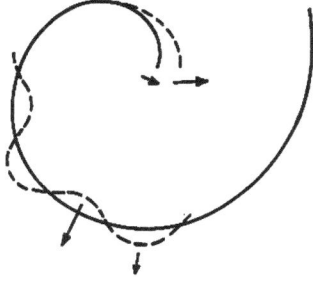

Fig. 4. Schematic illustration of the effect of curvature. Normal velocity (indicated arrows) decreases as curvature increases. As a result local or short-wavelength perturbations (dashed curves) that occur sufficiently far from the tip always decay, while those involving the tip may grow.

front interactions, on the other hand, act in the opposite direction. As the tip comes closer to the wavefront ahead of it, it is forced to slow down because of the highly refractory region it penetrates into. When the gain in normal velocity due to curvature exceeds the loss due to wavefront repulsion, tip perturbations, such as that illustrated in fig. 4, will grow. The parameter $\tilde{D}$ determines the strength of the curvature effect (see (27)). This explains why the transition to compound rotation occurs when $\tilde{D}$ is increased.

7. Discussion

The normal velocity expression (27) has been used in refs. [15, 16] to evaluate a relation among the parameters, ω, $c(\lambda)$, ρ_{tip} and $\tilde{D}$, for steadily rotating spiral waves. The spiral pitch, λ, was assumed to be a constant number. This relation together with the dispersion curve, $c = c(\lambda)$, were then used to determine the rotation frequency, $\omega = \omega(\rho_{tip}, \tilde{D})$, for a given core radius ρ_{tip} (and $\tilde{D}$). In the present approach, the two conditions are integrated together in (19). Stationary solutions of (19) select a particular radius, $\rho_{tip} = \rho_{tip}(\omega, \tilde{D})$, for a given rotation frequency, as shown in fig. 3a. The indeterminacy of ρ_{tip} (or ω) has recently been resolved, for a particular model system, by considering independently the shapes of the spiral wavefront and waveback [15, 17].

Transitions to compound rotation have been observed in numerical simulations [18–22] as well as experiments [8, 9, 20, 23]. In these studies various quantities have been used as control parameters and the question arises, what is the natural bifurcation parameter for this phenomenon? The transition mechanism proposed in the previous section points toward a possible answer. It suggests that factors that make the effect of curvature stronger relative to the effect of wavefront interactions, can lead to destabilization of steady rotation. These factors become apparent when considering the variation in normal velocity caused by tip perturbations of the form illustrated in fig. 4. Denoting the wavefront's radius of curvature by R, the velocity variation reads $\delta v_n \sim c'(\lambda)\delta\lambda + (\tilde{D}/R^2)\delta R$. For perturbations which are weakly nonuniform $\delta\lambda \propto -\delta R$ to leading order. Such perturbations therefore grow when the ratio $\tilde{D}/R^2$ is sufficiently large, or when the slope of the dispersion relation at the spiral pitch, $c'(\lambda)$, is sufficiently gradual. A dimensionless bifurcation parameter for the onset of meander can now be constructed as $R_M = \tilde{D}K_{tip}^2/c'(\lambda_{tip})$, where λ_{tip} and K_{tip} are the spiral pitch and curvature at the tip, respectively.

In the Belousov–Zhabotinsky reaction the transition to compound rotation has been observed by varying the potassium bromate concentration [8] and the acidity of the solution [9]. It may well be that $\tilde{D}$ does not remain constant when these quantities are varied. This possibility can be tested experimentally by measuring $\tilde{D}$ at various pH values or bromate-ion concentrations, using the method proposed in ref. [24]. Whether $\tilde{D}$ is sensitive to reaction parameters or not, can also be checked by estimating, analytically or numerically, the validity of (29) for a particular model system. In any event it is the parameter R_M that has to be studied as the instability point is approached.

A transition to compound rotation has also been obtained by Zykov [6]. In that work, the geometrical theory [1] has been extended to include a dependence of the local propagation speed on the time elapsed since the last excitation. This time interval has been evaluated numerically.

In integrating eq. (19), parameter regimes were found where neither steady rotation nor compound rotation were observed. Tip trajectories in these regimes were rather abrupt and inconsistent with the smallness assumptions $\partial_t \zeta \sim \partial_\sigma \zeta \sim \epsilon$. This behavior is attributed to the restrictive tip dynamics inherent in the present approach; the tangential velocity of the tip depends on the normal degree of freedom, $\zeta(\sigma_{\text{tip}}, t)$, and its derivatives, rather than being independent. A complete account of the two-dimensional spatial structure of the spiral tip is needed to relax this constraint.

Acknowledgements

The support of the Bantrell Career Development Fellowship for Scientific Research is gratefully acknowledged.

References

[1] V.S. Zykov, Simulation of Wave Processes in Excitable Media (Nauka, Moscow, 1984) [English transl.: Manchester Univ. Press, Manchester, 1987].

[2] P.K. Brazhnik, A.V. Davydov and A.S. Mikhailov, Teor. Mat. Fiz. 74 (1987) 440 [English transl.: Sov. Phys. Theor. Math. Phys.].

[3] P.K. Brazhnik, A.V. Davydov, V.S. Zykov and A.S. Mikhailov, Zh. Eksp. Teor. Fiz. 93 (1987) 1725 [English transl.: Sov. Phys. JETP].

[4] A.S. Mikhailov, in: Nonlinear Wave Processes in Excitable Media, eds. A. Holden, M. Markus and H. Othmer (Plenum Press, London, 1990), to appear.

[5] E. Meron and P. Pelcé, Phys. Rev. Lett. 60 (1988) 1880.

[6] V.S. Zykov, Biophysics 32 (1987) 365; J. Nonlinear Biol. 2 (1990).

[7] A.T. Winfree, When Time Breaks Down (Princeton University Press, Princeton, NJ, 1987).

[8] G.S. Skinner and H.L. Swinney, Physica D 48 (1991) 1–16.

[9] T. Plesser, S.C. Müller and B. Hess, J. Phys. Chem. 94 (1990) 7501;
S.C. Müller and B. Hess, in: Cooperative Dynamics in Complex Physical Systems, ed. H. Takayama (Springer, Berlin, 1989) p. 307.

[10] E. Meron, Phys. Rev. Lett. 63 (1989) 684.

[11] C. Elphick, E. Meron, J. Rinzel and E.A. Spiegel, J. Theor. Biol. 146 (1990) 249.

[12] A.T. Winfree, Science 175 (1972) 634.

[13] S.C. Müller, T. Plesser and B. Hess, Physica D 24 (1987) 87.

[14] C. Elphick, E. Meron and E.A. Spiegel, SIAM J. Appl. Math. 50 (1990) 490.

[15] J.J. Tyson and J.P. Keener, Physica D 32 (1988) 327.

[16] J.P. Keener and J.J. Tyson, Physica D 21 (1986) 307.

[17] P. Pelcé and J. Sun, Physica D 48 (1991) 353.

[18] V.S. Zykov, Biophysics 31 (1986) 940.

[19] E. Lugosi, Physica D 40 (1989) 331.

[20] W. Jahnke, W.E. Skaggs and A.T. Winfree, J. Phys. Chem. 93 (1989) 740.

[21] D. Barkley, M. Kness and L.S. Tuckerman, Phys. Rev. A 42 (1990) 2489.

[22] A. Karma, Phys. Rev. Lett. 65 (1990) 2824.

[23] K.I. Agladze, A.V. Panfilov and A.N. Rudenko, Physica D 29 (1988) 409.

[24] P. Foerster, S.C. Müller and B. Hess, Science 241 (1988) 685.

Physica D 49 (1991) 107–113
North-Holland

Vortex initiation in a heterogeneous excitable medium

A.V. Panfilov and B.N. Vasiev

Institute of Biological Physics, 142292, Pushchino, Moscow Region, USSR

We studied numerically the process of vortex initiation in the heterogeneous active medium which is described by a FitzHugh–Nagumo-type model. Vortex initiation results from interaction of two external stimuli with a stepwise inhomogeneity in refractoriness. The influence of distance between the place of stimulation and heterogeneity and geometrical sizes of the heterogeneity on the process of vortex initiation is examined. The drift and interaction of vortices is also studied.

1. Introduction

The existence of vortices is a general property of excitable media of different nature. Their appearance in physical, chemical and biological media leads to a special kind of chaos in these media – autowave chaos [1]. The initiation of vortices in cardiac tissue leads to a dangerous cardiac arrhythmias, in particular, to paroxysmal tachicardia and to heart fibrillation [2].

One of the most important mechanisms of the vortex initiation was proposed by Krinsky in 1968 [3], who studied an axiomatic model of a heterogeneous excitable medium (fig. 1). In this axiomatic model each point of the medium can be in the following three states: state of rest, excitation state and refractory state. The point at rest becomes excited if any of neighboring points is in the excited state. In this model, in contrast to the classical Wiener–Rosenbluth model [4], the excited state lasts for some finite period of time, usually referred as τ. In ref. [3] the following

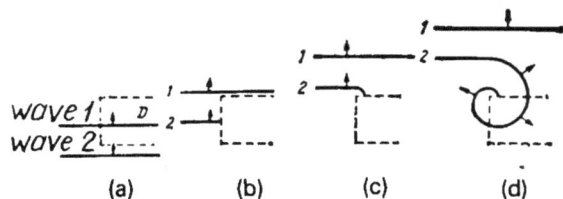

Fig. 1. Vortex initiation in an axiomatic model of an excitable medium.

process of vortex initiation was proposed. Assume that in an excitable medium with a refractoriness R_l a region D with a prolonged period of refractoriness R_h exists. If in this medium two waves propagate with an interval T_{st} (which has a value between R_l and R_h), then a second wave cannot penetrate into the region D, because its refractoriness R_h is higher than T_{st}. As a result, a wave break occurs. This break moves along the heterogeneity boundary, and when the refractoriness period in the region D ends, the break propagates into this region and a spiral wave is formed.

This mechanism has been confirmed in experiments on cardiac tissue [5], and it is perhaps one of the main mechanisms of the initiation of the cardiac arrhythmias.

In this paper we consider this process of vortex initiation in a reaction–diffusion model of an excitable medium.

2. Model, method of computation, and results

To represent the excitable medium the FitzHugh–Nagumo model was used:

$$\frac{\partial E}{\partial t} = \Delta E - f(E) - g, \quad \frac{\partial g}{\partial t} = \varepsilon(E)(E - g),$$

$$(1)$$

where Δ is the two-dimensional Laplacian, $f(E)$ is the nonlinear N-shaped function, $\varepsilon(E)$ is the parameter determining the temporal behavior of the slow variable g, i.e. the duration of the excited state and the duration of the refractory tail of the wave.

For calculations the following shapes of functions $f(E)$ and $\varepsilon(E)$ were used:

$$
\begin{aligned}
f(E) &= C_1 E, & E &< E_1, \\
&= -C_2(E-a), & E_1 &< E < E_2, \\
&= C_3(E-1), & E &> E_2, \\
\varepsilon(E) &= \varepsilon_1, & E &\leq 0, \\
&= \varepsilon_2, & 0 &< E < 1, \\
&= \varepsilon_3, & E &\geq 1, \quad (2)
\end{aligned}
$$

where ·the parameters determining the shape of the function $f(E)$ were as in a previous paper [6]: $E_1 = 0.018$, $E_2 = 0.94$, $C_1 = 4$, $C_2 = 1$, $C_3 = 15$, $a = 0.09$. These parameters specify the fast processes such as initiation and propagation of the wave front. The main parameter for the purpose of our paper is refractoriness, which is determined by function $\varepsilon(E)$. In $\varepsilon(E)$ (see eq. (2)) the parameter ε_1 specifies the duration of the refractory tail and ε_3 the duration of the excited state. To isolate the effects of refractoriness, the value of ε_1 was chosen smaller than ε_3; in particular, the values $\varepsilon_1 = 0.2$, $\varepsilon_2 = 0.05$ and $\varepsilon_3 = 2$ were used.

The inhomogeneity was preset by the parameter ε_1. In our calculations the value of ε_1 in region D changes from 0.2 to 0.03.

Neumann boundary conditions were used. Calculations were performed using the explicit Euler method with standard space step $h_x = 1.2$ and time step $h_t = 0.12$ [6].

2.1. Onset of the wave break

In the axiomatic model there are the following conditions for the break formation [3]:

$$
R_l < T_{st} < R_h - \tau, \quad (3)
$$

where T_{st} is the time interval between stimulations, R_l is the refractoriness period of the medium, R_h is the refractoriness in the D region, and τ is the duration of the excitable state. The width of the interval (3) in which the break could be formed is usally referred to as the width of the vulnerable phase.

In reaction–diffusion models where the velocity of the first wave is greater than that of the second one the condition of the break formation becomes

$$
R_l < T_{st} < R_h - \tau - (t_2 - t_1), \quad (4)
$$

where

$$
t_1 = \int_0^L \frac{dx}{V_1} = \frac{L}{V_1}, \quad t_2 = \int_0^L \frac{dx}{V_2(l)} \quad (5)
$$

are the times during which the first (t_1) or the second (t_2) wave propagates a distance L from the place of stimulation to the region D (fig. 2a), and x is the current distance. The $V_2(l)$ is the dispersion relation (the dependence of the velocity of the second wave on its distance l to the first wave).

The delay of the second wave from the first leads to a decrease in the width of the vulnerable phase, ΔT. In fact, the velocity of the second wave is smaller than that of the first one and the distance between them increases with time. This also formally follows from (4), (5), as the difference between the two integrals has constant sign and hence $t_2 - t_1$ increases with increasing L. At a sufficiently large L the width of the vulnerable phase, ΔT, can be equal to zero even if heterogeneity in refractoriness ΔR exists.

Fig. 3 shows the dependence of the width of the vulnerable phase, ΔT, on the distance L which was obtained numerically. It is seen that the value of ΔT decreases to zero with an increase of the value of L.

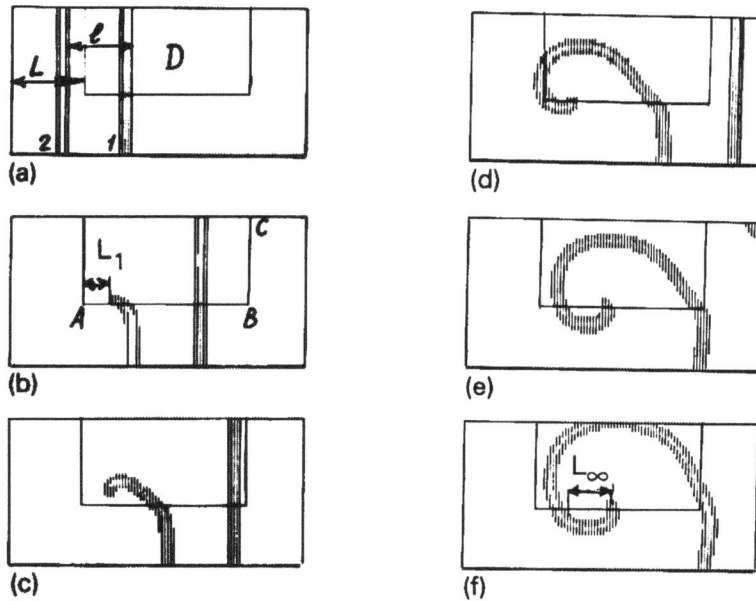

Fig. 2. Vortex initiation in a reaction–diffusion model. Numerical simulations. Model (1). In the dark regions $E > 0.8$, ε_1 in the medium is equal to 0.2; in the region D, $\varepsilon_1 = 1/17.5$. (At $\varepsilon_1 = 0.2$ the wavelength is $\lambda = 24$ and the period of the vortex is $T = 17$; at $\varepsilon_1 = 1/17.5$, $\lambda = 32$, $T = 43$.) The size of the heterogeneity is 36×66. In the successive figures the time is as follows: (a) $T = 43$, (b) $T = 71$, (c) $T = 86$, (d) $T = 100$, (e) $T = 215$, (f) $T = 532$.

2.2. Generation of a vortex

For the vortex to arise, the break of a front of the second wave has to penetrate into the region D. This penetration in the τ-model can happen only on the boundary BC (see fig. 2b) of region

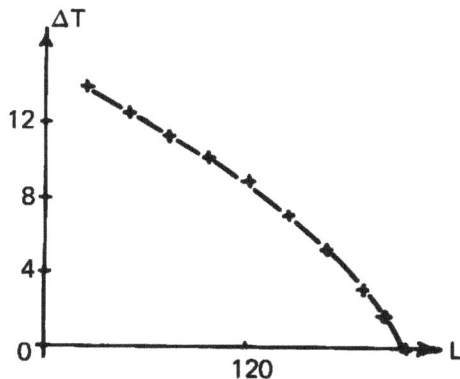

Fig. 3. The dependence of the vulnerable phase width, ΔT, on the distance L between the heterogeneity D and the region of stimulation (see fig. 2). $R_t = 12.6$, $R_h = 25.8$.

D, as the velocities of the first and the second waves are equal. In reaction–diffusion models, this can happen also on the boundary AB as the velocity of the second wave is smaller than that of the first one.

Fig. 2 shows this process. It is seen that the distance between the first and the second wave successively increases, and when it becomes more than the refractoriness in D, the second wave penetrates in region D and forms the vortex.

The distance L_1 on which the second wave propagates along the boundary before penetrating into the region D is determined at least by two factors: the lag of the second wave from the first one, and the bending of the front of the second wave. One can see this bend in fig. 2b. The bend of the second wave is associated with the fact that the wave near the break has a smaller velocity than the plane wave due to the leakage current at the break.

If the length of the boundary AB is less than the distance L_1, the excitation wave penetrates

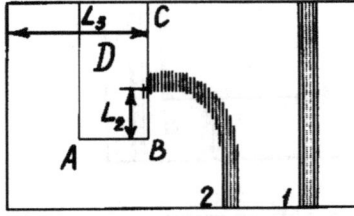

Fig. 4. Vortex generation in a dynamic model (1) in the case of $L_1 > AB$. $R_l = 12.6$; $R_h = 43.8$.

into the region D on the boundary BC (fig. 4). This case is similar to that in the axiomatic model (fig. 1) where the breaking wave propagates a distance $L_2 = (R_h - T_{st})V$ along the boundary BC and penetrates into region D and forms a spiral.

Let us estimate the path L_2 in the case of dynamic models. If the distance between the place of stimulation and the region D is equal to L_3 (fig. 4), then similar to (4), (5) we have:

$$L_2 = \left(R_h \cdot T_{st} + \frac{L_3}{V_1} - \int_0^{L_3} \frac{dx}{V_2(l)} \right) V_1. \qquad (6)$$

In (6) we take into account only one factor: the lag of the second wave from the first one on the path L_3; we do not consider the bend of the second wave and its dispersion on the boundary BC. If the length of the border BC is less than L_2 the vortex cannot arise and the breaking wave runs into the boundary of the medium and disappears.

2.3. Drift of the vortex

Numerical experiments have shown that the vortex arising on the stepwise inhomogeneity drifts (fig. 2). The velocity of the drift has two components: along the boundary of heterogeneity and transverse to it. Due to the transversal component the vortex shifts into the region D, its drift velocity decreases, and the vortex stops at some distance L_∞ from the point of origin.

The value of the distance L_∞ depends on the extent of inhomogeneity of the medium $\Delta R/R_l$.

Fig. 5 shows the dependence of L_∞ on $\Delta R/R_l$. It is seen that the distance L_∞ increases exponentially with an increase of the degree of inhomogeneity of the medium.

Calculations have shown that the direction of longitudinal drift of the vortex depends on the direction of its rotation and the direction of the transversal drift, and it is given by the vector:

$$U = [\omega \times T], \qquad (7)$$

where ω is the vector of angular velocity of the vortex and T is the vector in the direction of the transversal drift.

Interaction of two drifting vortices. In the case when the region D is within the medium (fig. 6), two vortices arise on the heterogeneity. These vortices rotate in opposite directions, and according to formula (7) will drift towards each other. Numerical experiments have shown that due to this drift the annihilation of the vortices can occur (fig. 6), i.e., vortices arising on heterogeneities of excitable media could have a finite lifetime.

Fig. 7 shows the dependence of the vortices' lifetime, T_l, on the initial distance between them. It is seen that the lifetime increases with an increase of the distance. When the initial dis-

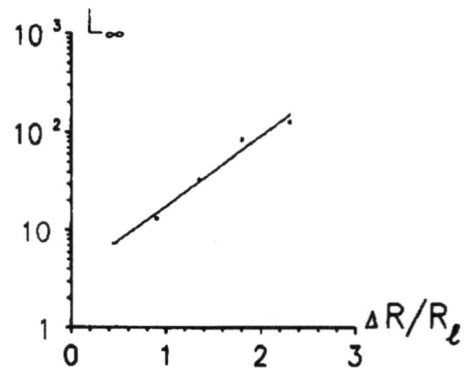

Fig. 5. The dependence of the distance of the vortex shift, L_∞, on the extent of inhomogeneity of the medium, $\Delta R/R_l$. $R_l = 12.6$, $R_h = 12.6$–43.8. L_∞ is given on a logarithmic scale.

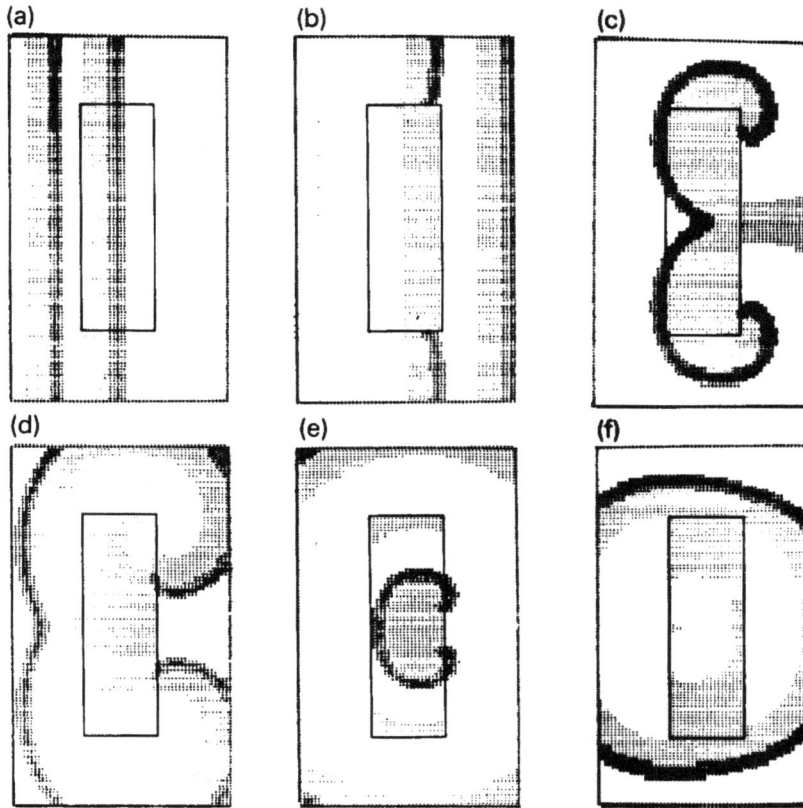

Fig. 6. The initiation of two vortices and their annihilation. $R_l = 12.6$, $R_h = 33.6$. (a) $T = 33$, (b) $T = 67$, (c) $T = 117$, (d) $T = 134$, (e) $T = 418$, (f) $T = 451$.

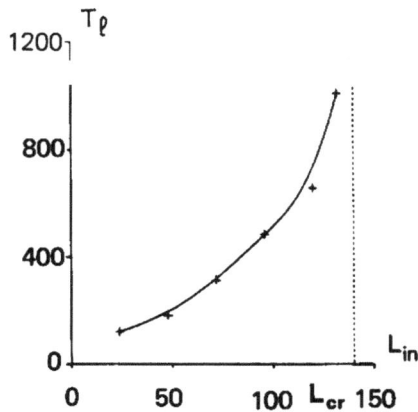

Fig. 7. The dependence of the lifetime of the vortices, T_l, on the initial distance between them, L_{in}. $R_l = 12.6$, $R_h = 30.0$.

tance is more than L_{cr} the lifetime becomes equal to infinity, i.e., due to the initial conditions two different regimes of vortex interaction could be realized: the regime of annihilation or the regime of the infinite lifetime.

In the absence of annihilation the vortices stabilize at some distance L_f between them. It is obvious that in the case of a large initial distance ($L_{in} \gg L_{cr}$) the following formula for the final distance L_f is valid:

$$L_f = L_{in} - 2L_\infty. \tag{8}$$

The dependence of L_f on L_{in} in the whole range is shown in fig. 8. It is seen that at large L_{in} the dependence is linear, in accordance with for-

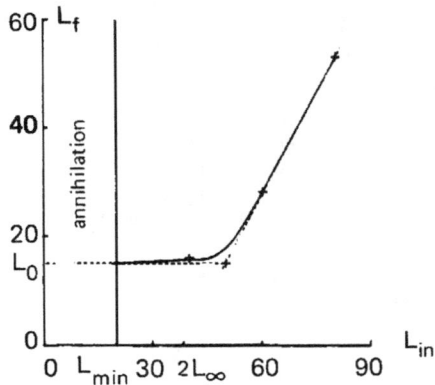

Fig. 8. The dependence of the final distance between vortices, L_f, on the initial distance between them, L_{in}. $R_t = 12.6$, $R_h = 22.8$.

mula (8). At small L_{in} the dependence departs from linearity and tends to some value L_0. If the initial distance between vortices is less than L_{min}, the vortices annihilate. Further investigations have shown that L_0 is the minimal possible distance between two vortices in homogeneous excitable media in which they can exist. The existence of this distance has been found in ref. [7], where it has been shown that two vortices rotating in opposite directions repel each other and diverge to some distance L_0.

3. Discussion

Some experimental results on vortex behavior in heterogeneous cardiac tissue [8] and in a temperature gradient in the Belousov–Zhabotinsky reaction [9] have been obtained recently. Let us compare them with the results of the present paper. The experimental conditions in ref. [8] are analogous to those in the present paper: a stepwise heterogeneity in refractory period, and a drift of the vortex along the boundary of heterogeneity does exist. The direction of this drift coincides with the direction of the longitudinal drift of the vortex shown in fig. 2, i.e., with that

given by formula (7). Unfortunately, the transversal component of the drift has not been observed in ref. [8]. This can be due to the small size of the fragment of cardiac tissue studied in that paper; the vortex ran into the wall and disappeared after two rotations.

In ref. [9] both components of the vortex drift were observed. The direction of the longitudinal component also coincides with that given by formula (7), but the direction of the transversal component is the opposite: in ref. [9] the vortex drifts so that its period decreases, whereas in the present paper the drift is directed to the increase of the vortex period. This difference is probably associated with the fact that the temperature gradient in the BZ reaction produces not only heterogeneity in refractoriness but also in the velocity of the wave propagation, while in the present paper heterogeneity in velocity is absent.

Finally, we point out one of the results which could be interesting for experiments in cardiac tissue. As shown in formulae (4), (5), the conditions of the break formation depend on the distance from the place of stimulation to the region of heterogeneity. This effect is due to the dependence of the wave velocity on the distance between the waves. This dependence is well known for cardiac tissue [10]. Therefore, we can assume that for cardiac tissue the width of vulnerable phase depends on the distance from the place of stimulation to the region of heterogeneity. This also means that the extent of heterogeneity of excitable medium obtained in experimental measurements will depend on the distance from the stimulating electrode to the boundary of heterogeneity.

References

[1] V.I. Krinsky, Autowaves: results, problems, outlooks, in: Self-Organization. Autowaves and Structures Far from Equilibrium, ed. V.I. Krinsky (Springer, Berlin, 1984) pp. 9–19.

[2] V.I. Krinsky, Pharmac. Theor. B3 (1978) 539–544.

[3] V.I. Krinsky, Problem. Kibern. 20 (1968) 59–80.

[4] N. Wiener and A. Rosenbluth, Arch. Inst. Cardiol. Mexico 16 (1946) 205–265.

[5] M.A. Allesie, F.I.M. Bonke and F.J.G. Schopman, Circ. Res. 39 (1976) 168–177.

[6] A.V. Panfilov and A.N. Rudenko, Physica D 28 (1987) 215–218.

[7] E.A. Ermakova, A.M. Pertsov and E.E. Shnol, Physica D 40 (1989) 185–195.

[8] V.G. Fast and A.M. Pertsov, Circ. Res. (1990), in press.

[9] K.I. Agladze, A.V. Panfilov and B.N. Vasiev, Physica D, submitted for publication.

[10] J.L.R.M. Smeets, M.A. Allessie, W.J.E.P. Lammers, F.I.M. Bonke and J. Hollen, Circ. Res. 58 (1986) 96–108.

Physica D 49 (1991) 114–124
North-Holland

Wave patterns in an excitable reaction–diffusion system

H. Sevcikova and M. Marek

Department of Chemical Engineering, Prague Institute of Chemical Technology, Technika 5,
166 28 Prague 6, Czechoslovakia

Simulations of a non-autonomous periodically forced excitable reaction–diffusion system described by two parabolic partial differential equations with SH kinetics reveal the existence of various types of periodic and aperiodic wave trains. The arrangement of the main entrainment regions in the forcing amplitude–forcing period plane reflects the existence of both dynamic and static thresholds. Bursting phenomena are observed when the effects of the sinusoidal forcing at the boundary are studied.

1. Introduction

Excitability is most often studied in the context of biological or mechanical systems [1, 2]. Here we discuss properties of concentration waves excited by periodic forcing applied at the boundary of a distributed, spatially one–dimensional excitable reaction–diffusion system. Our results are interesting not only for studies of chemical wave generation, but also for the interpretation of signals traveling in various excitable biological media [3–5]. For example, one of the theories of cooperative changes in the conductivity of Na^+ and K^+ channels of neural membranes in the course of excitation couples these changes to the changes in the concentration of –S–H and –S–S groups in proteins contained in the membrane [6]. The SH kinetic model used in the following study is based on the reaction mechanism that describes changes of –S–H and –S–S groups of proteins with low molecular thiols [7].

Both the lumped parameter and distributed excitable systems have been found to show entrainment under periodic stimulation, similar to forced oscillatory systems [1, 8–11]. In analogy with the introduction of rotation numbers for the description of the structure of resonance regions in forced oscillatory systems, the firing number has been defined [12, 13] and used to describe the entrainment behaviour in the parametric plane of amplitude versus period of forcing for excitable systems. The firing number, in principle, corresponds to the number of cycles of excitation (or evoked waves) per forcing period. Altogether, both experimental and mathematical studies show devil's staircase-like dependence of the phase-locked firing numbers on one or both forcing parameters and, in some cases [8, 11, 14, 15], aperiodic behaviour was also observed.

In the resonance diagram for forced oscillators resonance regions form so-called "Arnold's tongues" starting at the forcing period axis. The resonance regions in forced excitable systems are usually schematically presented in the shape of narrow strips bent along both axes [9, 10, 12].

In this paper we present results of a numerical study of a mathematical model of a periodically forced one-dimensional reaction–diffusion system where both a periodic pulse-like forcing and a sinusoidal periodic forcing are considered. Kinetic parameters used in the SH kinetic model correspond to the saddle–point type of excitability [16].

A resonance diagram exhibiting a discontinuous region of 0 : 1 resonances and a nonmonotonous shape of boundaries between individual resonance regions is presented and discussed. In addition to the parametric studies of the asymptotic behaviour (the properties of final solutions of the system of PDEs), the evolution of the

system after its initial rest state has been subjected to the stimulation is also discussed. The
generation of pulse wave trains by periodic subthreshold stimulus and the existence of aperiodic
solutions are documented. The sinusoidal periodic forcing is shown to support bursting-like
behaviour of the system.

2. Mathematical model

The mathematical model consists of a system
of two dimensionless partial differential equations:

$$\frac{\partial X}{\partial t} = \frac{D_X}{L^2}\frac{\partial^2 X}{\partial z^2} + f(X,Y),$$

$$\frac{\partial Y}{\partial t} = \frac{D_Y}{L^2}\frac{\partial^2 Y}{\partial z^2} + g(X,Y), \qquad (1)$$

describing the evolution of concentrations (X,Y)
of two reaction species in time (t) and space
(spatial coordinates $z \in \langle 0,1 \rangle$). L denotes the
characteristic length of the system and D_X and
D_Y diffusion coefficients of the related reaction
species. The following values of parameters were
used in computations: $L = 1$, $D_X = 0.008$ and
$D_Y = 0.004$. The nonlinear functions $f(X,Y)$ and
$g(X,Y)$ describing the reaction kinetics are given
by the SH kinetic scheme [7]:

$$f(X,Y) = \frac{\mathrm{d}X}{\mathrm{d}t} = \alpha\frac{\nu_0 + X^\gamma}{1 + X^\gamma} - X(1+Y),$$

$$g(X,Y) = \frac{\mathrm{d}Y}{\mathrm{d}t} = X(\beta+Y) - \delta Y, \qquad (2)$$

where α, β, γ, δ and ν_0 are dimensionless positive kinetic constants. By changing the values of
the kinetic parameters one can create different
types of dynamics of reaction [11, 17].

In this study, the values of the kinetic parame-
ters were set equal to: $\alpha = 12.0$, $\beta = 1.5$, $\gamma = 3.0$,
$\delta = 1.0$ and $\nu_0 = 0.01$. The phase portrait of this

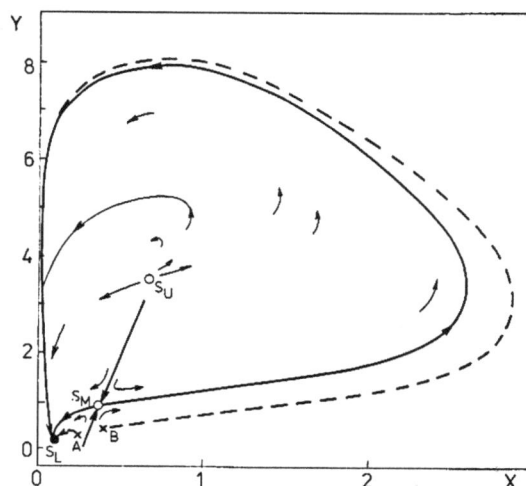

Fig. 1. Phase portrait of the excitable system (eqs. (2)); (●)
stable, (○) unstable stationary states, (A) subthreshold, (B)
superthreshold stimulus; (– – – –) excitable cycle evoked by a
superthreshold stimulus.

type of excitable system (eqs. (2)) is shown in fig.
1. There are three stationary state (S_L, S_M and
S_U), the first one being asymptotically stable. The
separatrix of the middle stationary state S_M (saddle point) defines the threshold of excitability.
The subthreshold perturbation of the stable stationary state S_L rapidly vanishes towards the
initial stable state while the superthreshold perturbation leads to a large excursion (excitation
cycle) before the system is restored to its initial
stable state again.

In a spatially distributed system (1) with
Neumann boundary conditions, the spatially homogeneous concentration profiles corresponding
to the lower stable state S_L,

$$X(z,t) = X_L, \quad Y(z,t) = Y_L \quad \text{for } z \in \langle 0,1 \rangle,$$
$$(3)$$

are also stable and were chosen as the initial rest
state. To simulate the wave initiation and propagation by periodic forcing, time-dependent
boundary conditions for variable X at $z = 0$ were

used:

$$X(0,t) = A_f \quad \text{for } t \in \langle nT_f, nT_f + \Delta\tau \rangle \,,$$

$$\frac{\partial X(0,t)}{\partial z} = 0 \quad \text{for } t \in \langle nT_f + \Delta\tau, (n+1)T_f \rangle$$

$$n = 0, 1, 2, \dots . \tag{4}$$

These conditions represent periodic pulse-like forcing, where $\Delta\tau$ denotes the length of the stimulus duration and A_f and T_f stand for the amplitude and period of periodic stimulation, respectively. The relation

$$X(0,t) = A_f + A_f \sin\left(\frac{2\pi}{T_f}t\right) \tag{5}$$

was used to represent the sinusoidal periodic forcing. In both cases the Neumann boundary conditions for variable Y at $z = 0$ and for both variables at $z = 1$ were preserved to complete the system of equations. The full mathematical model (eqs. (1)–(3) and (4) or (5)) was solved numerically using a three-point finite-difference scheme of the Crank–Nicolson type. The use of 128 spatial mesh points and a time step of 0.025 were found to be sufficient to obtain the required accuracy.

The numerical solutions were characterized by a "firing number" ν computed, by definition [12], as

$$\nu = \limsup_{q \to \infty} \frac{p}{q}, \tag{6}$$

where p is the number of traveling waves generated in the course of q applied stimuli.

Another characteristic we have studied was the sequence of mutual periods $\{T_n\}$, i.e., the time intervals elapsed when successive waves passed the same position. The concept of local characteristics (taken at certain chosen positions along the system), such as the time profiles of variables, phase portraits and stroboscopic maps, have been accepted to reflect properties of the wave solutions in studies of dynamical systems. The use of local characteristics corresponds to an experi-

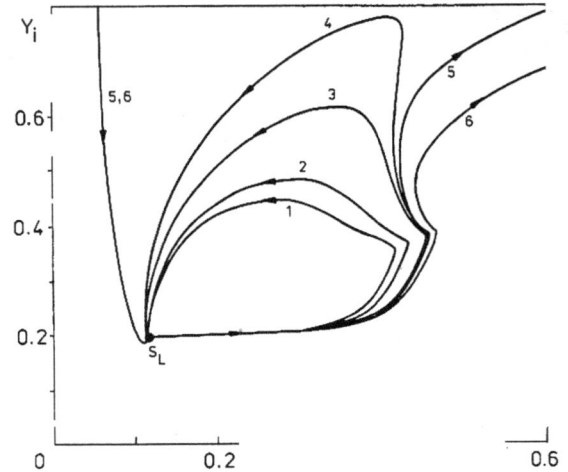

Fig. 2. Detail of the local phase portraits taken at the spatial point $i = 3$. The response of the system to several subthreshold (trajectories 1–4) and superthreshold (trajectories 5, 6) stimuli applied to a rest state (S_L). Trajectories 5, 6 follow the excitable cycle (cf. fig. 1). $A_f = 0.455$ (1), 0.46 (2), 0.48 (3), 0.482 (4), 0.483 (5) and 0.49 (6).

mental situation where, for example, the spreading of signals along nerve axons is often observed and recorded by inserting measuring electrodes at one or several specific points along the nerve fiber [14].

3. Results

3.1. Periodic pulse-like forcing

The excitable system under study is characterized by a "true" threshold behaviour [18] where the threshold is sharp and the differences between all or none responses to the stimulus are clearly defined (cf. fig. 2). To find the threshold value of the pulse stimulus A_{th}, the model equations (1)–(4) were solved for $T_f \to \infty$. The threshold value was found to lie in the interval $A_{th} \in \langle 0.482, 0.483 \rangle$.

By applying pulse-like periodic stimulation, three main types of solutions of the system of eqs. (1)–(4) have been found, depending on the values of forcing parameters (A_f, T_f):

(i) periodic solutions without traveling waves;

(ii) periodic solutions with traveling wave trains;

(iii) aperiodic solutions with traveling wave trains.

(i) *Periodic solutions without traveling waves* have firing numbers equal to zero since there are no waves traveling through the system under periodic forcing. The asymptotic solution of PDEs is in the form of spatially nonhomogeneous profiles of X and Y, periodically oscillating in time in the vicinity of the stable steady state S_L [19]. The amplitude of local oscillations decreases with the distance from the point of stimulation and approaches zero for $z > 0.4$, where the system variables preserve their initial steady state values (X_L, Y_L).

(ii) *Periodic solutions with traveling wave trains* are of two different types. The so-called "equally spaced wave trains" form the first group of solutions. Firing numbers of these wave solutions can generally be expressed as the ratio $\nu = 1 : q$, where q is a finite integer denoting the periodicity of the solution. These solutions are formed by a train of equal waves generated with a mutual period $T = qT_f$.

The wave solutions of the second group have firing numbers that can be expressed as the ratio of two finite integers: $\nu = p : q$, $p, q > 1$. We will call them "unequally spaced wave trains". The periodicity of a given solution is again the q multiple of the forcing period $(T = qT_f)$, but in this case the wave trains are composed of a periodic sequence of wave packets consisting of p different waves. The periods between the waves are different and fulfill the relation

$$\sum_{n=1}^{p} T_n = T = qT_f, \qquad (7)$$

where T is the period of the wave packet repetitions.

Fig. 3 illustrates the unequally spaced wave train solution with $\nu = 2 : 5$ and the method of firing number evaluation. The number of large

loops in the local phase space corresponds to the number of different waves in the packet, and the number of points on the stroboscopic map reflects the periodicity of the given solution.

(iii) *Aperiodic solutions with traveling wave trains* are solutions for which the firing numbers could not be expressed as the ratio of two finite integers. Local characteristics of two qualitatively different aperiodic solutions are shown in fig. 4 (each column describes one of these two aperiodic solutions).

In the first case the $A_+ = 0.494$ is less than the threshold value A_{th}, and thus several stimuli are necessary to sum up in time (see later) before the excitation occurs. The time course of the X variable observed near the point of stimulation (fig. 4a) shows several high peaks of excitation and several small peaks between the high ones, illustrating the summation of subthreshold stimuli. As a result there is only one traveling wave generated for every fifth or sixth applied stimulus. Since the small peaks do not propagate along the system, only high peaks corresponding to the

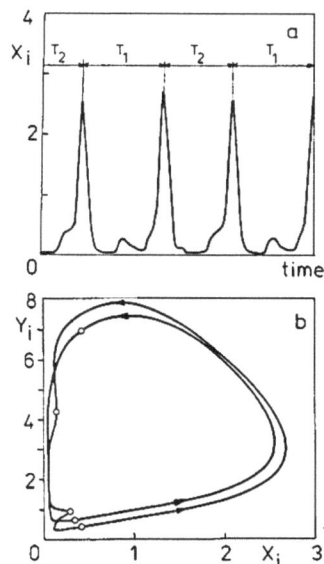

Fig. 3. Unequally spaced periodic wave solution ($A_f = 0.6$, $T_f = 2.75$, $\nu = 2:5$). (a) The local time course of X_i taken at $i = 10$, $T_1 = 7.423$, $T_2 = 6.337$. (b) The local phase portrait (solid line) and the local stroboscopic map (circles) taken at $i = 10$.

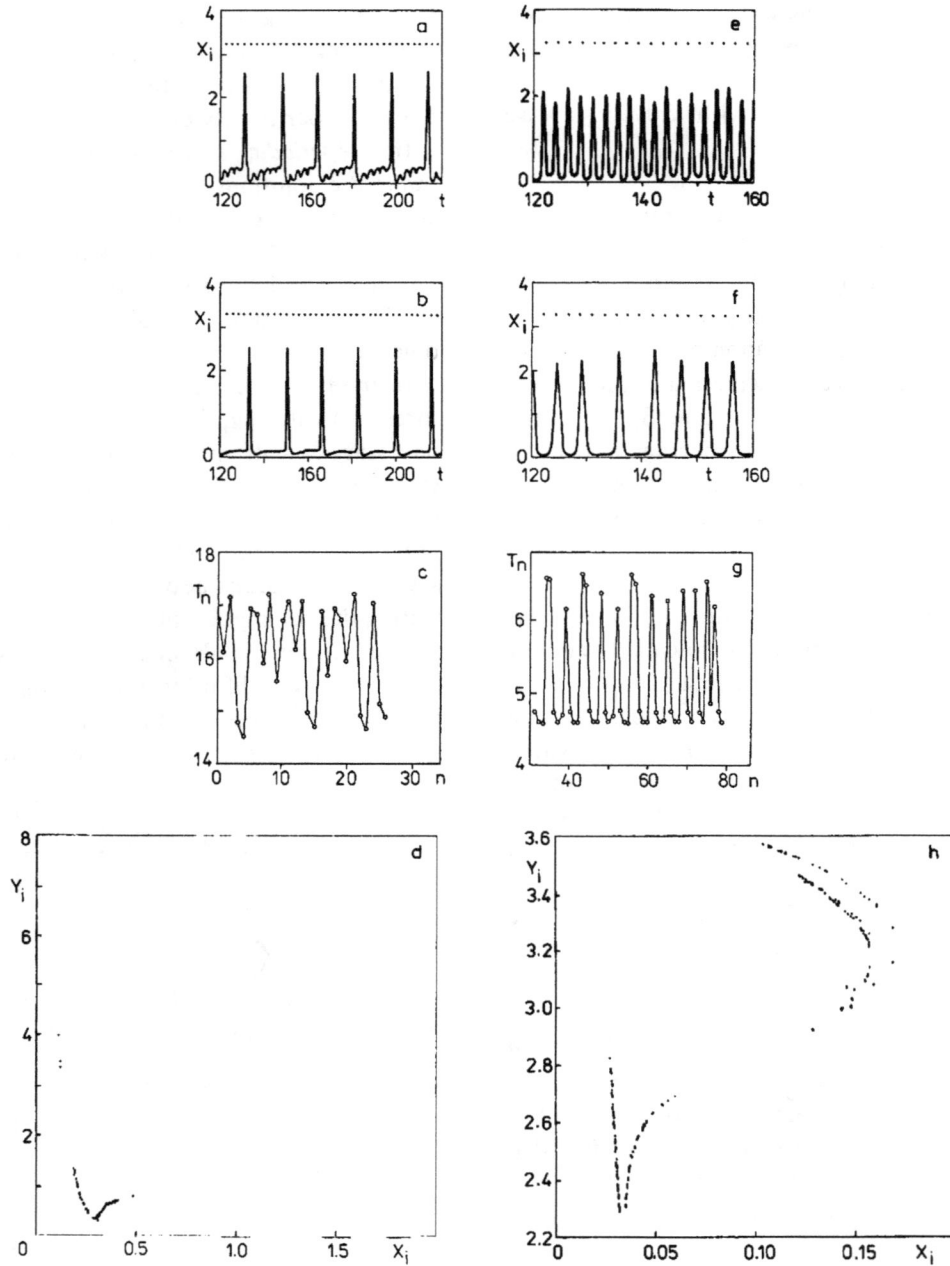

Fig. 4. Aperiodic behaviour observed for: (a–d) $A_f = 0.494$ and $T_f = 2.75$, (e–h) $A_f = 7.0$ and $T_f = 2.25$. (First and second row) Local time profiles of X_i taken at $i = 10$ (a, e) and $i = 77$ (b, f); the dots denote the time of pulse stimulus application. (c, g) Sequences of mutual periods measured at $z = 0.6$. (d, h) Local stroboscopic maps.

traveling waves are recorded in the time course of the X variable at the more distant positions (cf. fig. 4b). Mutual periods between waves are quite long and alternate irregularly in the range $T_n \in (14.5, 17.3)$, cf. fig. 4c. The aperiodicity of the observed behaviour is also illustrated by the stroboscopic map in fig. 4d, where the points randomly fill a closed curve.

The aperiodic solution in the right column is somewhat different. The forcing amplitude $A_f = 7.0$ is greater than the threshold A_{th}, and thus every applied stimulus excites a peak, cf. fig. 4e. However, due to the medium refractoriness not every excited peak gives rise to a traveling wave. This fact is seen in fig. 4f, where the peak appears once within two or three applied stimuli. The mutual periods between waves are shorter in this case and vary in the range $T_n \in (4.6, 6.8)$. The distribution of points on the stroboscopic map in fig. 4h also confirms the deterministic nature of the observed aperiodic behaviour.

We speculate that the wave aperiodicity in the first case (figs. 4a–4d) originates from the non-linear dynamics of the process of the wave generation at the boundary. Every fifth or sixth stimulation randomly elicits the wave, and then the aperiodic wave pattern in principle preserves its qualitative local characteristics when traveling along the system. We call this case a "pattern generation aperiodicity". In the second case (figs. 4e–4h) each stimulation generates the wave, but due to the interaction of generated waves with the refractoriness of the medium, a qualitatively different type of aperiodic pattern is developed in the course of propagation. We call this situation a "pattern propagation aperiodicity".

(iv) *The structure of resonance regions* in the plane of forcing parameters is represented in fig. 5. Contrary to the expected shape [9, 10, 12], we have found the region with $\nu = 0:1$ to be interrupted with a strip of repetitively fired wave trains. The zero wave region is thus divided into two

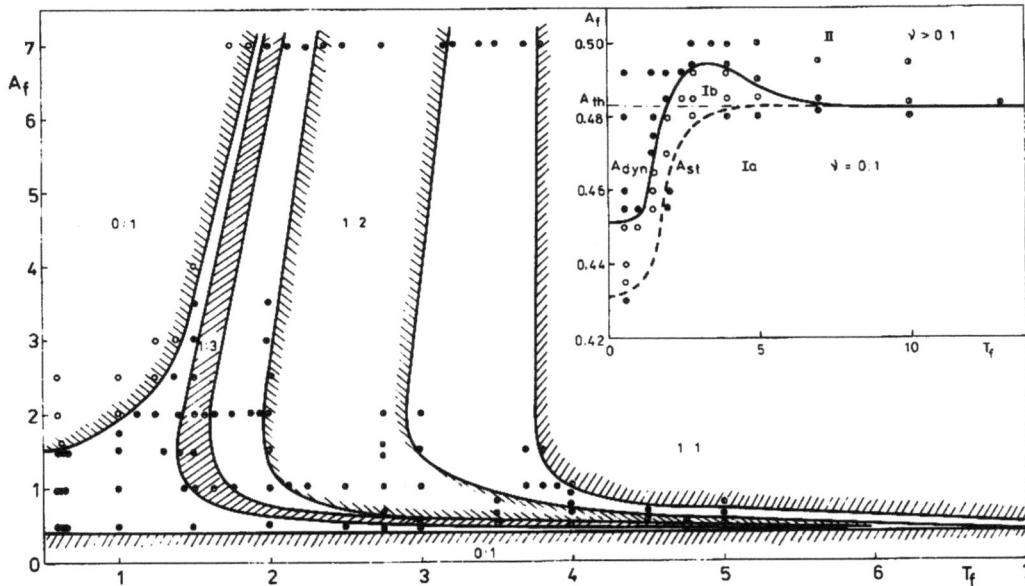

Fig. 5. The resonance diagram for the system of PDEs (1)–(4) in the plane of the forcing amplitude (A_f) versus the forcing period (T_f). The main resonances found have firing numbers as follows: (◐) 1:3, (◑) 1:2, (◒) 2:4, (◓) 1:1; (●) denotes other solutions with repetitive wave trains for which the firing numbers are different from all mentioned above. The inset: The detail of the boundary between the lower region of $\nu = 0:1$ resonances (Ia and Ib) and the region of repetitively fired wave trains (II, $\nu > 0:1$). The full line denotes the dependence of the dynamic threshold on the forcing period; the dashed line denotes the dependence of the static threshold on the forcing period.

parts, the lower one being naturally associated with the threshold value A_{th}.

It is important to point out that the value A_{th} denotes the threshold value for a single stimulus applied to a system in the rest state. The results obtained with the periodic forcing suggest that the threshold value of periodic stimulus differs from the value A_{th} and appears to be a complicated function of the forcing period. This fact is illustrated in the inset of fig. 5, where the boundary between the lower region of periodic solutions without traveling waves ($\nu = 0:1$) and the region of solutions with traveling wave trains ($\nu > 0:1$) is shown (full line). Moreover, the lower zero wave region is divided into two subregions differing in behaviour of the system during the transition from the original rest state to the final periodic solution without waves. While the transition in the subregion Ia is "straightforward", the system behaviour in the subregion Ib is characterized by the appearance of a single traveling wave elicited shortly after the beginning of the periodic stimulation.

Here we distinguish between two different types of the threshold magnitude of periodic stimulation. The one that excites a traveling wave in the rest state we call the "static" threshold (A_{st}), and the one responsible for the repetitive wave firing we call the "dynamic" threshold (A_{dyn}). Both thresholds are functions of the period of forcing (cf. the inset of fig. 5). In the limit of $T_f \to \infty$ both threshold values coincide and are equal to A_{th}:

$$\lim_{T_f \to \infty} A_{st}(T_f) = \lim_{T_f \to \infty} A_{dyn}(T_f) = A_{th}. \qquad (8)$$

When $T_f \to \infty$, the system returns to its original rest state. When the period of forcing shortens, both thresholds diverge. The dynamic threshold at first grows with the decreasing period and then decreases. The static threshold was found to decrease with decreasing period of forcing.

The decrease of both threshold magnitudes below A_{th} at very short periods of forcing is due to the ability of an excitable medium to integrate various stimuli in time. The mechanism is documented in fig. 6, where the effects of one, two and three stimuli of the magnitude $A_f = 0.46 < A_{th}$ applied with period $T_f = 1.5$ are illustrated. Curve 1 represents the phase trajectory after a single stimulus. Since its value is less than the threshold A_{th}, no excitation could be evoked. When the second stimulus of the same amplitude is applied, the effects of both stimuli are summed; then the system crosses the threshold boundary, and the excitation cycle (and the wave) is initiated (cf. curve 2 in fig. 6). Notice that the third applied stimulus causes a certain time delay in the initiation of the excitation cycle, and thus has an inhibitory effect, temporarily freezing the present state of the system.

As mentioned above, the existence of static and dynamic thresholds is reflected in two different types of transient behaviour observed in the lower zero resonance region. Stimulations with magnitudes lower than both thresholds do not evoke any wave (region Ia in fig. 5). Stimulations with magnitudes $A_{st} < A_f < A_{dyn}$ (region Ib in fig. 5) are (due to a summation effect) sufficient to overcome the static threshold, but they are insufficient to overcome the dynamic threshold and thus only a single traveling wave is generated before the system approaches the final solution without traveling waves.

This type of transition to the final solution is characteristic of the upper zero wave region where $A_f > A_{th}$, and thus one wave is always generated when the rest state is stimulated. While the failure of the repetitive wave firing in the lower zero region originates from the complex threshold behaviour of the excitable system discussed above, the failure of the wave generation in the upper zero region is associated with the existence of absolute refractoriness of the medium. Even though every applied stimulation evokes an excitation cycle at the point of its application (see e.g. fig. 4e), the neighbouring medium does not restore its excitable state as the diffusion flux from the point of initiation pulls the system back to its refractory, nonexcitable state; thus the medium is

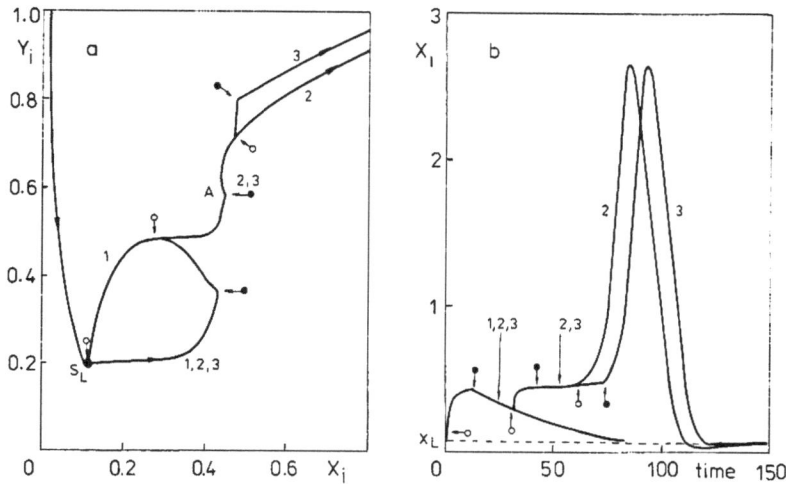

Fig. 6. Summation effect of periodically applied stimulation. ($A_f = 0.46$, $T_f = 1.5$). $\bigcirc\!\!\rightarrow$ ($\bullet\!\!\rightarrow$) denotes the beginning (or end) of stimulus application. Curve numbers correspond to a number of applied stimuli. (a) Trajectories in the local phase space, $i = 3$. (b) Time profiles of X_i taken at $i = 3$.

unable to support propagating waves. This diffusion flux increases with increasing stimulation amplitude and frequency, and thus the upper zero wave region enlarges with growing A_f.

The compounded action of the medium refractoriness, threshold behaviour, and summation effects gives rise to a quite complicated structure of resonance regions in the plane of forcing parameters. The non-monotonic shapes of the boundaries between several resonance regions shown in fig. 5 suggest that the dependence of the firing number on either of both forcing parameters does not always have to be a growing stepwise function similar to the devil's staircase, but rather it can exhibit local maxima.

On the other hand, some parametric dependences of ν have the devil's staircase-like structure, fulfilling the Farey tree rule for the ordering of periodic solutions (cf. fig. 7). The periodic solutions with higher resonances are condensed in narrow intervals of forcing amplitudes, while the main resonances ($\nu = 1:3$, $1:2$, $1:1$) cover large intervals of the values of A_f (cf. also fig. 5).

All solutions with firing number $\nu = 1:2$ observed for the forcing period T_f have the same periodicity ($T = 2T_f = 5.5$ in the case shown in fig.

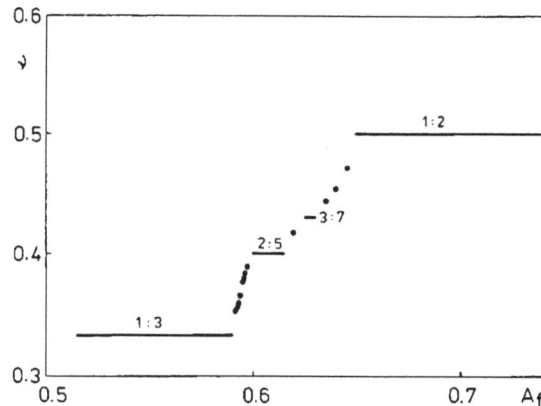

Fig. 7. The dependence of the firing number ν on the forcing amplitude for $T_f = 2.75$. The points denote calculated periodic wave solutions with higher resonance ratios which are (from left to right): $5:14$, $4:11$, $7:19$, $3:8$, $20:53$, $8:21$, $5:13$, $7:18$, $5:12$, $4:9$, $5:11$ and $8:17$.

7), and waves which are elicited and propagate along the system are the same in the sense that far away from the point of stimulation the local phase portraits coincide for all A_f. The difference arises in the phase shift relating the position on the local excitation cycle to the time of the stimulus application. The phase shift, expressed as the Y coordinate of points in the local strobo-

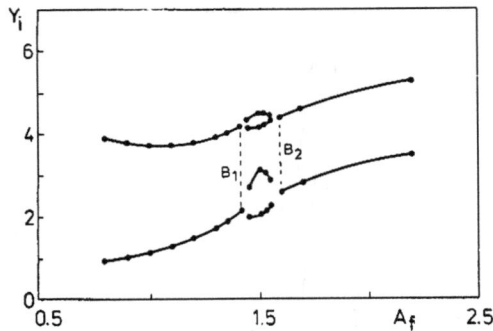

Fig. 8. The dependence of the stroboscopic maps on the forcing amplitude. At points B_1 and B_2 the period-two solutions bifurcate to period-four solutions.

scopic maps taken at the time instance corresponding to the end of the stimulating pulse, is plotted against the forcing amplitude in fig. 8. One can see that the changes of the phase shift with the amplitude of forcing lead to period-doubling bifurcations. Here the branch of solutions with the periodicity $T_2 = 2T_f$ having one wave generated with period T_2 bifurcates into the branch of solutions with periodicity $T_4 = 4T_f$, where two different waves are generated over the period T_4. Correspondingly, the two-point stroboscopic map bifurcates to the four-point one. The period-doubled solutions with $\nu = 2:4$ were found in other parts of the resonance region with $\nu = 1:2$ (see fig. 5), but no further subsequent bifurcations have been observed.

The boundary between the resonance region with the firing number $\nu = 1:1$ and regions with lower firing numbers has two significant limits, which are characteristic for an excitable medium. For $T_f \to \infty$ the boundary coincides with the upper boundary of the lower zero region, converging to the threshold value A_{th}. The other limit point, the value of T_f for which $A_f \to \infty$, corresponds to the so-called absolute refractory period, i.e., the shortest time required for the medium to restore its excitability. The value of the absolute refractory period also determines the lowest permissible value of the mutual period between two successive waves. The boundary itself determines the threshold value for the excita-

tion when the system was allowed to recover from the preceding stimulation for the time T_f. As we can see, the threshold is a decreasing function of the forcing period. Similar dependences were found for the measured recovery curves for the action potential propagation in nerve fibers [4] and were also confirmed by the measurements of phase excitation curves in the excitable Belousov–Zhabotinsky reaction in a CSTR [20].

3.2. Sinusoidal forcing

If a pulse-like forcing is used, then the $1:1$ synchronization region is found for high values of T_f and $A_f > A_{th}$, i.e., the stimulus of a short duration ($\Delta \tau = 0.6$ in our case) is able to evoke at most one propagating wave. Another situation occurs when the external forcing is of a sinusoidal form where long intervals of the superthreshold stimulation alternate with long intervals of the subthreshold stimulation. It can be basically expected that during the superthreshold phase of the forcing cycle several waves will be excited depending on the mutual ratio between the length of the superthreshold phase and the refractory period of the excitable medium.

Fig. 9 shows periodic wave patterns with firing numbers $\nu = 1:1$, $4:1$ and $6:1$ computed for the sinusoidal forcing. Periodic wave solutions with firing numbers $n:1$, for $n > 1$ are in the form of bursting oscillations, where n waves initiated with quite short mutual periods are followed by a long quiescent period. The longer the period of forcing, the higher the number of initiated bursts. Bursting phenomena have been observed in many cells of invertebrates, mammalian smooth muscles, etc. [21, 22]. In many studies, bursting is assumed to occur due to an interaction of slow subcellular oscillations with a fast spiking mechanism based on the existence of limit cycle oscillations in the fast subsystem, cf. e.g. refs. [23, 24]. Slow oscillations then act as a switch between stationary and limit cycle behaviour of the fast subsystem, thus causing switching between the passive and the active phases of the bursting

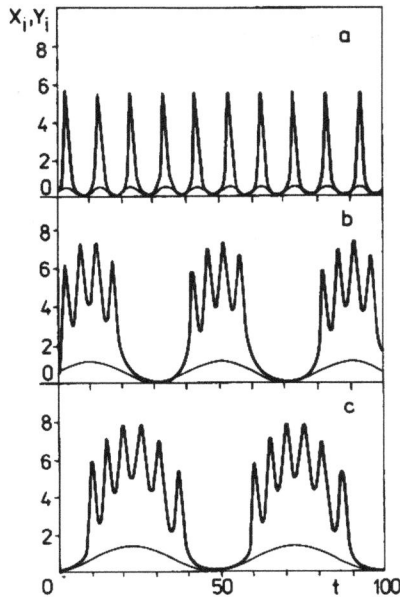

Fig. 9. Time profiles of X_i (thin line) and Y_i (thick line) at $i = 1$. Forcing parameters: (a) $A_f = 0.5$, $T_f = 10$, (b) $A_f = 0.6$, $T_f = 40$ and (c) $A_f = 0.7$, $T_f = 50$.

cycle. In our studies the slow processes are represented by sinusoidal external forcing, and the spiking mechanism stems from the excitability of the medium. A threshold phenomenon is then responsible for switching on and off the spiking mechanism.

4. Conclusion

A relatively simple nonautonomous periodically forced excitable reaction–diffusion system exhibits a number of phenomena often experimentally studied and modeled in more complicated biological systems. This study supports the idea of the use of models based on long-range interactions for the description of wave patterns generated in a squid giant axon [25, 26]. The simplicity of the system studied here permits clear interpretation and suggests possibilities of experimental verification in quasi-one-dimensional systems [10, 27].

References

[1] A.V. Holden, M. Markus and H.G. Othmer, eds., Nonlinear Wave Processes in Excitable Media (Plenum Press, London, 1990).

[2] W. Schiehlen, ed., Nonlinear Dynamics in Engineering Systems (Springer, Berlin, 1990).

[3] A.T. Winfree, The Geometry of Biological Time (Springer, New York, 1980).

[4] J. Field, H.W. Magoun and V.E. Hall, eds., Handbook of Physiology. Neurophysiology, Vol. 1 (Am. Physiol. Soc., Washington, DC, 1959).

[5] H. Sevcikova and M. Marek, in: From Chemical to Biological Organization, eds. M. Markus, S.C. Müller and G. Nicolis (Springer, Berlin, 1988) pp. 103–113.

[6] V.J. Geletjert and V.N. Kazachenko, Cluster Organization of Ionic Channels (Nauka, Moscow, 1990) [in Russian].

[7] E.E. Selkov, Biofizika 15 (1970) 1065–1075.

[8] D.R. Chialvo and J. Jalife, Nature 330 (1987) 24–31.

[9] H.G. Othmer, Resonance in oscillatory and excitable systems, Ann. NY Acad. Sci., to appear.

[10] J.C. Alexander, E.J. Doedel and H.G. Othmer, On the resonance structure in a forced excitable system, SIAM J. Appl. Math., to appear.

[11] H. Sevcikova and M. Marek, Physica D 26 (1986) 61–77.

[12] M. Marek, I. Schreiber and L. Vroblova, in: Structure, Coherence and Chaos, eds. R.D. Parmentier and P.L. Christiansen (Manchester Univ. Press, Manchester, 1988) pp. 233–244.

[13] M. Marek and I. Schreiber, in: Bifurcation: Analysis, Algorithms, Applications, eds. T. Küpper, R. Seydel and H. Troger (Birkhauser, Basel, 1987) p. 201.

[14] G. Matsumoto, K. Aihara, Y. Hanyu, N. Takahashi, S. Yoshizawa and J. Nagumo, Phys. Lett. A 123 (1987) 162–166.

[15] K. Aihara, M. Kotani and G. Matsumoto, in: Structure Coherence and Chaos, eds. R.D. Parmentier and P.L. Christiansen (Manchester Univ. Press, Manchester, 1988) pp. 613–617.

[16] G.B. Ermentrout and N. Kopell, SIAM J. Appl. Math. 46 (1986) 233–253.

[17] H. Sevcikova, M. Kubicek and M. Marek, in: Mathematical Modeling in Sciences and Technology, eds. X.J.R. Avula, R.E. Kalman, A.I. Ljapis and E.Y. Rodin (Pergamon Press, New York, 1984) pp. 477–482.

[18] W.D. McCormick, Z. Noszticzius and H.L. Swinney, Excitability and quasi-excitability: Definitions and experiments, in Proceedings of the International Conference on Dynamics of Exotic Phenomena in Chemistry, Hajduszoboszlo, Hungary, 1989.

[19] L. Ketnerova, H. Sevcikova and M. Marek, Periodic forcing of a spatially one-dimensional excitable reaction-diffusion system, in: Nonlinear Wave Processes in Excitable Media, eds. A.V. Holden, M. Markus and H.G. Othmer (Plenum Press, London, 1990).

[20] J. Finkeova, B. Hrudka, M. Dolnik and M. Marek, J. Phys. Chem. 94 (1990) 4110–4115.

[21] N. Kopell and G.B. Ermentrout, Math. Biosci. 78 (1986) 265–291.

[22] D.M. Himmel and T.R. Chay, Biophys. J. 51 (1987) 89–107.

[23] J. Rinzel, in: Proceedings of the International Congress for Mathematicians, Berkeley (1986) pp. 1578–1593.

[24] J. Rinzel and G.B. Ermentrout, in: Methods in Neural Modelling. From Synapses to Networks, eds. C. Koch and I. Segev (MIT, Cambridge, MA, 1989) pp. 135–169.

[25] Y. Hanyu and G. Matsumoto, Spatial long-range interactions in squid giant axons, Physica D 49 (1991 198–213, these Proceedings.

[26] N. Takahashi, Y. Hanyu, T. Musha, R. Kubo and G. Matsumoto, Global bifurcation structure in periodically stimulated giant axons of squid, Physica D 43 (1991) 318–334.

[27] M. Dolnik, J. Finkeova, I. Schreiber and M. Marek, J. Phys. Chem. 93 (1989) 2764–2774.

Physica D 49 (1991) 125–140
North-Holland

Alternative stable rotors in an excitable medium

A.T. Winfree[1]

326 Biological Sciences West, Department of Ecology and Evolutionary Biology, University of Arizona, Tucson, AZ 85718, USA

A given excitable medium (continuous, uniform, isotropic, simply connected, two- or three-dimensional) may support alternative types of vortex-like solution radiating from a 'rotor'. These have distinct periods, wavelengths, behaviors near interfaces, and, in rotationally symmetric vortex rings, different contraction or expansion rates and drift rates. Initial conditions select among the discrete alternatives. Rotors of different types may coexist within one medium, until the interface between their domains encroaches upon the rotor of longer period. Examples are given using the FitzHugh–Nagumo model. They are interpreted in terms of the measured dispersion curve for one-dimensional propagation and a hypothetical 'curvature relation' with one undetermined parameter.

1. Introduction

Any continuous, simply connected two-dimensional excitable medium is believed to have a unique rotor solution under given conditions in the laboratory: Whether in the Belousov–Zhabotinsky reagent, in layers of slime mold amoebae, in the retina, or in heart muscle, the isolated rotor's period and wavelength are the same regardless of the manner of its creation, and are believed to depend only on the medium's local, time-independent parameters.

Observation of numerical solutions of the parabolic partial differential equations of reaction and diffusion commonly used to represent such media leads to the same conclusion: Whatever initial conditions may be contrived, they create some field of propagating waves that eventually radiate away, leaving some residual assortment of isolated left- and right-handed single-armed rotors, all of which move in the same way (if at all) and radiate waves of exactly the same period and wavelength. Minor exceptions to these dogmatic generalizations are known only in the following cases:

(1) There is usually a range of parameters in which the rotor rotates rigidly about a fixed pivot,

and another in which it spontaneously moves along a flower-like 'meander' trajectory [1–3]. Like its spatial position and its phase during single-period rotation, its stage in this doubly periodic dance is established not by the medium's intrinsic parameters but by initial conditions.

(2) The rotor might not be isolated: Rotors close enough together to interact, or equivalently, half that far from a no-flux boundary, generally move and possibly change their period sligtly [4–6]. Such interaction has never been observed between counter-rotating vortex centers more than 2 diameters apart (taking the diameter as the wavelength/π); that record is held by the Oregonator rotor, its center standing off about 1 diameter from a no-flux boundary [7]. Interaction between co-rotating rotors is first observed at about 1 wavelength range in several excitable media.

(3) The rotor might have 'topological charge' other than 1, i.e. might have 2, 3, or more spiral arms [4]. The existence and stability of such rotors, as entities distinct from several adjacent 1-armed rotors with negligible interaction, remains somewhat uncertain or depends markedly on quantitative details of the particular medium. All '2-armed rotors' constructed in my laboratory in a diversity of excitable media have promptly resolved themselves into two adjacent 1-armed rotors.

---

[1]Internet: ATWINFREE@RVAX.CCIT.ARIZONA.EDU.

(4) The medium might not be uniform: Given a gradient of any parameter affecting rotor behavior, the rotor's period, wavelength, etc. obviously become continuous functions of position, thus rotors of distinct character may coexist until waves from the shorter-period rotor reach the longer-period source (but at the same level in the parameter gradient, any and all rotors have the same properties).

(5) The medium might not be simply connected: Given a hole in the continuum, a spiral wave can circulate around it in distinct alternative modes distinguished by the tip of the wave front being attached to or detached from the boundary of the hole [8], or characterized by the presence of 1, 2, 3, or more separate wave fronts around a large enough hole, or even characterized by distinct pulse profiles and speeds in media whose dispersion curves are multiple-valued, and rotors created near such a hole tend to become centered on it.

(6) The medium might not be two-dimensional: Given a third dimension, the rotor appears in every perpendicular cross-section of a vortex filament, but its exact period and wavelength depend on the curvature and twist associated with that filament [7, 9, 10].

Otherwise, the period, wavelength, and possible motion of the rotor are universally acknowledged to be unique functions of the medium's parameters, independent of initial conditions. So far as I am aware the results of every computation and every laboratory experiment in any uniform medium have been compatible with this generalization. However early thinking in this field contemplated a continuum of rotor solutions in any given excitable medium, and later when they did not appear, a discrete spectrum. The latter also did not appear, but one always remained free to doubt dogma, there apparently being no mathematical proof or reason of fundamental physical chemistry why alternative rotors should not exist. To test the viability of this doubt I undertook to deliberately contrive a counterexample to the dogma, starting in media with

non-monotonic dispersion curves. If proved unnecessary to look further. The basic rationale follows.

2. Dispersion curves and spiral waves

If there were alternative rotors of distinct period, then the periodic wave trains radiating from them would have different periods, and presumably different speeds and wavelengths. A periodic wave train can have any period, within limits, and to each period corresponds some wavelength (and speed = wavelength/period); the set of allowed combinations defines the dispersion curve. The dispersion curve (in the representation used in this paper) plots the speed, $c(\lambda)$, of a periodic wave train against the wavelength, λ. (Other familiar representations plot period versus wavelength, reciprocal period versus reciprocal wavelength, etc.) In media that recover monotonically the longer the interval between pulses (period or wavelength), the more thoroughly the medium has recovered its pre-pulse excitability, the lower is the threshold for contagious excitation, and the greater the speed of a wave front propagating into such medium: speed is a monotone increasing function of wave spacing. This feature lends stability to the uniform spacing in a wave train. But why is one such combination picked out for the rotor? Evidently there is some other constraint, some curve intersecting the dispersion curve at the particular period of the rotor's periodic wave train. One was proposed [1, 11] on grounds that the physico-chemical structure of the pivoting wavetip (the rotor) must be structured by diffusion, and diffusion only effectively covers an area $2\pi D\tau_0$ during one rotation period, τ_0. Taking the perimeter of that circular area as the wave source and wavelength, λ_0, we have a hyperbolic constraint $\lambda_0^2/D\tau_0 = c_0\lambda_0/D \approx 8\pi^2 \cong 79$. This argument fails in the case of rotors in which pivoting occurs about a point remote from the wavetip, in media of marginal excitability (Pertsov and Panfilov [12]): in such cases $c_0\lambda_0/D \gg 8\pi^2$.

Mathematical reasoning in the singular perturbation limit led Keener [13] and Tyson and Keener [14] to derive an alternative constraint from the assumed linear dependence of wave speed on the curvature, H, of the two-dimensional wave front (not to be confused with the curvature, k, of the one-dimensional vortex filament): $c(H) = c(0) - DH$. They argue that the 'curvature relation' for the case of a spiral wave rotating about a circular hole of substantial radius, r, is independent of aspects of such a medium's kinetics other than the multiplicative rate constant $1/\varepsilon$ specified below (at least, in the limit $\varepsilon \to 0$, and supposing alteration of other aspects leaves the medium adequately excitable to support propagation):

In dimensionless form, for $a = cr/D \geq 1$ [14, eq. (22)],

$$Q = \frac{c\lambda(c)}{D}$$

$$= \frac{8\pi a(1+a)}{1 + 4a - \sqrt{1 + 8a}}$$

$$\cong 8\pi \text{ to } 50\pi \cong 25 \text{ to } 157 \quad \text{for } a = 1 \text{ to } 20 \quad (1a)$$

or

$$\lambda(c) = \frac{8\pi Db(1+cb)}{1 + 4cb - \sqrt{1 + 8cb}} \quad \text{with } b = \frac{r}{D}, \quad (1b)$$

and in special application to the spiral wave radiated by a rotor in a medium with no hole,

$$\lambda_0(c_0) = \frac{8\pi Db(1 + c_0 b)}{1 + 4c_0 b - \sqrt{1 + 8c_0 b}}, \quad (1c)$$

where r now represents the radius of some 'ef-

fective hole' surrounding the pivot point, and turns out to be somewhat less than $\lambda_0/2\pi$. I use natural units of space and time here, not scaled by kinetic parameters as in refs. [13–15].

This formulation is suggested by the reasonable guess that vorticity in a simply connected medium entails some finite 'effective' hole near the wave's pivot. This would be exactly true if all iso-concentration contours, not only those at the wave front, were to pass perpendicularly through the boundary of some disk around the center of rotation (for then there would be no diffusion across that boundary); this is topologically impossible, but one still needs to check whether the idea may provide a useful approximation anyway. If the rotor does indeed contain a functional hole of radius r and if $c(H) \cong c(0) - DH$ over the range of small curvatures H encountered, then eq. (1) results. Unfortunately a criterion anticipating r and its dependence on kinetic parameters is still lacking. In the case of wavetip and center of rotation overlapping in a 'rotor', r values might be expected to be no larger than the wavelength $/2\pi$, or perhaps several times D/c as a rough estimator of the wave front thickness or the radius of the critical nucleus for spreading excitation (and this will be confirmed in numerical experiments below). Thus $b =$ several times $1/c$. Along curves of form (1) with such values of b, the dimensionless parameter $Q = c_0\lambda_0/D = \lambda_0^2/D\tau_0$ does indeed cover roughly the range observed in numerical experiments and laboratory experiments (about 20 to 150, independent of the choice of units for space and time).

Computed rotors in biochemical models of *Dictyostelium* give $Q \approx 16$ [16] or ≈ 49 [17], straddling the **observed** value ≈ 25 [18]. Rotors **computed using** modified versions of the **Beeler–Reuter electrophysiological model give 144–400** [19–22]; **Zykov's simpler model with less contrast between** excitation and recovery time constants gives 20–70 [2]. Actual ventricular myocardium gives 100–300 [20–22]. Rotors computed from the Oregonator model of the Belousov–Zhabotinsky reagent give 64 to 121, depending on chemical parameters [1]. Taking D to be 1.5×10^{-5} cm^2/s independent of temperature, we find $Q =$ rising from 13 to 73 in experiments using the Belousov–Zhabotinsky reagent, as the temperature rises from 6 to 45°C [23, 24]. Keener and Tyson [15] tabulate experiments with various recipes from which we find ratios ranging from 36 to 100. The spiral observed by Muller et al. [25] gave 56. Tam et al. [26] reported wavelength and velocity from spirals over a range of bromate concentrations, from which I find $Q \approx 25$ to 30.

There seems to be a minimum $6\pi \approx 20$ in the dependence of Q on kinetic parameters, but the value becomes arbitrarily large as parameters approach a locus beyond which front propagation is not sustainable. This limit is typically reached by increasing the excitation time constant from very small to less small values, and failure typically occurs with propagation speed only slightly diminished. Approaching this locus, front propagation speed, whether $c(0)$ into virgin territory or c_0 in a spiral's wavetrain, stays about the same while the spiral wavelength, λ_0, becomes arbitrarily large. Wave fronts in this parameter range experience conduction block when curved even slightly, and fail even at curvature 0 as the parameter-space locus is crossed at which plane wave propagation fails. The fall-off of speed with curvature, $H = 1/(\text{radius of curvature})_1 + 1/(\text{radius of curvature})_2$, presumably follows the familiar rule $c(H) = c(0) - DH$ at very small H, but propagation fails while DH is still $\ll c(0)$. Such wave fronts are unable to bend into tight spirals, and so they are compelled to rotate about enormous disks of quiescence, making λ_0 arbitrarily large, as first noticed by Pertsov and Panfilov [12]. Thus $c_0\lambda_0$ blows up as the time constant of excitation increases toward the critical value at which one-dimensional propagation fails. In such cases parameter r in eq. (1) would have to be very much larger than D/c.

With theory in this promising but quantitatively incompletely form (for want of knowing how r depends on the kinetics), it may still be worthwhile to try to measure the putative r in numeri-

cal experiments, and to contemplate possible consequences of the existence of a curvature relation. For example, could the corresponding curve intersect the dispersion curve more than once? The chances of such multiplicity would be enhanced were the dispersion curve non-monotonic. What kind of kinetics would favor that? Suppose the threshold of excitation were to oscillate in the wake of a pulse, falling from initially high values during the immediate refractoriness that gives the excitation front a direction, but then rising back through the initial low threshold. Then a wave front following a pulse during intervals of relatively high threshold would be relatively slow, and if it fell back further into any interval of lower threshold, it would propagate faster. The dispersion curve might have maxima of wave speed at the intervals of low threshold. In what kind of kinetics would threshold fluctuate in this way? Any kinetics would suffice in which the return trajectory approaches equilibrium in an oscillatory way, with wide gentle turns below threshold, i.e. threshold not too low and excitation not too abrupt [27–29]. The more nearly such kinetics, linearized about equilibrium, has purely imaginary eigenvalues, the more persistent are the ripples in the dispersion curve; and the lower is threshold at equilibrium, and the slower is excitation after threshold has been transgressed, the more dramatic is the effect of such ripples. Kinetics with relatively gentle excitation, such as the Oregonator model of Belousov–Zhabotinsky kinetics with parameter ε about as large as possible and parameter f near the Hopf bifurcation point, $f = 1 + \sqrt{2}$, have non-monotone dispersion curves due to the vanishing of the eigenvalue's real part [27]. There are abundant other examples, e.g. the piecewise-linear kinetics first used to demonstrate the stability of computational rotors [30–32]. In the latter case the speed is multiple-valued over a range of wavelengths that happens not to be close to the rotor wavelength, but the possibility of alternative rotors in principle is clearly suggested.

I went looking for some, hill-climbing in the three-parameter space of a generic excitable medium to enhance non-monotonicity in the dispersion curve:

$$\partial A/\partial t = (A - A^3/3 - B)/\varepsilon + D\nabla^2 A$$

'propagator' variable,

$$\partial B/\partial t = \varepsilon(A + \beta - \gamma B)$$

'controller' or 'recovery' variable   (2a)

A nullcline for $\nabla^2 A = 0$:

$$dA/dt = 0 = A - A^3/3 - B,$$
S-shaped between $A = \pm 1$,
symmetric through $(0,0)$ at unit slope.   (2b)

B nullcline in any case:

$$dB/dt = 0 = A + \beta - \gamma B,$$
line with slope $\gamma > 0$ and
intercept $0 > A = -\beta$ at $B = 0$.   (2c)

Spatially uniform equilibrium:

$$0 = A^* - (A^* + \beta)/\gamma - A^{*3}/3,$$
chose $\beta > 0$ with $\gamma < 1$ to create
a unique equilibrium (A^*, B^*).   (2d)

Diffusion coefficient D has units (space units^2/time units, su^2/tu) and merely determines the spatial scale of any non-uniform solutions; we choose a unit of distance to make $D = 1$ su^2/tu. Parameters β and γ determine the intersection between the cubic nullcline $dA/dt = 0$ of the 'propagator' variable and the linear nullcline $dB/dt = 0$ of the 'controller' variable (fig. 1). The unique attracting equilibrium point of the stable spatially uniform solution is thus determined, and with it the 'excitability' of this medium, meaning the ratio of the pulse amplitude to the threshold displacement ΔA required to initiate a pulse.

In this paper $\beta = 0.7$, $\gamma = 0.5$: see nullclines in fig. 1. Parameter ε (equivalent to $\sqrt{\varepsilon}$ in refs. [13–15] and proportional to $\sqrt{E}$ in eq. (3) below) is varied between 0.28 and 0.35, entirely above the meander range that begins below $\varepsilon = 0.27$ at this (β, γ). The maximum ε compatible with

one-dimensional or plane-wave propagation is 0.40 at this (β, γ), empirically defined very nearly by $4u_0 + 5\varepsilon = 2$ in eqs. (2) and (3). There is no minimum ε, and wave speed in fully recovered medium increases in proportion to $1/\sqrt{\varepsilon}$. This particular parameter range was chosen following the rationale suggested above: its excitation threshold at equilibrium is just slightly positive, and its eigenvalues near equilibrium have a substantial imaginary component. Zero threshold requires $3\beta + 2\gamma = 3$. Single rather than triple intersection of nullclines (unique equilibrium) is guaranteed for $0 < \gamma < 1$ and $0 > \beta > \sqrt{3}$. I chose $\beta = 0.7$, $\gamma = 0.5$ within these limits simply because other published studies at the same parameters provide a cross-check on the numerical consistency of new programs written for this study.

Papers on this kinetics commonly use eq. (2) or the re-parameterized equivalent:

$$\partial u/\partial \tau = -u(u - u_0)(u - 1) - v + D\sqrt{E}\,\nabla^2 u$$

'propagator' variable,

$$\partial v/\partial \tau = E(u - \alpha v)$$

'controller' or 'recovery' variable.   (3a)

u nullcline for $\nabla^2 u = 0$:

$$du/d\tau = 0 = u(u - u_0)(u - 1) + v,$$
S-shaped with zeros at $u = 0, 0 > u_0 > 1, 1$.
    (3b)

v nullcline in any case:

$$dv/d\tau = 0 = u - \alpha v,$$
line with slope $\alpha > 0$ through the origin.   (3c)

Spatially uniform equilibrium:

$$u = v = 0.$$ (3d)

In representation (2) the threshold is u_0 and pulse amplitude is 1, so 'excitability' is $1/u_0$. Formulae for conversion of E, u_0, α, and τ in eq. (3) to ε, β, γ, and t in eq. (2) are given in ref. [6]. In this paper $u_0 = 0.0217$ (essentially 0, but kept positive), the nullclines in fig. 1 intersecting just below the lower-left turn-around of the cubic.

B

0.667

-2.   -1.   1.   2.
                                              A

-0.667

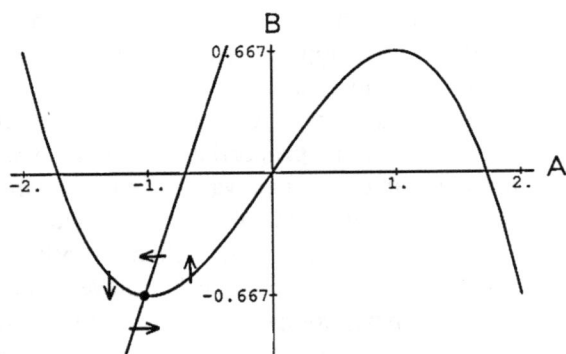

Fig. 1. On the (A, B) coordinate plane of eq. (2), nullcline loci are plotted along with $\mathrm{d}A/\mathrm{d}t = 0$ (trajectories vertical in the spatially uniform case, $\nabla^2 A = 0$: the S-shaped curve) or along which $\mathrm{d}B/\mathrm{d}t = 0$ (trajectories horizontal, even in spatially structured contexts: the straight line). They cross at equilibrium, close to $A = -1$, $B = -2/3$. Altering ε alters rates, but not these zero loci.

Under condition $u_0 = 0$, the eigenvalues of the uniform kinetics around equilibrium are $-(\varepsilon\gamma/2) \pm \mathrm{i}\sqrt{1 - (\varepsilon\gamma/2)^2}$. Slope $\alpha = 1.53$, and E will be varied between 0.0084 (corresponding to $\varepsilon = 0.28$) and 0.013 (corresponding to $\varepsilon = 0.35$). When I refer to 'space units' and 'time units' in this paper, I mean those of eq. (2).

The parameter values used here ensure that rotation is indeed rigid about a fixed pivot, a condition assumed in most mathematical analyses of rotors. Many such analyses (e.g. refs. [13–15]) deploy singular perturbation methods in the limit $E \to \infty$ or $\varepsilon \to 0$, scaling space and time in an ε-dependent way to keep finite measures of wavelengths and wave speeds. Toward that limit rotors commonly meander, possibly affecting mathematical inferences based on the assumption of rigid rotation. The present work stays far from that limit.

Dispersion curves were determined as follows. A ring of initial length $L = 50$ space units (su), sampled as 100 gridpoints (2 gridpoints per space unit), was constructed for Euler integration according to eq. (2). This ring is intended to represent one wavelength of the periodic wave train. It imposes strict periodicity, thus exposing for ob-

servation segments of the dispersion curve in which simply periodic wave trains are not stable (where $\mathrm{d}c/\mathrm{d}\lambda < 0$). With appropriate initial conditions $\{A(i), B(i)\}$, a solitary impulse begins to circulate along the ring, each gridpoint following the same excitation–recovery loop $(A(t), B(t))$ at staggered intervals of time. The times are noted at which gridpoint 1 passes $\overline{B}$ with $A > \overline{A}$, $\overline{A}$ and $\overline{B}$ indicating mean values during the cycle. The interval between successive times divided by the ring's current perimeter is taken as the wave speed for that wavelength or pulse spacing. At intervals of one circulation period (long enough so that the resulting dispersion curve is not materially affected by waiting longer between deletions) a single gridpoint is removed from the array, shortening the ring. This continues until the ring is too short and propagation fails, all gridpoints converging to the same state, ultimately the equilibrium state. The resulting dispersion curves were spot-checked at finer discretization and found to be only trivially different, and were compared against those available from analytic solutions of other excitable media and/or from other numerical techniques, and found to be indistinguishable for our purposes. The same curves can alternatively be obtained by gradually increasing D, keeping the number of cells fixed, since the ring's perimeter is proportional to $1/\sqrt{D}$.

Fig. 2 shows numerically determined dispersion curves for eq. (2) at eight values of parameter ε, with $\beta = 0.7$ and $\gamma = 0.5$. As another check on the numerical procedures, to first order, the speed of a solitary front (large wavelength) should be inversely proportional to $\sqrt{\varepsilon}$, as can be seen merely from dimensional considerations applied to the propagator variable's rate law (the upper equation) with ε small enough that the recovery process (the lower equation) can be ignored for purposes involving only front propagation: Rename variables $e = \alpha^2\varepsilon$, $X = \alpha x$, $T = \alpha^2 t$ and an equation of the same form is obtained in which $x/t = \alpha X/T$. This scaling is not quite found with ε as large as 0.3 and wave fronts as close together as 50 su, but the products $0.99 < c(50)\sqrt{\varepsilon} \le 1.04$ do maintain rough constancy across the examined range $0.28 \le \varepsilon \le 0.35$ (and the computational result fits more closely to a known higher-order dependency on $\sqrt{\varepsilon}$ [33]). At shorter wavelengths, ripples are conspicuous; maxima of speed occur at wavelength intervals near 10 and wave speeds near 1.7, corresponding to temporal period $10/1.7 \approx 2\pi$ as expected from the imaginary part of the near-equilibrium kinetics' eigenvalues: $\sqrt{1 - (\varepsilon\gamma/2)^2}$, close to ± 1. On the negative-slope part of each ripple one might expect the measurement to be spoiled by an instability, successive pulse intervals tending to adjacent extrema of positive-slope segments. This would happen were the measurement conducted using pulses spaced along a line, but with a single pulse on a ring the instability is suppressed, as can be seen from the smoothness of the trail of circumference/interval dots outlining this curve.

With ε not too small at parameters $\beta = 1.0$, $\gamma = 0.0$ (where the real part $-\varepsilon\gamma/2$ of the near-equilibrium eigenvalues vanishes) the dispersion curve has even more pronounced and persistent ripples [34], potentially giving rise to a profusion of distinctly different rotors, depending on the shallowness of the putative intersecting curve's slope. However this report is limited to merely demonstrating the stable coexistence of two alternative rotors at $\beta = 0.5$, $\gamma = 0.7$.

3. Numerical experiments define a 'critical curve'

Two-dimensional calculations (explicit, Euler method again) used fixed square grids $L = 120$ su on edge with 2 gridpoints/su (thus $N = 240$ gridpoints in each direction), with no-flux boundary conditions. Initial conditions were constructed usually by either:

(a) as in ref. [32], replicating waves $\{A(i), B(i)\}$ from the one-dimensional calculation at perimeter L, into all columns of the array west of the midline, with the pulse moving north in the mid-

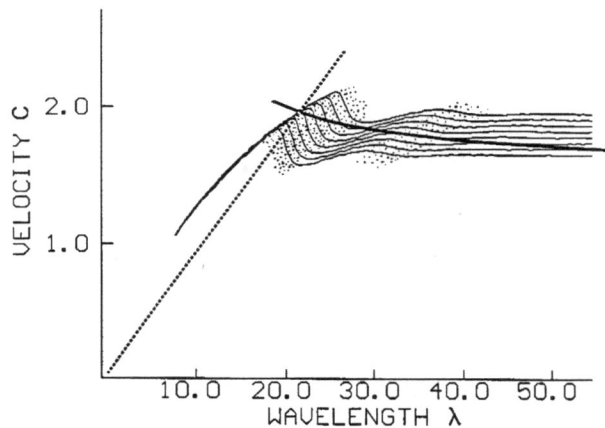

Fig. 2. For each of eight equispaced values of parameter ε (0.28, 0.29, 0.30, 0.31, 0.32, 0.33, 0.34, and 0.35 in order from bottom to top) the propagation speed of the impulse (in su/tu) is plotted against the distance (in su, proportional to the time) of the pulse behind a forerunner. The rotor period observed at each ε specifies a (dotted) radial line (e.g. period 11.2, dotted) and so specifies a unique (c, λ) point on the corresponding (solid) dispersion curve. These points can be threaded on a smooth 'critical curve' that might represent a locus of intersections with the 'curvature relation' corresponding to the same ε value [13–15]. The critical curve intersects curves $\varepsilon = 0.30$, 0.31, and 0.32 each twice (or more): inside that range of ε, rotors of two distinct periods are encountered. The bottom three dispersion curves ($\varepsilon > 0.32$) miss the period-11 intersection. The next three ($0.32 \ge \varepsilon > 0.29$) have two upslope intersections. The top two ($\varepsilon \le 0.29$) have only the period-11 intersection. Some of the intersections lie on downslope segments of the dispersion curve (shaded zones), where simple periodism is expected to be unstable; no rotors are observed on the most dramatic downslope.

dle of each column, and filling the rest of the array with medium at equilibrium; or

(b) as in refs. [30, 31], creating a linear gradient of A east–west, and a linear gradient of B north–south, in both cases spanning values found in the rotor: $A(J, K) = 2.5(K - 1)/(N - 1) - 1.25$ to initially span the range $-1.25 < A < 1.25$, and $B(J, K) = 1.4(J - 1)/(N - 1) - 1.00$ to initially span the range $-1.00 < A < 0.40$; or

(c) letting a spiral wave form after (a) or (b) then replacing a strip of the solution with medium at equilibrium, so that only one wave front is truncated: the neighborhood of its endpoint becomes a rotor.

A single rotor evolves from such initial conditions. Depending on ε, the rotor has period $\tau_0(\varepsilon)$, which we seek to determine. The intent is to traverse parameter space by varying any parameter or combination of parameters, seeking abrupt changes in $\tau_0(\varepsilon)$. Re-crossing any such boundary in the opposite direction, we hope to observe whether $\tau_0(\varepsilon)$ is a single-valued function, or perhaps part of a hysteresis loop. The asymptotic wavelength $\lambda_0(\varepsilon)$ corresponding to $\tau_0(\varepsilon)$ cannot be measured exactly from contour maps of $A(x, y)$ or $B(x, y)$, nor can the wave speed $c_0(\varepsilon)$ be obtained exactly by comparing successive contour maps, because both quantities are slightly decreased by wave front curvature, especially near the wave source along the rim of the rotor. But the period at each ε is clearly measurable and prescribes a unique radial line (dotted) in fig. 2, which intersects the dispersion curve for the same ε (solid) at speed c_0 and wavelength λ_0. These indeed turn out to be just a little larger than values estimated from contour maps. Such determinations of (c_0, λ_0) seem to outline a descending 'critical curve' (thicker line) similar to the curvature relation imagined above to be relatively independent of kinetic parameters other than multiplier ε. The dimensionless quantity $Q = c_0\lambda_0/D = \lambda_0^2/D\tau_0 = c_0^2\tau_0/D$ ranges from about 40 to about 90 in fig. 2: values typical of eq. (1) with reasonable values of a (or b). However within this range of $c\lambda$ our 'critical curve' does

not closely resemble the $\lambda(c)$ prescribed by eq. (1) with any plausible choice of fixed r, for two reasons:

(a) in the range of ε values used here (around 0.3, equivalent to $\sqrt{0.3} = 0.1$ in terms of the ε of refs. [13–15]) the singular perturbation limit required in refs. [13–15] is deliberately not approached, in order to retain an oscillatory character to the trajectories some distance from equilibrium, and in order to preserve rigid rotation of the vortex solution in physical space, avoiding meander, and

(b) the best-fitting r value proves to depend markedly on ε, ranging from about 2 to 6 su in this narrow range of ε, so the 'curvature relation' is deforming as we change ε, and the 'critical curve' is actually the locus of intersections of successive curvature relations with the corresponding dispersion curves.

4. Discovery of coexisting alternative rotors

Any excitable medium is characterized by (at least) two essential parameters: the threshold for excitation, normalized by the amplitude of full-blown excitation, and the ratio of the rate of excitation to that of recovery. The reciprocal of threshold/amplitude is often called the 'excitability'; in these numerical experiments it is held fixed ($1/u_0 = 46$). The recovery/excitation rate parameter is ε^2 in eq. (2) or E in eq. (3) and it is varied in these experiments. These two parameters suffice to describe the essential features of most excitable media. In eq. (2) at any ε, for example, any combination of β and γ that results in the same A^*, and therefore the same threshold/amplitude ratio, gives roughly the same dispersion curve and rotor behavior (though not exactly; and it is of course possible to build more complex excitable media, for example the Beeler–Reuter model of cardiac muscle membrane [19–22], which has two relatively independent mechanisms of excitation; then more parameters are needed to minimally characterize its main properties).

In a computation gradually varying ε, the dot constructed in section 3 may be described in retrospect as moving along the intersections of the (gradually changing) curvature relation with the (gradually changing) dispersion curve, perhaps jumping over descending segments of the latter. Wave trains cannot be stable on descending segments ($dc/d\lambda < 0$) of the dispersion curve, with speeds increasing as distance decreases behind the preceding wave. On each dispersion curve two or more narrow ranges of period can be identified by drawing radial lines through local extrema of wave speed, enclosing the negative-slope segments stippled in figs. 2 and 3.

Fig. 3 plots the observed period of the rotor, $\tau_0(\varepsilon)$, as ε is slowly increased from 0.25 to 0.34 and then slowly decreased back again from 0.34 to 0.25. The period in this experiment can be assayed with adequate precision by the usual technique of recording wave front arrivals at a fixed station. But a more refined technique might be wanted because:

(a) those times record something about events that transpired within the rotor some time ago (propagation delay), confounded by possible rearrangement of pulse spacings in the travelling wave train, and

(b) in three-dimensional experiments to follow, the rotor moves, so arrival times are confounded by Doppler shifts.

The more comprehensive technique of contouring $A(x, y)$ at two successive times, then fitting the two maps by a rotation about the rotor's center [32] only evades distortion (b) since (a) still spoils the map's rotational symmetry. Instead, at intervals of 10 dt (about 4/9 tu, thus 55 times in period 10 tu) I first located the inner tip of the spiral wave front as the geographical centroid of those ten or so (depending on ε) gridpoints in which $-0.6 < A(x, y, t) < 0.6$ and $-0.3 < B(x, y, t) < 0.3$. These combinations are not found during wave propagation, but only in the wave source near the 'phase singularity'. This centroid rotates along a tiny circle (radius a few space units, depending on ε) with readily observable period. This is taken as the most accurate possi-

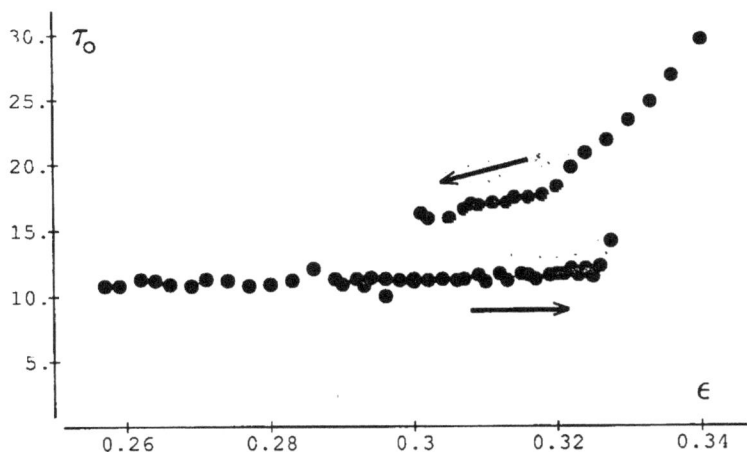

Fig. 3. The rotor's period evolves while parameter ε is slowly varied from 0.25 to 0.34, then back again. The stippled bands of fig. 2 (downslope segments of the corresponding dispersion curves) are mapped to this figure's coordinates and interpreted as zones of instability for simply periodic wave trains.

ble measure of the rotor's period and the source period of the spiral wave train radiating from the rotor, and reported here as $\tau_0(\varepsilon)$.

In some ranges of ε, τ_0 changes abruptly; these changes will be interpreted below as actual discontinuities corresponding to a leap across a range of unstable solutions. Most noteworthy is the conspicuous dissimilarity of $\tau_0(\varepsilon$ increasing) from $\tau_0(\varepsilon$ decreasing): The data present one large hysteresis loop. Starting from small ε, τ_0 remains essentially constant at 11 tu until the rotor abruptly shifts near $\varepsilon > 0.327$ to period $\tau_0 > 20$, and continues to climb, through period 56 at

$\varepsilon = 0.35$ and probably toward ∞ before ε approaches 0.40, the upper limit of ε compatible with planewave propagation. Starting from such a large, slow rotor and decreasing ε, τ_0 decreases past the former jump point, falling abruptly near $\varepsilon = 0.297$ from period near 17 tu, back to the period-11 plateau. Thus in the range $0.297 < \varepsilon < 0.327$, rotors may coexist at both periods, nominally 17 and 11 tu.

At $\varepsilon = 0.30$ there are two alternative rotors, one with period 11.2 (wave speed 1.9, wavelength 21 with a narrow excitable gap in which the recovered medium lingers near equilibrium await-

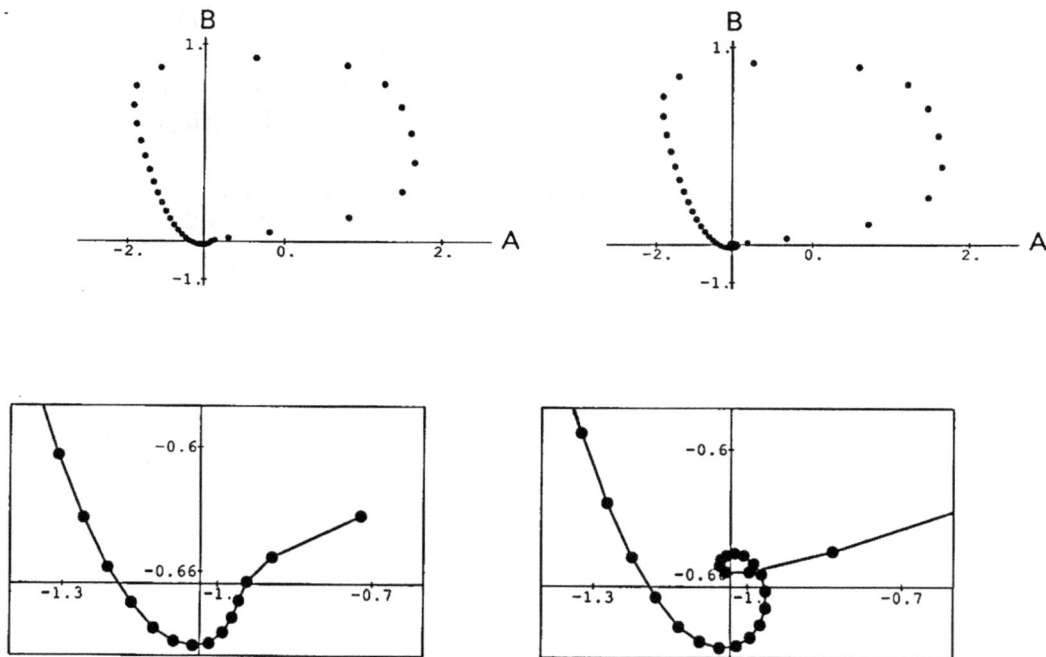

Fig. 4. Snapshots of $\{A(i), B(i)\}$ during propagation of a periodic trait of pulses on an open line, or of a single pulse around a closed ring, of eq. (2) medium with $\varepsilon = 0.30$, $\beta = 0.70$, $\gamma = 0.50$. The coordinate space is as in fig. 1 but the axes are drawn through the equilibrium state ($A = -1.033$, $B = -0.6655$) rather than through $(0,0)$, and the nullclines are deleted. The pulse has essentially the same shape regardless of interpulse spacing (and period), except for the 'excitable gap' of gridpoints near equilibrium: (Right) Ring perimeter $L = 30$ su, slack enough so that cells near equilibrium can follow at least one full loop of nearly uniform local kinetics; the region near equilibrium is enlarged below. (Left) Ring perimeter $L = 25$ su, tightened up a bit so that the former loop is no longer stable: cells approaching equilibrium are already too close to the next wave front, and as the enlarged region around equilibrium shows, they are re-excited without the delay of an intervening 'supernormal' period. This wavelength is in a downslope of the dispersion curve so the even spacing between consecutive pulses would be unstable but for the stabilizing constraint provided by running the single pulse on a closed ring; the map shown here is only slightly different at $L = 22$ su, well inside an upslope arc of the dispersion curve.

ing the next front) and one with period 16.5 (wave speed 1.7, wavelength 28 with a wide excitable gap but otherwise essentially the same pulse shape). Fig. 4 plots in (A, B) space two alternative one-dimensional wave trains at $\varepsilon = 0.30$, dramatizing the physical basis of their distinction:

In one case (on the right, with longer period and slightly higher wave speed, so pulses are farther apart) gridpoints follow the local kinetics relatively faithfully through one loop approaching equilibrium, before excitation from attached neighbors pulls them across threshold and into the next excitation; while in the other case (on the left, with shorter period and lower wave speed, so pulses are closer together), even that one loop is bypassed by the local levelling effect of diffusion. This effect becomes more dramatic as the eigenvalues of uniform near-equilibrium kinetics become purely imaginary; at $\beta = 1.0$, $\gamma = 0.0$ plots like fig. 4 show many loops at long inter-pulse intervals.

This is the same mechanism as was first seen in similar plots in the similar piecewise-linear kinetics mentioned above, but in this case it occurs close enough to the descending curvature relation to affect the options for stable rotors. It is a mechanism common to all excitable media in which equilibrium is approached through a damped oscillation. In such media the local threshold for excitation oscillates after a prior excitation; in cardiology the recurrent intervals of diminished threshold (when the local state between equilibrium and threshold on a looping trajectory in fig. 4) are called 'supernormal'; supernormality might be connected with vulnerability to alternative modes of re-entrant action potential (thus distinct tachycardias), as in these alternative rotors.

Each of the four successive panels of fig. 5 show the two alternative rotors at $\varepsilon = 0.30$ cohabiting a 240×240-gridpoint square ($L = 120$ su) of uniform equation (2) medium with $\beta = 0.7$, $\gamma = 0.5$. All four frames are contoured for $A(x, y)$, the 'propagator' variable. The excitable gap (wide

Fig. 5. Propagator variable A is contoured on snapshots from numerical solution of eq. (2), with $\varepsilon = 0.30$, $\beta = 0.70$, $\gamma = 0.50$ and no-flux boundary conditions, in a box 120 su on edge, following initial conditions suitable to conjure one rotor of each of the two kinds simultaneously viable in this medium. Most of the variation in A is confined to a narrow pulse, but the excitable gap separating successive pulses is different in the wave trains radiating form these two rotors in a uniform medium. During the first interval (34 tu, (a) to (b)) both rotors execute an integer number of turns. The collision boundary between them (curving locus of cusps, unmarked) advances 12 su toward the longer-period, clockwise source in the lower right of each panel. The second interval (16.5 tu from (b) to (c)) was chosen to illustrate that the quicker, anti-clockwise rotor in the upper left of each panel executes more than one turn while the lower right rotor does exactly one turn. The third interval (11.2 tu from (c) to (d)) was chosen to illustrate that the lower right rotor in each panel executes less than one turn while the upper left rotor does exactly one turn.

region near equilibrium) is conspicuous for its sparse contours. The rotor in the lower right of each panel turns clockwise and the other turns anticlockwise, but this is not the basis of their distinctness. The final frame, (d), follows $T_1 = 11.5$ tu (one period of the shorter-period rotor in the upper left) after frame (c). Frame (c) follows $T_2 = 16.5$ tu (one period of the longer-period rotor in the lower right) after frame (b). Frame b follows 34 tu after frame (a), 34 being nearly $3T_1$

or $2T_2$. During this time the collision boundary has moved visibly. It moves at speed $(1/T_1 - 1/T_2)/(1/\lambda_1 + 1/\lambda_2) = 0.34$ su/tu (about 20% of wave speed, 'Mach 0.2') toward the longer-period source in the lower right. Until it reaches the longer-period rotor, both behave as though isolated. But when it does, the latter switches to the shorter period (not shown). Then there are two rotors of equal period, so the boundary moves no further. If the originally shorter-period source is now erased, the originally longer-period phase singularity is still impacted by the several wave fronts that had already radiated from the erased source, but thereafter it restructures its environment in the form of a rotor of the original handedness and resumes its own radiation ... as it happens, in this case, at period 11.2.

The range of ε admitting such bistability is narrow in this particular model. As can be seen from fig 2, the dispersion curve's left-most segment of negative slope moves from wholly below the critical curve to wholly above it as ε is changed from 0.297 (where the $T_2 \approx 17$ rotor loses stability) to 0.327 (where the $T_1 \approx 11$ rotor loses stability). This range depends slightly on the time and space grids used in the Euler integration. The numbers given were obtained with $dt = 0.04$ tu and $dx = 1/2$ su gridpoints. With smaller increments the results scarcely differ, but with larger $dx = 2/3$ su, the bistable range moves over slightly so that 0.30 is no longer included. The relative stability of the period-11 rotor vis-a-vis the period-17 rotor in these numerical experiments is probably dominated by our choice of $\varepsilon = 0.300$, in retrospect seen to be near the edge of the stability range of the longer-period rotor. Near the opposite edge of the bistable interval ($\varepsilon = 0.325$) both of our standard initial conditions (truncated waves and crossed gradients) evolve spontaneously to the *longer*-period rotor.

These rotors are repelled from no-flux boundaries and from mirror-image rotors whenever they get too close. For example, the period-11 rotor at $\varepsilon = 0.30$ stays a minimum of 10–12 su (about $\lambda_0/2$) from other rotors and stands off from a wall motionlessly at half that distance, slightly less than one nominal rotor radius of 22 su/2π [6]. The period-17 rotor's motion shows no effect at wall-to-pivot distances exceeding 8 su (about $\lambda_0/4$), but pushed closer than 6 su it converts to a period-11 rotor, and rebounds slightly. The period 56 rotor at $\varepsilon = 0.35$ also stands off from no-flux walls, at range 10 su, somewhat less than the nominal radius 90 su/2π, and creeps around the wall in a direction opposite to its rotation. (This 'skidding', rather than 'rolling', reaction to a wall was also observed in the Oregonator model of the Belousov–Zhabotinsky chemically excitable medium [35] but this is the first such instance noticed in FitzHugh–Nagumo kinetics.)

What initial conditions are required to initiate the distinct rotors at $\varepsilon = 0.30$ in the bistable range? A sufficiently faithful facsimile of either rotor suffices as initial condition, but among less faithful initial conditions tried, all led to the period-11 solution. For example, constructing a stable periodic plane-wave field at either period, then replacing half of it by medium near equilibrium to truncate the wave fronts, leads each wave front to coil up around its end, creating period-11 rotors. Period-17 rotors were obtained only by gradually decreasing ε from 0.35 then saving that solution to be replicated as initial condition in various later experiments. Replacing a region of that solution by medium at equilibrium, a period-11 rotor is created wherever a wave front is interrupted; this is how fig. 5 was initiated. No spontaneous conversion from shorter to longer period has been observed, nor, in terms of fig. 4, is it easy to see how this might come about.

The stability of these rotors is attested not only by convergence to these two forms (only) from diversely imperfect initial conditions, but also by their persistence without noticeable change during dozens of rotations while eccentrically situated in a square box, the four corners of which recurrently perturb the passing wavefront. Recomputing with time or space grid steps altered two-fold, essentially the same results are obtained.

5. Three-dimensional vortex rings

The vortex ring is a three-dimensional self-sustainingly periodic solution to the reaction–diffusion equations; it resembles the two-dimensional solution rigidly rotated about a distant axle [1, 6, 9, 11, 21, 22, 30, 36, 37]. Eq. (2) with parameters $\varepsilon = 0.30$, $\beta = 0.70$, $\gamma = 0.50$ is the medium in which stable vortex rings were first demonstrated [6, 36] using the shorter-period alternative. In retrospect, this turns out to be a doubly peculiar parameter combination. First of all, it lies very close to the locus $u_0 = 0$: the nullclines cross almost at a vertical segment of the cubic. And fig. 3 shows that it lies very close to the edge of the stable range for the longer-period rotor. Rotors started by truncating a wave front all evolved to the period-11 solution. Rotationally symmetric computations, started from a *disk* of wave front, evolved to vortex rings with period only slightly longer than 11, in proportion to the ring's curvature, $k = 1/$radius of curvature, r. These rings all contracted at a speed depending parabolically on k: very slowly at very small k, faster at $k \approx 1/$(10 su), and again very slowly near the attracting stable curvature $k = 1/$(4.8 su).

In contrast, in the same medium the period-17 vortex ring of radius exceeding 35 su (a little more than λ_0) *expands*. Over the range of radii observed ($r = 40$ to 55 su), it expands at rate $dr/dt = 1.3/r$, nearly the opposite of the period-11 vortex ring with comparable curvature [6]. It also drifts at speed $0.2/r$, about $1/5$ the speed of the period-11 vortex ring, also in the opposite direction. As it comes closer than one rotor radius (5–6 su) from the no-flux boundary, it abruptly converts to a period-11 vortex ring, which then turns around to contract in the familiar way from that large radius. In other words, 'expansion' or 'contraction' in scroll rings is not so much a property of the *medium* as of the particular solution inhabiting the medium.

Initiated at radii smaller than 35 su (i.e. with more curvature), the period-17 vortex ring proves unstable: it immediately converts to a period-11

vortex ring and then contracts in the familiar way.

Had this experiment been done in a rectangular rather than a cylindrical box, parts of the period-17 ring would have grazed the walls and converted before other parts, presenting us with the spectacle of a vortex ring composed of alternating segments of period ≈ 17 and of period ≈ 11 moving in opposite directions. The interfaces would rapidly become twisted; this might result in explosion of helicity [7], possibly nucleating turbulence in a uniform excitable medium; or the period-17 segments might be forced promptly to convert to period ≈ 11. In general, the apparition of alternative rotor solutions potentially presents us with the spectacle of vortex filaments of hybrid character, and new questions about dynamics near the juncture, e.g. conversion from one period to the other. How fast does this disturbance move along the filament, relative to collision boundaries and relative to wave propagation? If linked rings were of period-11 filament and period-17 filament, presumably the former would overwhelm the latter from a distance, via collision boundary migration at Mach 0.2. There could be no change in the topological necessities of this kind of organizing center [37–39], but a solitary period-11 twisted vortex ring would be left without a linking mate, linked only by a ring-shaped singularity that acts as a sink for period-11 wave fronts: would this change not affect the stability of the object?

6. Conditions for the existence of rotors in an excitable medium

The 'critical curve' estimated in fig. 2 might have turned out to represent a single relatively ε-independent condition on wave speed and wavelength that must be satisfied by any rotor, as originally suggested in ref. [15]. But it now seems better thought of in terms of the theory of rotors advanced by Keener and Tyson [13–15] as a locus of intersections between curve pairs parametrized by ε, not closely delineating any single curvature relation. Given the observed rotor period at any ε, one can choose a value of undetermined parameter r in eq. (1) to force $\lambda(c)$ to intersect the corresponding dispersion curve at the observed period. The best-fitting r values turned out to range from 2 to 6 su as ε ranged from 0.28 to 0.34 (roughly $62\varepsilon - 16$): always about $\lambda_0/10$, somewhat smaller than the nominal core radius $\lambda_0/2\pi$ based on the 'rule of thumb' that wave speed is independent of curvature [23]. (In the Oregonator model, r is always about $\lambda_0/(7$ to $10)$ [27]; were this empirical rule exact, it would provide the missing key to a quantitative theory of rotors.) With these r values, the best-fitting curvature relations of form (1) necessarily cut the corresponding dispersion curves at the dots determined from the corresponding rotor periods in fig. 2.· Their (negative) slopes turn out to be nearly equal to the steepest downslope of these dispersion curves, so none of them can provide the two alternative rotor solutions observed. This curve-fitting exercise thus does not satisfactorily account for our numerical experiments, but it was not expected to since the theory pertains to a singular perturbation limit ($\varepsilon \rightarrow 0$), and these experiments were performed close to the opposite limit of ε so large that propagation fails (0.4). The point of the present exercise was not to check the singular perturbation limit, but to use that idealization as a compass for seeking alternative rotor solutions in practical contexts. This was successful.

Some of these curves intersect *only* in a downslope region of the dispersion curve, or at such a shallow angle that the point of intersection may be labile. In such a case the simply periodic wave train would probably not be stable and more complex behavior would develop [28, 29], e.g. pulses bunching together in pairs as the wave train radiates from its source in the rotor, so that all inter-pulse intervals come to lie in upslope arcs of the dispersion curve. In the present case, $0.32 < \varepsilon < 0.33$ presents such intersections. There is no particular reason to think this would preclude the existence of a rotor, even a strictly

periodic one, and nothing particularly peculiar appeared during simulations as ε passed through this range, but the passage could have been slower, and only wave-tip centroid motion was assayed, and regularity of the wave train radiating away was not assayed at any large distance from its source. The question warrants analytical attention and more refined numerical experiments.

The existence of alternative rotors in a given medium provides an opportunity to experimentally answer another fundamental question: Is 'the rotor' really the causal source of the waves radiating away from it (as I have generally assumed), or is it merely a dislocation imposed and maintained by the surrounding wave field (as seems generally assumed elsewhere)? Perturbing a period-17 rotor near its nominal edge (radius about $\lambda_0/2\pi$ from the pivot, using a small region of excitation, $\Delta A = 0.25$, applied wherever the medium is not still too refractory) causes it to convert to period ≈ 11, which causes the surrounding medium soon to be filled with period-11 wave trains. Similar perturbations elsewhere have little lasting effect, except for those not too far away in the 'excitable gap' that start a propagating circular front timed to arrive at the rotor's edge when it is vulnerable. Thus it seems that the medium is indeed driven by the rotor, the active part of which may be its edge. If so then dynamics near the pivot may merely passively supported by wavefront kinematics outside, along the active edge of the rotor, but in the case at hand the entire rotor is so small (2 to 6 space units in radius, about equal to the pulse thickness) that it is hard to make an experimentally meaningful distinction between 'the edge' and 'the pivot'. Suppose the opposite had been found, viz., that perturbations near the rotor accomplish little, but remote perturbations just ahead of the wave front locally iron-out the loop shown in fig. 4, and that this transition spreads along the moving front both outward and inward. Then it would have seemed that the 'rotor' is not the active source of waves radiating outward but a mere name for the region around the moving wave tip, a reification the topological singularity implicit in any spiral wavefield. This appears not to be the case.

Acknowledgements

Programming utilities constructed by Chris Henze graphed the dispersion curves and iso-concentration contours. John Tyson caught some of my errors in struggling with eq. (1); careful reading by three referees improved the presentation throughout. This project was supported by the National Science Foundation (Chemical Dynamics and Applied Mathematics).

References

[1] W. Jahnke, W.E. Skaggs and A.T. Winfree, Chemical vortex dynamics in the Belousov–Zhabotinskii reaction and in the 2-variable Oregonator model, J. Chem. Phys. 93 (1989) 740–749.

[2] E. Lugosi, Analysis of meandering in Zykov kinetics, Physica D 40 (1989) 331–37.

[3] G. Skinner and H. Swinney, Periodic to quasiperiodic transition of chemical spiral rotation, Physica D 48 (1991) 1–16.

[4] K.I. Agladze and V.I. Krinsky, Multi-arm vortices in an active chemical medium, Nature 296 (1982) 424–426.

[5] E.A. Ermakova, A.M. Pertsov and E.E. Schnol, On the interaction of vortices in two-dimensional active media, Physica D 40 (1989) 185–195.

[6] M. Courtemanche, W.E. Skaggs and A.T. Winfree, Stable three-dimensional action potential circulation in the FitzHugh–Nagumo model, Physica D 41 (1990) 173–182. [There is a typographical error on page 174: $w = 3/(u_0^2 - u_0 + 1)^{0.5}$, and not $3/(u_0^2 - u_0 - 1)^{0.5}$.]

[7] C. Henze, E. Lugosi and A.T. Winfree, Stable helical organizing centers in excitable media, Can. J. Phys. 68 (1990) 683–710.

[8] A.M. Pertsov, E.A. Ermakova and A.V. Panfilov, Rotating spiral waves in a modified Fitzhugh–Nagumo model, Physica D 14 (1984) 117–124.

[9] A.T. Winfree, Stable particle-like solutions to the nonlinear wave equations of three-dimensional excitable media, SIAM Rev. 32 (1990) 1–53.

[10] A.T. Winfree, Multiple stable solutions to the kinetic equations of an excitable medium, in: Integral Methods in Science and Engineering – 90, eds. F.R. Payne and A. Haji-Sheikh (Hemisphere Publ., Washington, DC, 1991) pp. 172–183.

[11] A.T. Winfree, Spatial and temporal organization in the Zhabotinskii reaction, Adv. Biol. Med. Phys. 16 (1973) 115–136.

[12] A.M. Pertsov and A.V. Panfilov, Spiral waves in an active medium, in: Autowaves in Reaction–Diffusion Systems (Gorky, 1981).

[13] J.P. Keener, A geometrical theory for spiral waves in excitable media, SIAM J. Appl. Math. 46 (1986) 1039–1056.

[14] J.J. Tyson and J.P. Keener, Singular perturbation theory of traveling waves in excitable media, Physica D 32 (1988) 327–361.

[15] J.P. Keener and J.J. Tyson, Spiral waves in the Belousov–Zhabotinskii reaction, Physica D 21 (1986) 307–324.

[16] P.B. Monk and H.G. Othmer, Relay, oscillations and wave propagation in a model of *Dictyostelium discoideum*, Lect. Math. Life Sci. 21 (1989) 87–122.

[17] J.J. Tyson, K.A. Alexander, V.S. Manoranjan and J.D. Murray, Spiral waves of cyclic AMP in a model of slime mold aggregation, Physica D 34 (1989) 193–207.

[18] K.J. Tomchik and P.N. Devreotes, Adenosine 3′,5′-monophosphate waves in *Dictyostelium discoideum*: A demonstration of isotope dilution-fluorography, Science 212 (1981) 443–446.

[19] M. Courtemanche, W.E. Skaggs and A.T. Winfree, Two-dimensional rotating depolarization waves in a modified Beeler–Reuter model of cardiac cell activity, in: Science at the John von Neumann National Supercomputer Center, Vol. 3, ed. G. Cook (Consortium for Scientific Computing, Princeton, 1990) pp. 79–86, reprinted as: A two-dimensional model of electrical waves in the heart, Pixel 1(3) (1990) 24–31.

[20] A.T. Winfree, Electrical instability in cardiac muscle: Phase singularities and rotors, J. Theor. Biol. 138 (1989) 353–405.

[21] A.T. Winfree, Ventricular reentry in three dimensions, in: Cardiac Electrophysiology, from Cell to Bedside, eds. D.P. Zipes and J. Jalife (Saunders, London, 1989).

[22] A.T. Winfree, Vortex action potentials in normal ventricular muscle. Ann. NY Acad. Sci. 591 (1990) 190–207.

[23] A.T. Winfree, Spiral waves of chemical activity, Science 175 (1972) 634–636.

[24] R. Haan and A.T. Winfree, unpublished 1973 lab books using the recipe given in ref. [23].

[25] S.C. Muller, T. Plesser and B. Hess, Two-dimensional spectrophotometry of spiral wave propagation in the Belousov–Zhabotinsky reaction I. Experiments and digital data representation, Physica D 24 (1987) 71–86; II. Geometric and kinematic patterns, Physica D 24 (1987) 87–96.

[26] W.Y. Tam, W. Horsthemke, Z. Noszticzius and H.L. Swinney, Sustained spiral waves in a continuously fed unstirred chemical reactor, J. Chem. Phys. 88 (1988) 3395–3396.

[27] W. Jahnke and A.T. Winfree, A survey of spiral wave behaviors in the Oregonator model, Int. J. Bifurcations Chaos, in press.

[28] E. Meron, Nonlocal effects in spiral waves, Phys. Rev. Lett. 63 (1989) 684–687.

[29] C. Elphick, E. Meron and E.A. Spiegel, Spatiotemporal Complexity in Travelling Patterns, Phys. Rev. Lett. 63 (1986) 496–499.

[30] A.T. Winfree, Rotating chemical reactions, Sci. Am. 230(6) (1974) 92–95.

[31] A.T. Winfree, Rotating solutions to reaction/diffusion equations, SIAM/AMS Proc. 8 (1974) 13–31.

[32] A.T. Winfree, Stably rotating patterns of reaction and diffusion, Prog. Theor. Chem. 4 (1978) 1–51.

[33] A.C. Scott, The electrophysics of a nerve fiber, Rev. Mod. Phys. 47 (1975) 487–533.

[34] A.T. Winfree, Discrete spectrum of rotor periods in an excitable medium, Phys. Lett. A 149 (1990) 203–206.

[35] C. Henze, unpublished observations.

[36] W.E. Skaggs, E. Lugosi and A.T. Winfree, Stable vortex rings of excitation in neuroelectric media, IEEE Trans. Cir. Svs. 35 (1988) 784–787. [There is a typo in the equation: parameters 0.7 and 0.5 are interchanged.]

[37] A.T. Winfree and S.H. Strogatz, Singular filaments organize chemical waves in three dimensions: I. Geometrically simple waves, Physica D 8 (1983) 35–49; II. Twisted waves, Physica D 9 (1983) 65–80; III. Knotted waves, Physica D 9 (1983) 335–345; IV. Wave taxonomy, Physica D 13 (1984) 221–233.

[38] A.T. Winfree and S.H. Strogatz, Organizing centers for three-dimensional chemical waves, Nature 311 (1984) 611–615.

[39] A.T. Winfree, E.M. Winfree and H. Seifert, Organizing centers in a cellular excitable medium, Physica D 17 (1985) 109–115.

Physica D 49 (1991) 141–160
North-Holland

Chapter 3. Fronts and Turing patterns

Instabilities of front patterns in reaction–diffusion systems

A. Arneodo[a,1], J. Elezgaray[b,1], J. Pearson[c] and T. Russo[a]

[a]*Center for Nonlinear Dynamics, Department of Physics, The University of Texas, Austin, TX 78712, USA*
[b]*Center for Applied Mathematics, 306 Sage Hall, Cornell University, Ithaca, NY 14583, USA*
[c]*Center for Studies in Statistical Mechanics and Complex Systems, The University of Texas, Austin, TX 78712, USA*

Recent experiments in chemical reaction–diffusion systems with externally imposed concentration gradients may provide access to a host of spatio-temporal pattern formation phenomena. These systems tend to form steep reaction fronts in response to the external gradient. We use singular perturbation techniques, normal form calculations and numerical simulations to investigate the existence and the stability of such sustained fronts. In one-dimensional systems, the theoretical predictions are found in quantitative agreement with direct simulations of the Hopf bifurcation from steady to periodically oscillating front structures observed in the Couette flow reactor. Also conditions are found under which oscillations of the spatial structure become chaotic. In two-dimensional systems, we address the issue of realizing an experimental situation hitherto unattained: a one-dimensional chain of coupled oscillators at the onset of the Hopf destabilization of the front structure. We point out the intimate relationship between the frequency of oscillation, ω, of the homogeneous front pattern and the characteristic wavelength, λ, of the Turing pattern that can develop along the front; $\lambda \approx 2\pi(D/\omega)^{1/2}$. We comment on subsequent bifurcations that may result from the nonlinear interaction between Hopf and Turing instabilities as the precursors to spatio-temporal chaos. In conclusion, we emphasize the possibility of probing the transition to "mediated defect turbulence" in thin film gel reactors.

1. Introduction

Pattern formation phenomena and self-organization processes have become a topic of great importance in many different scientific fields as diverse as materials science, hydrodynamics, chemistry, biology and plasma physics [1–10]. In particular, a great deal of attention has been paid to the formation of dissipative structures in chemically reacting and diffusing systems. In well-mixed (homogeneous) media, the nonlinear aspect of chemical kinetics has provided a convenient experimental support for the study of low-dimensional dynamical systems [11–13]. When conducted in a continuously stirred tank reactor (CSTR), chemical reactions were shown to exhibit a transition from coherent temporal patterns to chemical chaos [11]. Among these chemical oscillators, the Belousov–Zhabotinsky (BZ) reaction [14] has revealed most of the well-

known scenarios to chaos including period-doubling, intermittency, quasiperiodicity, frequency locking, collapse of tori and crisis phenomena [11–13]. In contrast, there has been only a little progress in the experimental research on sustained spatial or spatio-temporal chemical dissipative structures where, in addition, a diffusive transport process competes with the local chemical kinetics.

In 1952, Turing [15] established the theoretical possibility that time-independent spatial patterns could organize in an initially homogeneous reaction–diffusion system. Since this pioneering work, the role of diffusion on the stability of steady states has been extensively studied from a theoretical point of view, both in discrete coupled cell systems and in extended continuous systems [1, 10, 16–22]. Diffusive instabilities have also been proposed to account for propagating patterns in homogeneously oscillating systems [4]. Near the oscillatory onset, the Ginzburg–Landau [23] amplitude equation (Hopf normal form for extended systems) can be reduced to the Kuramoto–

[1]Permanent address: Centre de Recherche Paul Pascal, Avenue Schweitzer, 33600 Pessac, France.

Sivashinsky equation [24, 25], which describes slow spatial and temporal variations of the phase of the oscillators. This equation has been the center of increasing interest during the past few years [26]. Besides regular cellular and propagating solutions, this equation was shown to exhibit chaotic solutions [26–28], a form of weak turbulence usually called "phase turbulence" [4, 24]. More recently, numerical simulations with the two-dimensional complex Ginzburg–Landau equation have provided a very appealing interpretation of weak turbulence in terms of the creation of spiral defects whose interactions and motion lead to a complex spatio-temporal behavior termed "topological turbulence" [29–33]. Ultimately, when more delocalized amplitude modes are awakened, "defect mediated turbulence" transforms itself into amplitude turbulence [24, 34].

Despite the great deal of theoretical and numerical effort, the experimental situation is much less satisfactory. Unlike heterogeneous systems where chemical stationary patterns are easily observed [35], there has been for years no clear experimental evidence for time-independent Turing patterns in single phase systems. A few experiments have yielded stationary patterns, but the interpretation of the origin of these structures is unfortunately made ambiguous by the presence of convective or interfacial effects [36–38]. The main drawback of past experiments on chemical pattern formation is that they were conducted in closed reactors (petri dishes) without feed, which prevented the observation of sustained patterns. In closed systems, only transient phenomena are accessible, since the system uncontrollably and irreversibly relaxes to thermodynamic equilibrium. Therefore the applicability of these experiments was limited to the study of patterns developing in a very short time, in practice those resulting from excitability phenomena such as the so-called target patterns and spiral waves commonly observed with the BZ reaction [5–7, 39–45].

Besides the lack of open reactors, severe hindrances have delayed experimental research of sustained nonequilibrium patterns. According to

theoretical work on model systems [1, 10, 16–18, 46], Turing instability to steady cellular patterns requires the diffusion coefficients of the different chemical species to be significantly different. These conditions are met in systems that are governed by the competition between an activator and an inhibitor, when the inhibitor diffuses faster than the activator. This is a common situation in biological systems [10], where many processes are activated by enzymes immobilized in a matrix. This is a generally unrealistic situation for chemical reactions, such as the much studied BZ reaction [14], involving small size molecules in aqueous solutions with comparable diffusion coefficients $D \sim 10^{-5}$ cm^2 s^{-1}.

As far as propagative structures are concerned, they are a common feature in excitable systems [5–7, 40–45]. However, these systems do not meet the prerequisite for the Kuramoto picture [4] to apply. The Hopf bifurcations identified in chemical systems are for the most part subcritical and lead to large-amplitude relaxation oscillations [5]. Relatively few supercritical Hopf bifurcations have been observed [47]. Therefore, wave patterns observed thus far in chemical systems are not the paradigm for Kuramoto propagative structures; they require a mathematical description specific to spatially distributed excitable media [48–52].

While the CSTR stirred up experimental studies of sustained temporal patterns in the late seventies [11–13], open spatial reactors have been devised only very recently. For a recent review of these experiments see ref. [53]. The difficulty in designing a reactor to obtain a genuine reaction–diffusion system comes essentially from (i) the feeding procedure: one has to ensure a spatially uniform feeding without introducing any convective motion; (ii) the transport process: as previously mentioned, molecular diffusion coefficients are very small and somewhat experimentally inconvenient; there is thus a need to control to some extent the transport process. Basically, three types of open reactors are currently operating: the Couette flow reactor [54–57], the continuously fed unstirred reactor [58], and the linear [59] or annular [53, 60] gel reactors. In this introduction, we will briefly describe the corresponding experimental systems with special emphasis on the observed sustained reaction–diffusion patterns rather than on technical details that can be found in the original publications.

The spatially extended open Couette flow reactor [53–57] provides a practical implementation of an effectively one-dimensional reaction–diffusion system with (i) well-defined boundary conditions: the fresh chemicals are fed at the two boundaries of the Couette reactor, and (ii) controlled turbulent diffusion process: the effective diffusion coefficient D is a tunable parameter that depends mainly on the rotation rate of the inner cylinder. The accessible D values range from 10^{-2} to 10 cm^2 s^{-1}, i.e., several orders of magnitude larger than molecular diffusion coefficients [61]. Two different reactions have presently been studied in the Couette flow reactor, namely the variants of the BZ [54–57] and chlorite–iodide [56–57] reactions. The BZ reaction has revealed a rich variety of steady, periodic, quasi-periodic, frequency-locked, period-doubled and chaotic spatio-temporal patterns [54], well described in terms of diffusive coupling of oscillating reactor cells, the frequency of which changes continuously along the Couette reactor as the result of the imposed spatial gradient of constraints [62]. With the chlorite–iodide reaction, steady and oscillatory (single- and multi-) front patterns have been observed when varying the control parameters [56, 57, 63]. This reaction provides a remarkable illustration that spatio-temporal patterns can organize in a chemical system from the diffusive coupling of (nonoscillating) steady state reactor cells [64, 65].

The continuously fed unstirred reactor and the gel reactors are two-dimensional systems where the transport process is essentially natural molecular diffusion [53]. The continuously fed unstirred reactor [58] is a two-dimensional thin film in which reaction and diffusion take place excluding convective processes. The gel support medium is uniformly fed through a thin membrane which discards turbulent transport effects, by contact with a large reservoir of homogeneous nonequilibrium reacting medium. This reactor has been applied mainly to study detailed motions of sustained spiral waves in excitable media [66]. The linear and annular gel reactors are strips of gel that are fed from the lateral boundaries. These reactors have a natural tendency to produce narrow front structures away from the boundaries where the perturbations associated to the feed may disturb the dynamics. In the annular gel reactor, rotating wave patterns have been observed travelling along the front [60, 67, 68]. Unlike the propagating structures in petri dishes, these travelling waves can be sustained indefinitely, since the medium is constantly maintained out of equilibrium. Very recently, the Bordeaux group has reported the observation of a symmetry-breaking instability leading to a stationary Turing structure along the front pattern in a very promising experiment in the linear gel reactor [59].

The aim of the present study is to provide theoretical and numerical support for the recent experimental observations of sustained dissipative structures in open spatial reactors. We will mainly focus our attention on the front structures which develop in the system when a concentration gradient is imposed from the boundaries [16, 53, 64]. These chemical fronts correspond to localized structures of strong chemical activity. Since instabilities can only develop inside the active region, the theoretical understanding of the existence and stability of front patterns is thus of fundamental importance for further progress in the comprehension of self-organization phenomena in reaction–diffusion systems. Our approach will be based on singular perturbation techniques and normal form calculations. For most of the oscillating reactions, the knowledge of the reaction mechanisms and rate constants is generally very sketchy. Since the details of a particular kinetic model are not relevant close to bifurcation conditions, we will consider only simple models that retain the main ingredients required for pattern formation. In section 2, we report on a numerical and theoretical study of spatio-temporal pattern formation in a one-dimensional reaction–diffusion system with equal diffusion coefficients that mimics the Couette reactor experiments performed with the chlorite–iodide reaction [64, 65, 69, 70]. In section 3, we generalize this study to two-dimensional reaction–diffusion systems that model sustained front patterns observed in gel reactors. We pay special attention both to the Hopf instability leading to oscillating fronts [71] and to the Turing instability that can develop along the front pattern [22]. In particular, we emphasize the intimate relationship between the characteristic wavelength of the Turing pattern and the frequency of the (damped) oscillation of the homogeneous front pattern [71] as a possible experimental test of the intrinsic nature of the observed spatial structure. In section 4, we comment on secondary bifurcations that can result from the interaction of these two instabilities, as the precursors to spatio-temporal complexity in confined geometries. In spatially extended geometries we emphasize the possibility of studying the transition from phase turbulence to topological turbulence with the specific goal of elucidating the creation mechanism of spiral defects. It is likely that theoretical and numerical studies currently in progress will indicate new directions for experimental research.

2. Front patterns in a one-dimensional chemical system with equal diffusion coefficients

2.1. The reaction–diffusion model

With a few exceptions [16, 22, 53, 54, 62], the crucial role of external gradients has not been considered in previous work, and the stability and dynamics of sustained steady front patterns is very poorly documented in the literature. The study reported in this section is a step in that direction [64, 65, 69, 70]. The reaction–diffusion

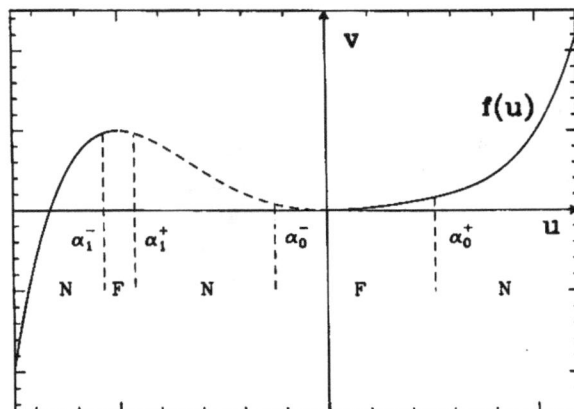

Fig. 1. Sketch of the slow manifold $v = f(u) = u^2 - u^3 + u^5$. The unique steady state: $u = \alpha$, $v = f(\alpha)$, is a focus (F) for $\alpha_1^- < \alpha < \alpha_1^+$ or $\alpha_0^- < \alpha < \alpha_0^+$, and a node (N) elsewhere. A solid line indicates a stable steady state ($\alpha < \alpha_L$ or $\alpha > \alpha_U$); a dashed line an unstable steady state ($\alpha_L < \alpha < \alpha_U$).

model presented below is a formal model that does not claim to describe faithfully the experimental situation encountered in the Couette flow reactor. It does not completely meet the experimental conditions and the specific requirements of chemical kinetic laws. However, it does retain the minimal ingredients necessary to reproduce most of the phenomena associated with the observed front patterns in the chlorite–iodide reaction.

Our model of the reaction term is a two-variable Van der Pol-like system [72, 73]

$$\frac{du}{dt} = \varepsilon^{-1}[v - f(u)], \quad \frac{dv}{dt} = -u + \alpha, \tag{1}$$

where ε is a small positive parameter and α is a free parameter. These equations ensure the existence of a pleated slow manifold $v = f(u)$, on which all trajectories are attracted in a time $\sim \mathcal{O}(\varepsilon)$. The "S" shape of this manifold accounts for the excitable character of the dynamics. The only steady state of the reaction term ($u_s = \alpha$, $v_s = f(\alpha)$) is necessarily located on the slow manifold; this steady state is stable for $\alpha < \alpha_L$ (lower branch), or $\alpha > \alpha_U$ (upper branch), while it is unstable for $\alpha \in [\alpha_L, \alpha_U]$, as sketched in fig. 1. The critical values α_L and α_U correspond to a

Hopf bifurcation leading to oscillatory behavior. According to the specific shape $f(u)$ of the slow manifold, this bifurcation can be either supercritical or subcritical [73]. When adding a flux term to this Van der Pol-like equation, bistability can also be recovered. Despite the fact that model (1) does not have all the properties required in a chemical scheme, u and v play the role of concentration variables and we will refer to the upper and lower branches of the slow manifold as the analogues of the reduced and oxidized state branches of the chlorite–iodide reaction[#1].

When taking into account the diffusive transport process, the reaction–diffusion equations read

$$\frac{\partial u}{\partial t} = D\frac{\partial^2 u}{\partial z^2} + \varepsilon^{-1}[v - f(u)],$$

$$\frac{\partial v}{\partial t} = D\frac{\partial^2 v}{\partial z^2} - u + \alpha, \qquad z \in [0,1], \quad (2)$$

where z is the single space variable; the length of the Couette reactor has been rescaled to unity for convenience. The diffusion coefficients are set equal to $D_u = D_v = D$, in order to mimic the turbulent mass transport that drives pattern formation in the Couette flow reactor.

In most experimental runs, the volume and feeding flows of the two CSTRs at both ends of the Couette flow reactor were large enough for their internal state not to be significantly influenced by the dynamics inside the Couette reactor. This corresponds mathematically to imposing Dirichlet boundary conditions on our model reaction–diffusion system (2). (We refer the reader to refs. [64, 70] for a similar study with Neumann boundary conditions.) The left-end CSTR ($z = 0$) is set in a (reduced) upper-branch state while the right-end CSTR ($z = 1$) is maintained in an

---

[#1]In our reaction model (1), the variables u and v can take on negative values. The fact that our concentration variables are not necessarily positive is irrelevant for the physics of the patterns described in this article.

(oxidized) lower-branch state:

$$v(z = 0) = f(u_0), \quad u_0 = u(z = 0) > \alpha_U,$$
$$v(z = 1) = f(u_1), \quad u_1 = u(z = 1) < \alpha_L. \quad (3)$$

For the sake of simplicity, the value α in the system (2) is set independent of z, $\alpha > \alpha_U$, so that when switching off the diffusion process, all the intermediate cell points evolve asymptotically to the same reduced steady state. A more realistic model should take into account a spatial dependence of α, with $\alpha(z = 0) = u_0$ and $\alpha(z = 1) = u_1$; but the practical implementation of this spatial constraint does not have much meaning in this formal approach since it should strongly depend on the specific kinetics of the reaction [56, 57, 63]. For further discussion of this issue we refer the reader to ref. [69], but let us mention that the spatio-temporal patterns reported in this numerical study are robust against smooth perturbations of the $\alpha = $ constant working hypothesis.

2.2. Numerical simulations

The partial-differential equations (2) have been integrated numerically by the method of lines (reduction to an ODE system by space discretization) and integrated with a stiff ODE solver [69, 70]. Care has been taken to vary the spatio-temporal resolution in order to check the reliability of the reported phenomena.

The numerical patterns shown in fig. 2 have been obtained with the following form of the slow manifold: $f(u) = u^2 - u^3 + u^5$; this choice makes the Hopf bifurcation in the reaction term subcritical [73], as is the case in most experimental situations. The parameters u_1 and ε are fixed to the values $u_1 = -1.5$, $\varepsilon = 0.01$, whereas u_0, D and α are taken as control parameters. In fig. 2, we use a space–time representation with a concentration coding similar to the one used to visualize the change of colors in the experimental study. Fig. 2 has to be compared with fig. 2 in ref. [56].

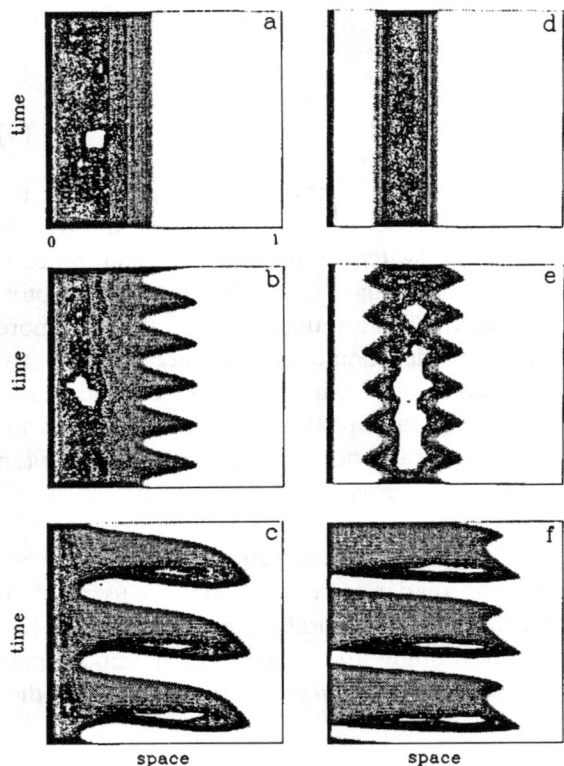

Fig. 2. The spatio-temporal variation of the variable $u(z,t)$ is coded in order to mimic the spatial color profile observed in the Couette flow reactor [56, 57]; 32 shades are used from the left-end upper branch (reduced) state (black) to the right-end lower branch (oxidized) state (white). The numerical spatio-temporal patterns in (a)–(f) were computed with the reaction–diffusion system (2) with the model parameters: $u_1 = -1.5$, $\varepsilon = 0.01$, $f(u) = u^2 - u^3 + u^5$. (a) Stationary single-front pattern ($u_0 = 1.1$, $D = 0.1$, $\alpha = 0.01$); (b) periodically oscillating single-front pattern ($u_0 = 1.1$, $D = 0.045$, $\alpha = 0.01$); (c) periodic alternation of a single-front and a three-front pattern ($u_0 = 1.1$, $D = 0.01$, $\alpha = 0.01$); (d) stationary three-front pattern ($u_0 = 0.5$, $D = 0.08$, $\alpha = 0.2$); (e) periodically oscillating three-front pattern ($u_0 = 0.5$, $D = 0.06$, $\alpha = 0.2$); (f) periodic alternation of a single-front and a three-front pattern ($u_0 = 0.5$, $D = 0.02$, $\alpha = 0.2$).

For $D \gg \varepsilon^{-1}$, the predominance of the diffusion term drives all the trajectories of the system towards a trivial steady state solution. This solution is merely a linear spatial concentration profile linking the two ending concentrations. For values of $u_0 \gg \alpha$, when decreasing the transport rate, the reaction term eventually becomes of the order of the diffusion term, and the former "dif-

fusion-like" solution develops a sharp front, whose spatial extent behaves like $\sim \mathcal{O}(\sqrt{D\varepsilon})$. This steady single front pattern corresponds to a spatial switching between the two attracting branches of the slow manifold (fig. 2a). When D is decreased the transition front becomes sharper and sharper, until this stationary pattern loses its stability and starts to oscillate as shown in fig. 2b. To gain some understanding of this instability, one may consider a spatially discretized version of our original continuous reaction–diffusion system (2). One can show that the coupled elementary reactor cells located at the front zone (where a steep gradient of concentration exists) can be driven by the diffusion coupling from a steady state to a limit cycle behavior via a Hopf bifurcation [69, 70]. The physical mechanism underlying this oscillatory instability has been clearly identified in the direct diffusive coupling of two unsymmetrically constrained CSTRs [74] (the simplest two-point reaction–diffusion system). The soundness of an extrapolation to the continuous limit is supported by analytical studies as discussed in section 2.3. When D is further decreased, the amplitude of the oscillations increases until a qualitative change occurs in the spatio-temporal evolution of the system (fig. 2c). A three-front profile alternates periodically with a single front proceeding from the periodic appearance and coalescence of two additional travelling fronts. A similar pattern where the period is about twice the period of the previous one is shown in fig. 2f. For $u_0 \sim \alpha$, we were able to freeze a pattern involving three spatial switches between the two branches of the slow manifold (fig. 2d). Again, a decrease of the diffusion coefficient induces a transition to a periodically oscillating structure (fig. 2e).

The patterns collected in fig. 2 do not constitute a bifurcation sequence since they were not obtained by a continuous variation of a single control parameter. However, they present some striking similarities [63] with the experimental patterns in fig. 2 of ref. [56]. In particular, the initial and final stages of the experimental sequence are identical to the numerical sequences in figs. 2a–2c and figs. 2d–2e, both obtained by a continuous change of the parameter D. Nevertheless, in the numerical simulations, the stationary multi-peaked structure coexists with the oscillating single front pattern, in contrast with the experimental situation, where these two patterns apparently take place in two different regions of the constraint space. We have strong indications that these multi-peaked structures arise from saddle–node bifurcations [70], i.e., at some distance from the "natural" steady state of the system, namely the stationary single front pattern. As pointed out in ref. [63], one cannot exclude the

presence of multiple stable states in the experiments.

In fig. 2, we have only illustrated a small sample of the results obtained with the reaction–diffusion model (2). A detailed description of the entire zoology of patterns observed both in the simulations and the experiments is reported in refs. [56, 57, 69, 70]. In particular, we have found conditions numerically where the oscillating front patterns undergo secondary instabilities leading to more complicated spatio-temporal behavior. In fig. 3a, we show a chaotically oscillating single front structure computed with the following slow

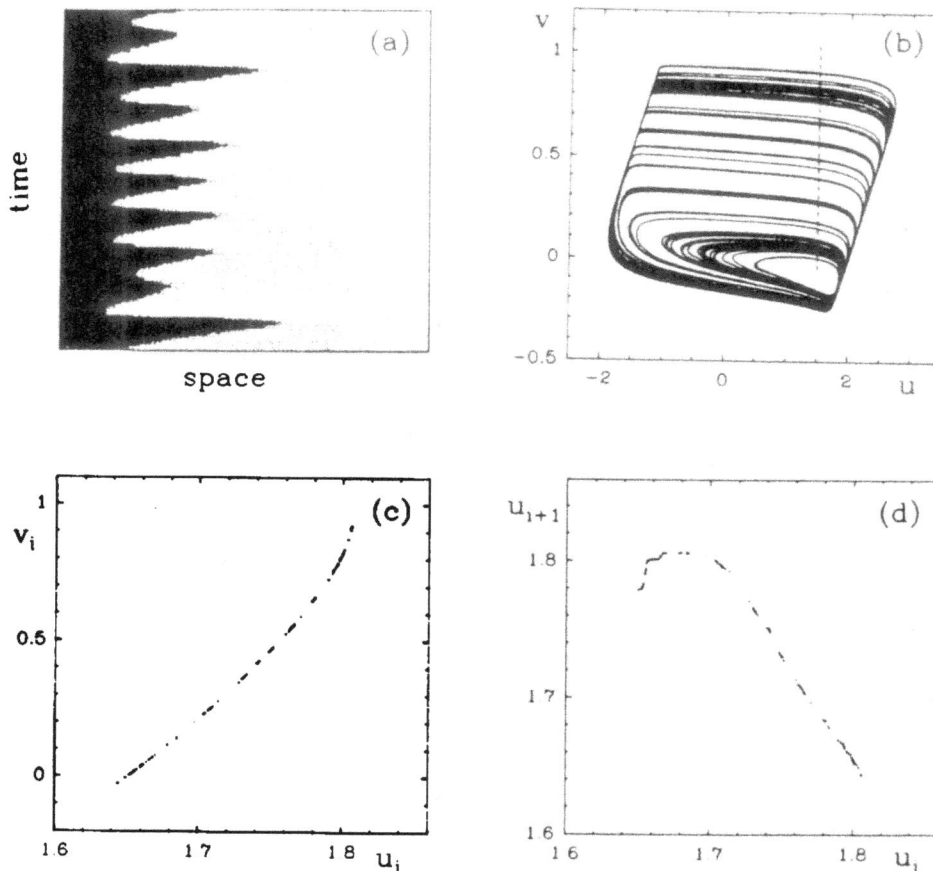

Fig. 3. Diffusion-induced chaos obtained when integrating the reaction–diffusion system (2) with the slow manifold $f(u)$ defined in eq. (4) ($\delta = 10^{-2}$). The model parameters are: $u_0 = 2.5$, $u_1 = -3.1$, $\varepsilon = 0.01$, $\alpha = 1.1$, $D = 0.05$. (a) Spatio-temporal variation of the variable $u(z, t)$ coded as in fig. 2; (b) phase portrait; (c) Poincaré map; (d) 1D map.

manifold ($\delta = 0$ corresponds to a piecewise linear function f which will be used in section 2.3 to derive analytical results):

$$f(u) = -f(-u)$$
$$= -u, \quad 0 \le u \le 1 - \delta,$$
$$= -(1-u)^4/2\delta^3 + 3(1-u)^2/2\delta - 1,$$
$$1 - \delta < u < 1 + \delta,$$
$$= u - 2, \quad u \ge 1 + \delta. \qquad (4)$$

The phase portrait reconstructed from the temporal evolution of the variables u and v recorded at an intermediate spatial cell point is shown in fig. 3b. The corresponding Poincaré map and 1D map are illustrated in figs. 3c and 3d, respectively. The fact that the Poincaré map is not a scattering of points but that all the points lie to a good approximation along a smooth curve indicates that the trajectories lie approximately on a (multi-folded) two-dimensional sheet in the phase space. The single humped shape of the 1D map in fig. 3d is a clear signature of the low-dimensional chaotic nature of these oscillations. It is somewhat puzzling that the phase portraits obtained in our simulations (fig. 3b) are strikingly similar to the strange attractors observed in the BZ reaction when conducted in a CSTR [11, 13]. Moreover, as in the homogeneous BZ reaction, period-doubling bifurcations are observed as precursors to these chaotic behaviors [55, 56]. The chaotic spatio-temporal patterns have been obtained when using a spatial discretization (~ 100 intermediate reactor cells) compatible with the number of characteristic diffusion lengths in the Couette reactor. We have checked that these chaotic patterns are preserved when increasing spatial resolution.

We have thus shown that bistability is the main ingredient required for our model reaction–diffusion system to reproduce the sustained spatial and spatio-temporal chemical patterns recently observed in the Couette reactor with asymmetric feeding. Our numerical results provide a remarkable illustration that coherent front patterns can organize in a one-dimensional extended medium in the limit of equal diffusion coefficients, provided a concentration gradient is imposed on the system. These sustained patterns organize due to the interaction of the diffusion process with a chemical reaction which itself would proceed in a stationary manner if diffusion was negligible [64, 65]. These front patterns can display diffusion-induced chaotic dynamics of the same mathematical nature as the chaos arising from the nonlinear complexity of the chemical kinetics of the BZ reaction.

2.3. Theoretical analysis

In this section, we study the existence and stability of the stationary states of the reaction–diffusion model (2) with Dirichlet boundary conditions (3). Our purpose here is not to give a full mathematical description of all the possible cases, but rather to emphasize some properties that are common to a large class of reaction–diffusion systems determined by a set of equations similar to (2). The existence problem for steady front patterns requires the use of a perturbative approach [69, 70]. Singular perturbation techniques [75] show that the equilibrium solutions of (2) display two characteristic length scales: one large scale, which is arbitrarily chosen to be $\sim \mathcal{O}(1)$ ($\sim$ the size of the Couette reactor), and a small scale $\sim \mathcal{O}(\sqrt{D\varepsilon})$. Generically, one of these solutions consists of N "outer" regions, where the solution is smooth and approximately satisfies the relation $v = f(u)$, separated by $N - 1$ narrow transition zones, or "fronts". In each of the N "outer" regions, the solution belongs alternatively to the upper or lower branch of the slow manifold. The transition between these two branches takes place at the front zones; in each of these zones, v is almost constant, and u varies abruptly as shown in fig. 4a. The value v_* of v at the front locations is a constant determined by

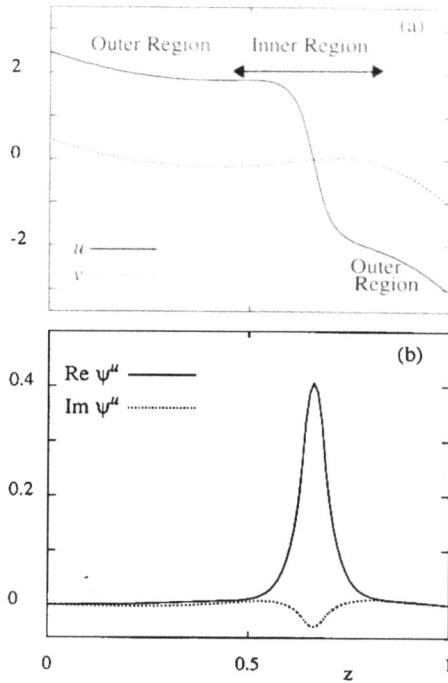

Fig. 4. (a) Spatial profile of a single-front solution: u (—) and v (...) variables. (b) Spatial profile of the u-component of the critical Hopf mode ψ^u: Re ψ^u (—) and Im ψ^u (...). Model parameters: $f(u)$ defined in eq. (4) with $\delta = 0$, $u_0 = 2.5$, $u_1 = -3.1$, $\alpha = 1.1$, $\varepsilon = 0.01$, $D_{\text{Hopf}} = 0.011$.

the condition

$$\int_{h_-(v_*)}^{h_+(v_*)} [v_* - f(u)] \, du = 0, \tag{5}$$

so that, in the limit $\varepsilon \to 0$, all the stationary solutions satisfy $v(z_i) \sim v_*$, where z_i is the location of the ith front. The two functions h_- and h_+ are the inverses of f: $f(h_-(v)) = f(h_+(v)) = v$. In the limit $\varepsilon \to 0$, it can be proved [70, 75] that, provided some mild conditions are satisfied, the v component of the stationary solutions of (2) is given by the reduced problem

$$D\frac{\partial^2 v}{\partial z^2} - h_\pm(v) + \alpha = 0, \quad v(0) = v_0, \quad v(1) = v_1, \tag{6}$$

with the additional requirement that, at the front

location where h_+ switches to h_- (or vice versa), $v(z)$ takes on the value v_* and is differentiable. In the particular case of piecewise linear kinetics (eq. (4) with $\delta = 0$), this theoretical analysis can be carried out analytically and it can be shown that any N-front solutions with $N \geq 2$ is created through a saddle–node bifurcation [69, 70]. In addition, for each $N \geq 1$, only one N-front solution persists in the limit $D \to 0$. We believe that these results can be generalized to a more general class of S-shaped slow manifolds.

As far as the linear stability of these front pattern solutions is concerned, we have established the existence of a Hopf bifurcation for single front solutions [69, 70]. Our proof relies heavily on the techniques developed recently by Nishiura and Fujii [76] for solving a similar problem with Neumann boundary conditions. These techniques allow us to reduce to two nonlinear equations, for the real and imaginary parts of the eigenvalue λ, the following infinite dimensional eigenvalue problem obtained by linearization around the single-front solution [69, 70]:

$$D\varepsilon \frac{\partial^2}{\partial z^2}\psi^u + \psi^v - f'(u(z))\psi^u = \lambda\varepsilon\psi^u$$

$$D\frac{\partial^2}{\partial z^2}\psi^v - \psi^u = \lambda\psi^v \tag{7}$$

where $u(z)$ is the u component of our single-front solution and (ψ^u, ψ^v) are the (u, v) components of the Hopf eigenvector. As illustrated in fig. 4b, one can demonstrate that the critical eigenvector is strongly localized in the front zone and decreases exponentially away from it [69, 70]. The critical eigenvalue has unbounded imaginary part in the limit $\varepsilon \to 0$, corresponding to a frequency of oscillation $\omega \sim \mathcal{O}(\varepsilon^{-1})$.

Close to the bifurcation value, the dynamics of the single-front pattern is governed by the Hopf normal form equation [77, 78]

$$\dot{A} = (\mu + i\omega)A + \kappa A|A|^2 + \text{h.o.t.}, \tag{8}$$

where A is the complex amplitude of the critical

Hopf mode (ψ^u, ψ^v), μ defines the deviation from criticality $(\mu = 0)$, and κ is a complex number whose real part determines the subcritical $(\mathrm{Re}\,\kappa > 0)$ or supercritical $(\mathrm{Re}\,\kappa < 0)$ character of the Hopf bifurcation. We have numerically checked the nature of the Hopf bifurcation for several choices of the slow manifold $f(u)$. It turns out that the nature of the Hopf bifurcation of the reaction term greatly influences the character of the Hopf bifurcation of the single-front solution of the whole reaction–diffusion system (2). For instance, the Hopf bifurcation shown in figs. 2a and 2b with $f(u) = u^2 - u^3 + u^5$ was found to be subcritical, while supercritical bifurcations were identified with $f(u) = u^2 + u^3$. Using center-

manifold theorem and normal form techniques [77, 78], we have explicitly reduced our reaction–diffusion system (2) to the Hopf normal form (8) of the single-front solution. Then by calculating the coefficient κ on the critical surface $\mu = 0$, we have compared our theoretical prediction to direct simulations of the partial-differential system (2). As shown in fig. 5, a quantitative agreement has been found between the normal form prediction and the numerical results in the supercritical situation $f(u) = u^2 + u^3$. Fig. 5a illustrates the numerical determination of the scaling exponent $(\beta = 1/2)$ of the amplitude of the Hopf mode as a function of the distance μ to the bifurcation point: $|A| = (-\mu/\mathrm{Re}\,\kappa)^{1/2}$. Fig. 5b shows the numerical values of the coefficient $\mathrm{Re}\,\kappa$ extracted from the μ dependence of the amplitude of the limit cycle as compared to the theoretical normal form prediction. The numerical points range on a straight line whose intercept at $\mu = 0$ is in remarkable agreement with the theoretical value $\mathrm{Re}\,\kappa(\mu = 0) = -0.7301$. This theoretical analysis is likely to be tractable in the more general case of N-front solutions [69, 70].

3. Front patterns in two-dimensional reaction–diffusion systems

3.1. The reaction–diffusion model

The basic idea of the open gel reactors (both linear and annular) [59, 60, 68] is to localize all the significant dynamical phenomena inside a narrow region at the front zone in order to both decrease the effective dimensionality of the problem and to distance the active region from the boundaries where feed-induced perturbations may disturb the dynamics [53]. The design of the front reactors was pioneered by Boissonade in ref. [22]. The basic configuration is a rectangular strip of gel of length L $(0 \leq x \leq L)$ and with l $(0 \leq y \leq l)$. A concentration gradient is imposed on the system by a continuous feeding of the fresh reactants at the two opposite large edges

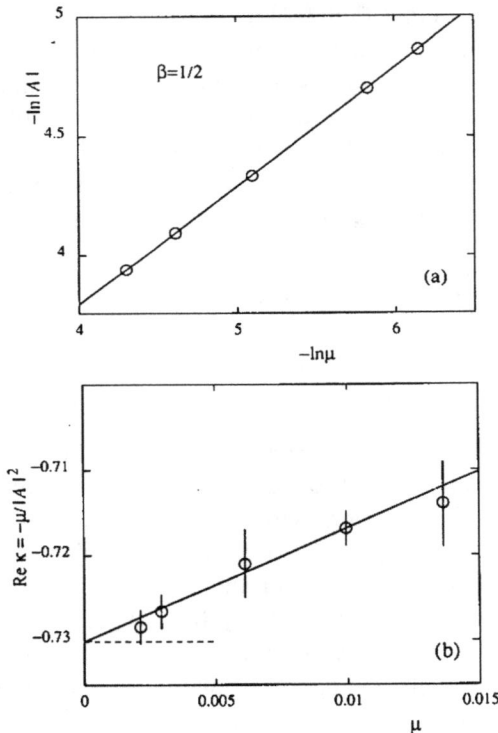

Fig. 5. Direct simulations of the reaction–diffusion system (2) versus Hopf normal form predictions. $|A| = (-\mu/\mathrm{Re}\,\kappa)^{1/2}$ is the amplitude of the Hopf critical model. (a) Determination of the critical exponent $\beta = 1/2$. (b) Measurement of $\mathrm{Re}\,\kappa$: the dashed line corresponds to the normal form predictions on the critical surface $(\mu = 0)$. Model parameters: $f(u) = u^2 + u^3$; $u_0 = 1.5$, $u_1 = -1.5$, $\alpha = +0.001$, $\varepsilon = 0.01$, $D_{\mathrm{Hopf}} = 1.537 \times 10^{-2}$.

(Ox direction) of the reactor. If one feeds asymmetrically with an oxidizing medium at one edge and a reducing medium at the other edge, one expects to recover in the direction Oy parallel to the gradient, the front pattern morphologies observed in the one-dimensional Couette reactor. These front patterns are likely to be translation invariant in the direction Ox transverse to the gradient. A direct consequence of the theoretical study carried out in section 2 is that these homogeneous (in the Ox direction) front patterns should destabilize into a periodically oscillating structure via a Hopf bifurcation. The aim of this section is to point out that the Turing symmetry-breaking bifurcation that can develop along the front (in the Ox direction) as originally described by Boissonade [22], is likely to occur in the neighborhood of the Hopf bifurcation of the homogeneous front pattern, and that under such conditions the characteristic wavelength of the Turing pattern is intimately related to the frequency of oscillation of the front structure [71].

Along the line of Boissonade's simulations [22], the reaction terms are modeled by the popular Brusselator model:

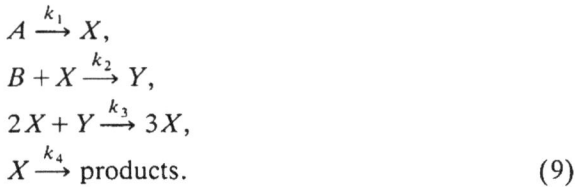

$$A \xrightarrow{k_1} X,$$

$$B + X \xrightarrow{k_2} Y,$$

$$2X + Y \xrightarrow{k_3} 3X,$$

$$X \xrightarrow{k_4} \text{products}. \tag{9}$$

When taking into account the diffusive transport process, the macroscopic rate equations are:

$$\frac{\partial A}{\partial t} = -k_A A + D_A \Delta A,$$

$$\frac{\partial B}{\partial t} = -k_B BX + D_B \Delta B,$$

$$\frac{\partial X}{\partial t} = A - (B+1)X + X^2Y + D_X \Delta X,$$

$$\frac{\partial Y}{\partial t} = BX - X^2Y + D_Y \Delta Y, \tag{10}$$

where A, B, X and Y are scaled variables. The first equation does not involve any species except A and thus can be integrated by hand. After the elimination of A, the system is reduced to three nonlinear partial differential equations which are solved numerically. At the boundaries $y = 0$ and $y = l$, the concentrations of the major species A and B are kept constant, whereas those of the intermediate species X and Y are zero. On the lateral sides ($x = 0$ and $x = L$) periodic boundary conditions are chosen, i.e. identical concentration profile on both sides. Under this working hypothesis, and neglecting curvature effects, the reaction–diffusion model (10) mimics the geometry of the annular gel reactor [60, 67, 68]. Inside the reacting strip the transport is controlled by molecular diffusion.

3.2. Numerical simulations

The two-dimensional reaction–diffusion equations (10) were first converted to a set of ordinary differential equations by using finite differences and integrated using the exponential propagation method of Friesner et al. [79]. The system was initialized in different randomly generated non-homogeneous configurations in order to let the different spatial modes grow. As previously emphasized, because of the concentration gradient the system evolves naturally toward a stationary single-front or multi-front structure which displays translation invariance in the direction transverse to the gradient. Those front structures can be more or less localized depending on the constraints.

Oscillatory instability of front patterns. According to the same mechanism reported in section 2, these front patterns can destabilize into periodically oscillating structures. The nature of the Hopf bifurcation can be either subcritical or supercritical depending on the working conditions and the model considered. This bifurcation can occur with *equal diffusion coefficients* for the different species. In fig. 6, we show snapshots of a periodically oscillating three-front pattern. The concentration of the intermediate species $X(x, y)$ was coded

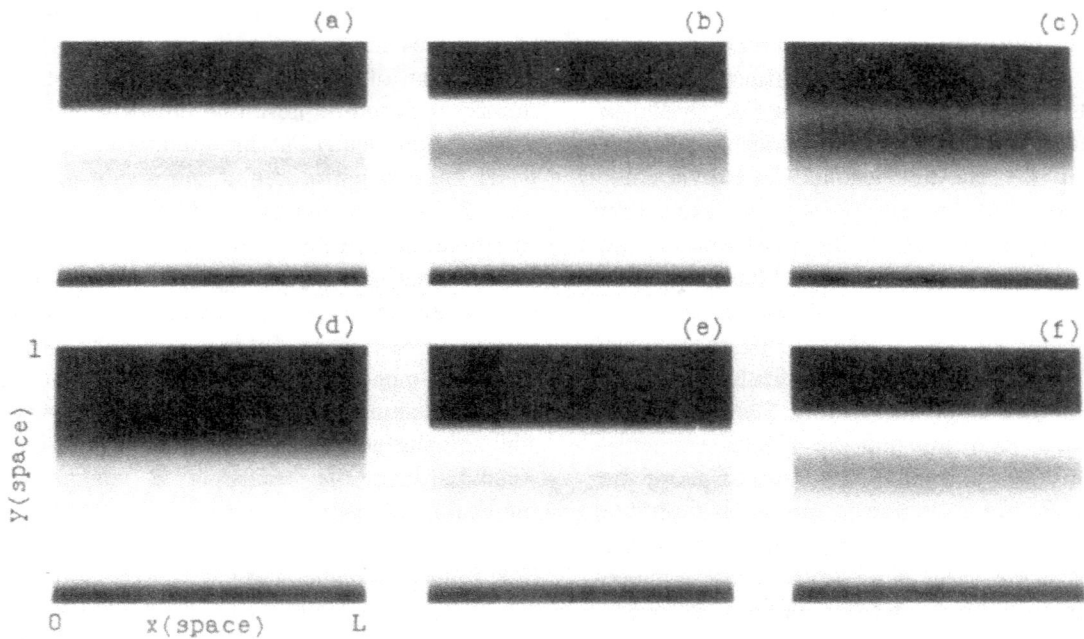

Fig. 6. Snapshots of a periodically oscillating three-front pattern computed with the reaction–diffusion model (10). The concentration of the intermediate species $X(x, y)$ is coded using 256 shades from black ($X = 0$) to white ($X = 2.65$). The model parameters are $k_A = k_B = 10^{-2}$; $D_A = D_B = 12$, $D_X = 4$, $D_Y = 6.5$. The boundary conditions on the opposite large edges of the strip were fixed to: $A = 3$, $B = 2$, $X = Y = 0$ for $y = 0$ and $A = 1$, $B = 6$, $X = Y = 0$ for $y = l$. The dimensions of the system are $l = 43$ and $L = 50$.

using 256 shades. This pattern was obtained for unequal values of the diffusion coefficients $D_X = 4$, $D_Y = 6.5$, which turn out to be close to the conditions for which a Turing instability can break the translation invariance in the direction perpendicular to the gradient. The three fronts in fig. 6 are not very sharp, but one can capitalize on the tendency of the system to provide narrow fronts, to realize a situation so far unattained in bench experiments: a nearly one-dimensional chain of coupled oscillators at the onset of a supercritical Hopf bifurcation of the front structure. As suggested by the theoretical analysis in section 2.3, the critical Hopf mode is condensed at the front zone with an exponential decrease away from it. The reaction–diffusion model (10) thus reduces to one or several very narrow strips of oscillating intermediate reactor cells coupled by diffusion and uniformly fed. In principle, all the phenomena found in the one-dimensional Ginzburg–Landau [23] and

Kuramoto–Sivashinsky [24] equations such as standing waves, travelling waves, phase turbulence, etc. should be observable in such a system. A systematic numerical investigation of these self-organized pattern formation phenomena is currently in progress. Let us point out that the wave patterns observed in these conditions are forced to travel along the front very much like the rotating waves observed in the annular gel reactor [53, 60, 67, 68]. However, there is a fundamental difference between the experimental waves which result from an external perturbation in an excitable regime and the Kuramoto-like travelling waves which originate from an instability induced by the interplay of the kinetics and the diffusion process. A direct consequence of this difference is that (i) excitation waves appear as sharp finite amplitude waves as the result of the relaxational character of local dynamics, while (ii) Kuramoto's travelling waves are phase waves that might be more difficult to detect, especially

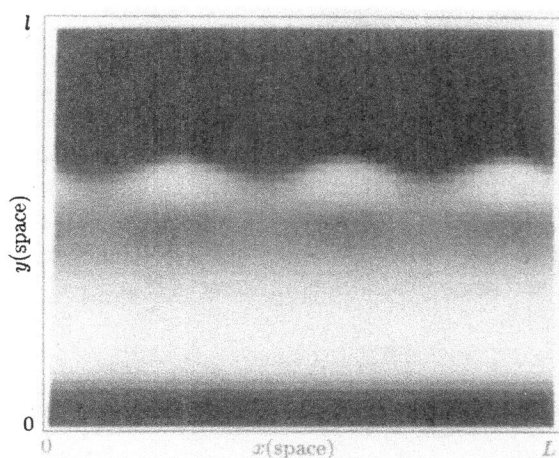

Fig. 7. Turing pattern developing along a front pattern as computed with the reaction–diffusion model (10) for parameter values identical to those of fig. 6, except $D_Y = 12$. The concentration of the X species is represented using 256 shades from black ($X = 0$) to white ($X = 3$).

near the onset of the Hopf bifurcation where the amplitude of oscillation is small.

Turing symmetry-breaking bifurcation. In fig. 7, we have reproduced a Boissonade–Turing pattern, but for conditions which are slightly different from those investigated in ref. [22]. The diffusion coefficients of the intermediate species $D_X = 4$, $D_Y = 12$ are closer to the Hopf bifurcation values of the three-front pattern described in fig. 6. When the diffusion of the inhibitor Y is faster than the diffusion of the activator X ($D_Y/D_X = 3$), modes transverse to the imposed concentration gradient can eventually destabilize and develop into a stationary, spatially modulated front pattern. When a bifurcation parameter is tuned such as [B] at one boundary, the amplitude of this modulation goes to zero at the bifurcation point with a characteristic square root law behavior [22]. This modulation is a perfect illustration of a (supercritical) symmetry-breaking bifurcation leading to the emergence of a genuine Turing cellular pattern from a uniform state.

Very recently, the Bordeaux group has reported the experimental observation of a sustained standing nonequilibrium chemical pattern

in a linear gel reactor under concentration gradient [59]. The observed pattern displays striking resemblance with the symmetry breaking Turing pattern shown in fig. 7. Besides the difficulty of identifying clearly the bifurcation threshold, the unambiguous evidence for Turing patterns requires the demonstration that the characteristic wavelength of the pattern is really intrinsic to the sole coupling of reaction and diffusion processes. The experimental wavelength $\lambda \sim 0.2$ mm is much smaller than any geometric size of the reactor and is very likely to be intrinsic. Our simulation of the reaction–diffusion model (10) suggests a demonstration which consists in checking the intrinsic interdependence of the characteristic wavelength λ of the Turing pattern (fig. 7) and the frequency ω of the periodically oscillating uniform front pattern (fig. 6). In the next section we elaborate on the intimate relation $\lambda \sim \omega^{-1/2}$ and we provide indications for practical measurement of this scaling law in the experiments.

3.3. Theoretical analysis

It has long been known that a necessary condition for the occurrence of Turing instability is inequality of diffusion coefficients [18, 46]. Only recently have sufficient conditions for the onset of symmetry-breaking bifurcations been found [80]. It has been shown that if the reaction kinetics are sufficiently close to a multicritical situation of Takens–Bogdanov type (simultaneous onset of Hopf and saddle–node bifurcations), then there exists a diffusion matrix arbitrarily close to a scalar times the identity such that a Turing bifurcation can be obtained [19]. Let us summarize the main steps of this demonstration when considering for the sake of simplicity, a two-variable system with a diagonal diffusion matrix. We will consider our steady front pattern as a uniformly fed one-dimensional spatial structure.

The stability of the homogeneous steady state to infinitesimal perturbations of characteristic wavevector k is determined by the dispersion

relation

$$\det\left[J - Dk^2 - Is_k\right] = 0, \tag{11}$$

where J is the 2×2 Jacobian matrix of the reaction term, D is the 2×2 diffusion matrix and I the identity matrix. For stability of the basic state in the well-mixed system, we have

$$\mathrm{Tr}\, J = J_{11} + J_{22} < 0, \tag{12}$$

and

$$\det J = J_{11}J_{22} - J_{12}J_{21} > 0. \tag{13}$$

A Turing bifurcation occurs when $\Delta(k^2)$ has a degenerate positive root, where

$$\begin{aligned}\Delta(k^2) &= \det J - (D_1 J_{22} + D_2 J_{11})k^2 + D_1 D_2 k^4 \\ &= 0.\end{aligned} \tag{14}$$

D_1 and D_2 are the diagonal diffusion coefficients. Note that if $\Delta(k^2)$ is to have a real positive root k^2, then we necessarily have

$$D_1 J_{22} + D_2 J_{11} > 0. \tag{15}$$

For conditions (15) and (12) to hold, then J_{11} and J_{22} must be of opposite sign. If J_{11} and J_{22} are of opposite sign, then J_{12} and J_{21} must also be of opposite sign in order to satisfy (13). The condition for a degenerate root is a null discriminant in (14). Thus,

$$(D_1 J_{22} + D_2 J_{11})^2 = 4D_1 D_2 \det J \tag{16}$$

at the Turing bifurcation, and the critical wavenumber is given by

$$k_c^2 = \sqrt{\det J / D_1 D_2}. \tag{17}$$

Since $\det J = \omega^2$ at the Hopf bifurcation ($\omega =$ frequency of oscillation) of the homogeneous state ($k = 0$), we see that one can approximate the critical wavelength by

$$\lambda \sim 2\pi[D_1 D_2]^{1/4}/\omega^{1/2}, \tag{18}$$

when the Turing bifurcation occurs sufficiently close in the parameter space to the Hopf bifurcation of the $k = 0$ mode. From (16), one can compute the critical ratio of diffusion coefficients at the Turing bifurcation in terms of the J_{ij}:

$$\frac{D_1}{D_2} = \frac{\det J - J_{12}J_{21} \pm 2\sqrt{-J_{12}J_{21} \det J}}{J_{22}^2}, \tag{19}$$

where $+$ $(-)$ corresponds to the case $J_{11} < 0$ (> 0), $J_{22} > 0$ (< 0). Near the double-zero Takens–Bogdanov situation, $\det J$ is negligible relative to $-J_{12}J_{21}$. Then

$$\frac{D_1}{D_2} \approx -\frac{J_{12}J_{21}}{J_{22}^2}, \tag{20}$$

which, by $\det J = J_{11}J_{22} - J_{12}J_{21} = 0$ and $\mathrm{Tr}\, J = J_{11} + J_{22} = 0$, goes to unity as the double-zero point is approached. A rigorous demonstration of this result was carried out in refs. [19, 80] and shown to extend to n-variable systems.

The location of the codimension-two Takens–Bogdanov bifurcation critical surface in the well-mixed system turns out, however, to be a very hard experimental task. Our point here is to show the robustness of the approximation (18) far from the multicritical Takens–Bogdanov situation for values of the ratio D_1/D_2 which are not unrealistically different from unity [71]. We consider the following simple model recently introduced in molecular dynamics simulations [81]:

$$U + W \xrightarrow{k_1} W + V,$$
$$V + V \underset{k_{-2}}{\overset{k_2}{\rightleftharpoons}} W,$$
$$V \xrightarrow{k_3} \text{products.} \tag{21}$$

The macroscopic rate equations for the above system in an isothermal continuous flow stirred

tank reactor are

$$\frac{du}{dt} = -k_1 uw + \alpha(u_f - u),$$

$$\frac{dv}{dt} = k_1 uw - 2(k_2 v^2 - k_{-2} w) - k_3 v + \alpha(v_f - v),$$

$$\frac{dw}{dt} = k_2 v^2 - k_{-2} w, \qquad (22)$$

where u, v and w are the mole fractions of U, V and W, $k_{\pm i}$ are rate constants, α is the inverse residence time or feed rate, and u_f and v_f are the feed concentrations of u and v. The system is assumed impermeable to W.

In a reaction–diffusion setting, we append diffusion terms to the above ODEs. We take molecular diffusion values $D_U = 4.5 \times 10^{-5}$ cm²/s and $D_V = D_W = 10^{-5}$ cm²/s. We fix the parameters $k_1 = k_2 = k_{-2} = 1$ and $v_f = 1/30$. k_3 and α are considered as free parameters. The system length is $L = 3$ cm. The boundary conditions are no flux to mimic the Bordeaux experiment in the linear gel reactor [59]. We perform numerical simulations for different sets of parameters. We begin with spatially uniform initial conditions to which we add one percent random noise. We then evolve the system deterministically via finite centered differencing using from 100 to 500 spatial grid points. For each of the control parameter values investigated beyond the Turing bifurcation threshold, the system settles at a stationary pattern as shown in fig. 8a. Before the transition, the time evolution in an elementary reactor cell produces a time series which displays an oscillatory behavior that corresponds to the oscillatory damping of the homogeneous $k = 0$ mode as shown in fig. 8b. We define the temporal period T from the first oscillations in the time series. We have plotted in fig. 9 the logarithm of the wave number (m) versus $\log T$ for different values of the control parameters. We note that while T varies continuously, m varies discretely. This is a consequence of the discrete spectrum of the Laplacian operator on a finite domain. The slope of the line through the data is $-1/2$ in accord

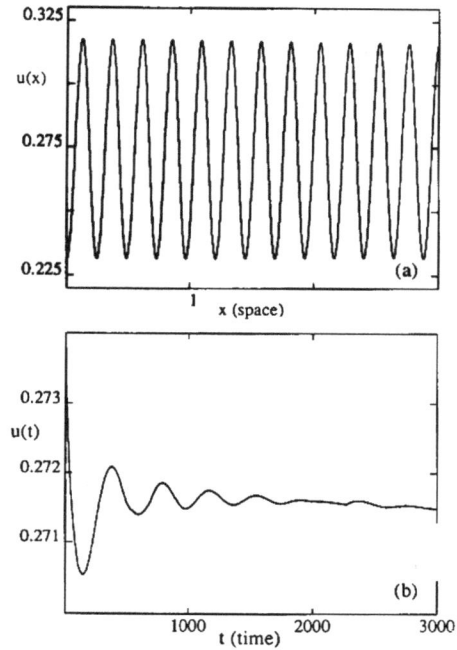

Fig. 8. Turing pattern as computed with the model (22) with diffusion terms. (a) Spatial concentration profile of the species U. (b) Oscillatory transients recorded in one elementary reactor cell before the onset of bifurcation. The model parameters are: $D_U = 4.5 \times 10^{-5}$ cm²/s, $D_V = D_W = 10^{-5}$ cm²/s, $k_1 = k_2 = k_{-2} = 1$, $v_f = 1/30$; the values of the free parameters are: $\alpha = 5.34919 \times 10^{-3}$, $k_3 = 3.0172 \times 10^{-2}$.

with the theoretical claim (18) that the wavelength $\lambda \propto \omega^{-1/2}$.

We have performed subsequent simulations that confirm that the intimate relation (18) between the characteristic wavelength of the Turing pattern and the Hopf frequency of the homogeneous mode can be observed in a rather wide range of parameters in the neighborhood of the locus of the oscillatory instability of the homogeneous front pattern. This can be understood since the Hopf frequency generally does not vary significantly in nearly critical situations. In ref. [19] a seven-variable model of the BZ reaction was investigated. The critical ratio of diffusion coefficients necessary to achieve a Turing instability was observed to remain reasonably close to unity over a wide range of parameters along a line of Hopf bifurcations, but increased extremely rapidly

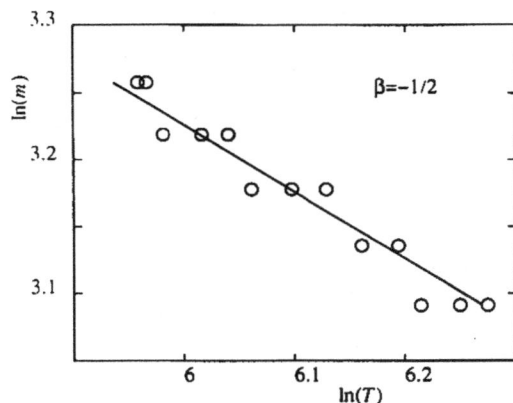

Fig. 9. The wavenumber m of the Turing pattern (fig. 8a) versus the period of oscillation T of the transients recorded in one elementary reactor cell (fig. 8b). The line of slope $\beta = -1/2$ corresponds to the theoretical prediction (18). The model parameters are the same as in fig. 8; the circles correspond to data points computed along a path near a line of Hopf bifurcations in the (α, k_3) plane.

with distance from the line of Hopf bifurcations. These results are likely to extend to a large class of reaction–diffusion systems that exhibit bistable or oscillatory behavior. Observation of the relationship between the wavelength of the Turing mode and the Hopf frequency of the $k = 0$ mode would be a strong indication that the patterns recently observed in the Bordeaux experiment [59] were indeed formed by Turing instability. The characteristic wavelength observed in this experiment $\lambda \sim 0.2$ mm corresponds to a period of oscillation $T \sim 10$ s when considering molecular diffusion characteristic values ($D \sim 10^{-5}$ cm^2/s) in the relation (18). This estimate is in quite good agreement with the frequency of oscillation of front patterns observed in the Couette flow reactor with the chlorite–iodide reaction [56, 57]. A strategy to extract the scaling law (18) in the linear gel reactor consists in playing on the control parameters in order to cross the Turing bifurcation threshold. Before the transition, at each step of the control parameter, the system settles to a homogeneous front pattern in a damped oscillatory way; one can thus estimate the frequency of oscillation in precritical situa-

tions (fig. 8b). Beyond the transition, the Turing pattern emerges and there is no difficulty in estimating the characteristic wavelength (fig. 8a). By reproducing this investigation in different experimental conditions, one can check whether or not relation (18) applies, taking advantage of the fact that one does not need a significant shift in the frequency of oscillation to induce a noticeable change in the wavenumber of the Turing structure in view of the smallness of λ ($L > 100\lambda$). In conclusion, relation (18) is likely to provide a decisive experimental test of the intrinsic nature of the steady cellular patterns recently observed in the linear gel reactor [59].

4. Future prospects

In sections 2 and 3, we have focused mainly on primary instabilities of reaction fronts in one- and two-dimensional reaction–diffusion systems that model recent experiments performed in open gel reactors. We have shown in section 3 that when the Hopf bifurcation of the homogeneous front pattern occurs in the neighborhood of the symmetry breaking Turing bifurcation, the characteristic wavelength of the steady cellular pattern which develops along the front behaves like $\lambda \propto \omega^{-1/2}$, where ω is the frequency of oscillation of the homogeneous front. We have emphasized this relation as a possible experimental test to provide unambiguous evidence that the stationary cellular patterns observed recently by the Bordeaux group [59] are genuine Turing structures originating from the interplay of the reaction kinetics and diffusion process.

The results of the present study naturally raise the issue of the interaction between these two primary instabilities. This question can be addressed theoretically by taking advantage of our control of the transverse concentration gradient imposed on the system from the boundaries, to confine the significant dynamical behavior of the system in a narrow strip located at the front zone. In this limit, the annular gel reactor can be modeled by a reaction–diffusion system posed on a circle. This system must be invariant under the group O(2) generated by rotations and reflections [82]. This symmetry, which is also present when using no-flux instead of periodic boundary conditions [83], is of fundamental importance as far as the classification of the dynamics in multiple bifurcation phenomena is concerned.

In the case of spatially confined geometry, i.e. when the length of the circle is small relative to the intrinsic chemical wavelengths, the mathematical technique of center manifold/normal form reduction [77, 78] can be used to calculate the normal form of this O(2) equivariant bifurcation problem. Similar approaches of steady-state mode interactions [84–86], Hopf-mode interactions [82, 87] and Takens–Bogdanov bifurcation [88] with O(2) symmetry have been carried out recently in the literature. The unfolded normal form for the interaction of a steady state mode ($k \neq 0$) and a Hopf mode ($k = 0$) in the presence of O(2) symmetry is likely to produce secondary bifurcations to standing waves (e.g., breathing of the Turing pattern) or travelling waves (e.g., drift of the Turing pattern). A detailed analysis of the secondary, tertiary, ... bifurcations that may result from this steady–Hopf mode interaction is currently in progress. Special attention is devoted to the dependence of the bifurcation behavior on the problem's parameters. In particular, the issue of the relative values of the diffusion coefficients of the different intermediate species is investigated with great care for experimental purpose.

In the case of spatially extended systems, i.e. when the circle is large as compared to the wavelength of the Turing pattern, which is the case in the Bordeaux experiment [59], then one can expect to encounter rich dynamics from the onset of either primary bifurcation. In extended systems, a continuum of modes becomes unstable instead of discrete ones. The marginal continuum may be viewed as a wave packet: the amplitude of the most unstable mode is slowly modulated in space. The dynamics of the system is then governed by an amplitude equation which is a partial differ-

ential equation that is invariant under O(2) symmetry.

Steady-state mode instability ($k \neq 0$) in extended systems in the presence of O(2) symmetry has been the focus of much recent interest [86, 89, 90]. Instabilities of one-dimensional steady cellular patterns are likely to break translation and reflection symmetries. On the basis of symmetry arguments [90], generic instabilities of such patterns have been shown to lead to a rich variety of patterns including breathing, vacillation and drift instabilities [86] of the basic cellular pattern. Because of the coupling with the phase of the underlying periodic pattern, localized solutions, defects and phase instabilities may result in the destruction of the spatial order [90].

It is well known that the Hopf instability of the homogeneous mode ($k = 0$) in spatially extended systems is described by the Ginzburg–Landau equation [23]. At the onset of the Hopf bifurcation, this amplitude equation can be reduced to the Kuramoto–Sivashinsky equation [24, 25] that accounts for the dynamics of the phase of the coupled oscillators. In the presence of O(2) symmetry, bifurcation sequences including stable standing waves, stable travelling waves, unstable modulated travelling waves and attracting heteroclinic cycles have recently been reported in a theoretical and numerical study of the Kuramoto–Sivashinsky equation [91]. This equation also possesses chaotic solutions that have been coined "phase turbulence" [4, 24, 26–28]. Ultimately, when the adiabatic elimination of the amplitude mode which leads to the Kuramoto–Sivashinsky equation breaks down, phase turbulence turns into amplitude turbulence [24, 34].

Those increasingly complex dynamics are thus very likely to be observed in the active region of the open gel reactors which is confined to the front, e.g. in a narrow circular strip in the annular gel reactor. In addition to nearly one-dimensional phenomena, the geometry of the open gel reactors may prove useful in investigations of the outstanding issue of the transition to chemical turbulence in two-dimensional reacting media. Starting from a phase turbulent state, one can progressively increase the width of the active region by playing on the control parameters in order to either decrease the sharpness of the front pattern or increase the amplitude of oscillation of the front. In other words, one has the possibility to continuously change the aspect ratio of the active zone from a one-dimensional chain of coupled oscillators to a finite width strip (annulus) of oscillators. Doing so, one can investigate the transition from phase turbulence [4, 24, 26–28] to topological turbulence [29–33], which is governed by the creation of topological defects. The understanding of this type of turbulence calls for the elucidation of the creation mechanism, the interactions and motion of topological defects. A possible mechanism from which spiral defects might originate would be the intrinsic breaking of the front pattern. Prospective numerical simulations are currently in progress. The preliminary results reported in this paper provide stimulating indications for future experimental research. The new set of recently designed open spatial reactors is likely to bring some answers to the most difficult questions concerning self-organization phenomena in chemical systems. We would like to thank the referee for bringing to our attention two recent articles on this subject [92, 93].

Acknowledgements

We are very grateful to J. Boissonade, P. DeKepper, Q. Ouyang and H.L. Swinney for several enjoyable discussions. Two of us (A.A. and J.E.) acknowledge the financial support of the Direction des Recherches Etudes et Techniques (DRET) under contract (No. 89/196). A.A. is also supported by the US Department of Energy under contract DE-FG05-88ER13821. J.E. is also partly supported by the US Army Research Office through the M.S.I. of Cornell University. One of us (J.P.) has been supported by the US Department of Energy under contract No. DE-FG05-88ER13897. T.R. is supported by the British Petroleum Venture Research program.

References

[1] G. Nicolis and I. Prigogine, Self-Organization in Non-Equilibrium Systems (Wiley, New York, 1977).

[2] H. Haken, Synergetics: An Introduction (Springer, Berlin, 1977); Advanced Synergetics (Springer, Berlin, 1983).

[3] A. Babloyantz, Molecules, Dynamics and Life (Wiley, New York, 1986).

[4] Y. Kuramoto, Chemical Oscillations, Waves and Turbulence (Springer, Berlin, 1984).

[5] C. Vidal and A. Pacault, eds., Nonequilibrium Dynamics in Chemical Systems (Springer, Berlin, 1984).

[6] V.I. Krinsky, ed., Self-Organization, Auto-Waves and Structures Far From Equilibrium (Springer, Berlin, 1984).

[7] R.J. Field and M. Burger, eds., Oscillations and Traveling Waves in Chemical Systems (Wiley, New York, 1985).

[8] A.R. Bishop, G. Gruner and B. Nicolaenko, eds., Spatio-Temporal Coherence and Chaos in Physical Systems, Physica D 23 (1986).

[9] M. Markus, S.C. Müller and G. Nicolis, eds., From Chemical to Biological Organization (Springer, Berlin, 1988).

[10] H. Meinhardt, Models of Biological Pattern Formation (Academic Press, New York, 1982).

[11] J.C. Roux and H.L. Swinney, in: Nonequilibrium Dynamics in Chemical Systems, eds. C. Vidal and A. Pacault (Springer, Berlin, 1984) p. 141.

[12] A. Arneodo, F. Argoul, P. Richetti and J.C. Roux, in: Dynamical Systems and Environmental Models, eds. H.G. Bothe, W. Ebeling, A.B. Kurzhanski and M. Peschel (Akademie Verlag, Berlin, 1987) p. 122.

[13] F. Argoul, A. Arneodo, P. Richetti, J.C. Roux and H.L. Swinney, Account. Chem. Res. 20 (1987) 436.

[14] B.P. Belousov, Ref. Radiats, Med. 1958, Medgiz, Moscow (1959) 145;
M. Zhabotinsky, Dokl. Akad. Nauk. SSR 157 (1964) 392.

[15] A.M. Turing, Phil. Trans. R. Soc. London B 237 (1952) 37.

[16] M. Herschkowitz-Kaufman and G. Nicolis, J. Chem. Phys. 56 (1972) 1890;
M. Herschkowitz-Kaufman, Bull. Math. Biol. 37 (1975) 585;
G. Nicolis, T. Erneux and M. Herschkowitz-Kaufman, Adv. Chem. Phys. 88 (1978) 263.

[17] A.M. Zhabotinsky and A.B. Rovinsky, J. Stat. Phys. 48 (1987) 959.

[18] J.A. Vastano, J.E. Pearson, W. Horsthemke and H.L. Swinney, Phys. Lett. A 124 (1987) 320; J. Chem. Phys. 88 (1988) 6175.

[19] J.E. Pearson and W. Horsthemke, J. Chem. Phys. 90 (1989) 1588.

[20] G. Dewel, D. Walgraef and P. Borckmans, J. Chim. Phys. (Paris) 84 (1987) 1335.

[21] G. Dewel and P. Borckmans, Phys. Lett. A 84 (1987) 1335.

[22] J. Boissonade, J. Phys. (Paris) 49 (1988) 541.

[23] L.D. Landau and V.L. Ginzburg, in: Collected Papers of L.D. Landau, ed. D. Ter Haar (Pergamon, New York, 1965) p. 217.

[24] Y. Kuramoto, Prog. Theor. Phys. 71 (1984) 1182; in Chemical Oscillations, Waves and Turbulence, ed. H. Haken (Springer, Berlin, 1984).

[25] G.I. Sivashinsky, Acta Astronaut. 4 (1977) 1177.

[26] J.M. Hyman, B. Nicolaenko and S. Zaleski, Physica D 23 (1986) 265.

[27] Y. Pomeau, A. Pumir and P. Pelce, J. Stat. Phys. 37 (1984) 39.

[28] A. Pumir, J. Phys. 46 (1985) 511.

[29] A.V. Gapanov-Grekhov and R.I. Rabinovitch, Usp. Phys. Nauk. 152 (1987) 159 [Sov. Phys. Usp. 30 (1987) 433].

[30] P. Coullet, L. Gil and J. Lega, Phys. Rev. Lett. 62 (1989) 1619.

[31] G. Goren, I. Procaccia, S. Rasenat and V. Steinberg, Phys. Rev. Lett. 63 (1989) 1237.

[32] S. Rica and E. Tirapegui, Phys. Rev. Lett. 64 (1990) 878.

[33] T. Bohr, A.W. Pedersen and M.H. Jensen, Phys. Rev. A 42 (1990) 3626.

[34] H.R. Brand, P.S. Lomdahl and A.C. Newell, Physica D 23 (1986) 345.

[35] R. Aris, The Mathematical Theory of Diffusion and Reaction in Permeable Catalysts, Vols. 1, 2 (Clarendon Press, Oxford, 1975).

[36] J.C. Micheau, M. Gimenez, P. Borckmans and G. Dewel, Nature 305 (1983) 43.

[37] D. Avnir and M. Kagan, Nature 307 (1984) 717.

[38] P. Borckmans, G. Dewel, D. Walgraef and Y. Katayama, J. Stat. Phys. 48 (1987) 1031.

[39] A.N. Zaikin and A.M. Zhabotinsky, Nature 225 (1970) 535.

[40] A.T. Winfree, The Geometry of Biological Time (Springer, Berlin, 1980).

[41] P. Fife, Mathematical Aspects of Reacting and Diffusing Systems (Springer, Berlin, 1979).

[42] C. Vidal and A. Pacault, in: Evolution of Order and Chaos, ed. H. Haken (Springer, Berlin, 1982) p. 74.

[43] C. Vidal, J. Stat. Phys. 48 (1987) 1017.

[44] A. Pagola, Thesis, Bordeaux (1987).

[45] S.C. Müller, T. Plesser and B. Hess, Science 230 (1985) 661; J. Stat. Phys. 48 (1987) 991.

[46] P. Hanusse, Compt. Rend. Acad. Sci. Paris C 274 (1972) 1245.

[47] F. Argoul, A. Arneodo, P. Richetti and J.C. Roux, J. Chem. Phys. 86 (1987) 3325.

[48] P. Ortoleva and J. Ross, J. Chem. Phys. 63 (1975) 3398.

[49] P.C. Fife, J. Chem. Phys. 64 (1978) 554.

[50] L.M. Pismen, J. Chem. Phys. 71 (1979) 462.

[51] J.P. Keener, SIAM J. Appl. Math. 46 (1986) 1039.

[52] J.J. Tyson and J.P. Keener, Physica D 32 (1988) 327.

[53] J. Boissonade, in: Dynamic and Stochastic Processes: Theory and Applications, eds. R. Lima, L. Streit and R. Vilela-Mendes, Lecture Notes in Physics (Springer, Berlin), to appear.

[54] W.Y. Tam, J.A. Vastano, H.L. Swinney and W. Horsthemke, Phys. Rev. Lett. 61 (1988) 2163.

[55] W.Y. Tam and H.L. Swinney, Physica D 46 (1990) 10.

[56] Q. Ouyang, J. Boissonade, J.C. Roux and P. De Kepper, Phys. Lett. A 134 (1989) 282;
Q. Ouyang, Thesis, Bordeaux (January 1989).

[57] Q. Ouyang, V. Castets, J. Boissonade, J.C. Roux, P. De Kepper and H.L. Swinney, J. Chem. Phys., to appear.

[58] W.Y. Tam, W. Horsthemke, Z. Noszticius and H.L. Swinney, J. Chem. Phys. 88 (1988) 3395.

[59] V. Castets, E. Dulos, J. Boissonade and P. De Kepper, Phys. Rev. Lett. 64 (1990) 2953.

[60] Z. Noszticius, W. Horsthemke, W.D. McCormick, H.L. Swinney and W.Y. Tam, Nature 329 (1987) 619.

[61] W.Y. Tam and H.L. Swinney, Phys. Rev. A 36 (1987) 1374.

[62] J.A. Vastano, T. Russo and H.L. Swinney, Physica D 46 (1990) 23.

[63] J. Boissonade, Q. Ouyang, A. Arneodo, J. Elezgaray, J.C. Roux and P. De Kepper, in: Nonlinear Waves in Excitable Media, eds. A.V. Holden, M. Markus and H.G. Othmer (Pergamon, New York, 1990), to appear.

[64] A. Arneodo and J. Elezgaray, in: Spatial Inhomogeneities and Transient Behavior in Chemical Kinetics, eds. P. Gray, G. Nicolis, F. Baras, P. Borckmans and S.K. Scott (Manchester Univ. Press, Manchester, 1990) p. 415.

[65] J. Elezgaray and A. Arneodo, in: New Trends in Nonlinear Dynamics and Pattern Forming Phenomena: The Geometry of Nonequilibrium, eds. P. Coullet and P. Huerre (Plenum, New York, 1990) p. 21.

[66] G.S. Skinner and H.L. Swinney, Physica D 48 (1991) 1–16.

[67] N. Kreisberg, W.D. McCormick and H.L. Swinney, J. Chem. Phys. 91 (1989) 6532.

[68] E. Dulos, J. Boissonade and P. De Kepper, in: Nonlinear Wave Processes in Excitable Media, eds. A.V. Holden, M. Markus and H.G. Othmer (Pergamon, New York, 1990), to appear.

[69] A. Arneodo and J. Elezgaray, Phys. Lett. A 143 (1990) 25; preprint (1990), submitted to J. Chem. Phys.

[70] J. Elezgaray, Thesis, University of Bordeaux (1989).

[71] A. Arneodo, J. Pearson and T. Russo, in preparation.

[72] B. Van der Pol, Philos. Mag. 3 (1927) 65.

[73] W. Eckhaus, Lect. Notes Math. 985 (1983) 449.

[74] M. Boukalouch, J. Elezgaray, A. Arneodo, J. Boissonade and P. De Kepper, J. Phys. Chem. 91 (1987) 5843.

[75] P.C. Fife, J. Math. Anal. Appl. 54 (1976) 497.

[76] Y. Nishiura and H. Fujii, SIAM J. Math. Anal. 18 (1987) 1726.

[77] V.I. Arnold, Supplementary Chapters to the Theory of Differential Equations (Nauka, Moscow, 1978).

[78] J. Guckenheimer and P. Holmes, Nonlinear Oscillations, Dynamical Systems and Bifurcations of Vector Fields (Springer, Berlin, 1984).

[79] R.A. Friesner, L.S. Tuckerman, B.C. Dornblaser and T.V. Russo, J. Sci. Comp. 4 (1989) 327.

[80] J.E. Pearson, Ph.D. Thesis, The University of Texas (1988).

[81] F. Baras, J.E. Pearson, M. Malek-Mansour, J. Chem. Phys. (1990), to appear.

[82] W.W. Farr and M. Golubitsky, preprint (1990).

[83] J.D. Crawford, M. Golubitsky, M.G.M. Gomes, E. Knobloch and I.N. Stewart, preprint (1989).

[84] G. Dangelmayr, Dyn. Stab. Syst. 1 (1986) 159.

[85] D. Armbruster, J. Guckenheimer and P. Holmes, Physica D 29 (1988) 257.

[86] S. Fauve, S. Douady and O. Thual, J. Phys. (Paris), to appear.

[87] P. Chossat, M. Golubitsky and B. Keyfitz, Dyn. Stab. Sys. 1 (1986) 255.

[88] G. Dangelmayr and E. Knobloch, Phil. Trans. R. Soc. London A 322 (1987) 243.

[89] P. Coullet, R.E. Goldstein and G.H. Gunaratne, Phys. Rev. Lett. 63 (1989) 1954.

[90] P. Coullet and G. Iooss, Phys. Rev. Lett. 64 (1990) 866.

[91] D. Armbruster, J. Guckenheimer and P. Holmes, SIAM J. Appl. Math. 49 (1989) 676.

[92] T. Ohto, M. Mimura and R. Kobayashi, Physica D 34 (1989) 115.

[93] D.A. Kessler and H. Levine, Phys. Rev. A 41 (1990) 5418.

Physica D 49 (1991) 161–169
North-Holland

Turing-type chemical patterns in the chlorite–iodide–malonic acid reaction

P. De Kepper, V. Castets, E. Dulos and J. Boissonade

Centre de Recherche Paul Pascal, Université Bordeaux I, Avenue Schweitzer, 33600 Pessac, France

We describe experimental observations of symmetry breaking stationary patterns. These patterns are interpreted as the first unambiguous evidence of Turing-type structures in a single-phase isothermal chemical reaction system. Experiments are conducted with the versatile chlorite–iodide–malonic acid reaction in open spatial reactors filled with hydrogel. A phase diagram gathering the domain of existence of symmetry breaking and no-symmetry breaking standing patterns is discussed.

1. Introduction

Recently several types of open spatial reactors have been designed to produce sustained spatio-temporal structures [1–7]. In these reactors the experiments are no longer limited to the rapid determination of some transient triggered wave properties in excitable media [8–12]. Many other asymptotic spatio-temporal behaviors can be reproducibly attained and controlled. Furthermore, these reactors make a larger number of chemically different oscillating and bistable reactions [13, 14] amenable for spatial studies. This tremendously increases **the likelihood of uncovering,** in the field of chemistry, many new types **of** spatial self-organization phenomena [15, 16] besides the classic triggered waves. A rejuvenation of experimental studies of spatial self-organization phenomena in chemical systems has just started. Though a number of these studies are still based on the classical excitable properties of the popular Belousov–Zhabotinsky (BZ) reaction [2–4], other previously undocumented wave-like [5–7] and standing [6, 7] concentration patterns have been discovered.

In a previous paper [17], we reported the first clear evidence of a Turing structure [18] in an isothermal single phase reaction system. Here, we shall further elaborate on these observations and show how these Turing structures are closely associated with the emergence of standing front patterns organized parallel to the feed boundaries.

Let us first make clear what we mean by the term "Turing structures" and briefly enumerate the main properties of these structures: (a) They are *stationary* concentration patterns, originating solely from the coupling of *reaction* and *diffusion* processes. In particular, patterns associated with any type of hydrodynamic motion are excluded. Seemingly unmoving patterns, called "mosaic structures", were reported to develop in a number of different oscillating and even monotonic reactions [19, 20] when the reacting solutions were spread in a thin layer of free fluid. At one time thought of as examples of standing patterns resulting from a reaction–diffusion instability, they are now considered merely the result of convective instabilities induced by adverse density gradients of different origins, i.e., surface evaporative cooling of solvent, specific partial volume changes between reagents and products, since convective motions have actually been observed in a few cases [21, 22]. (b) The patterns result from *spontaneous symmetry breaking* phenomena associated with steady state instabilities. The symmetry breaking is defined in comparison with the boundary condition geometry. (c) The patterns are characterized, at the transition, by an *intrinsic* wavelength which does not depend on the geometrical parameters of the system but only on the concentrations or input rates of the reactants, the diffusion coefficients, and the macroscopic reaction rates. This distinguishes the Turing patterns from other well known nonequilibrium structures like the convective Bénard cells or the Taylor vortices in Couette flows, which depend on some geometric length of the system [23]. In Turing patterns, the size of the system is important only when the system is smaller than a few wavelengths – due to the necessary fit with the boundary conditions – but disappears for larger systems. Geometry, nonetheless, can play an important role in the pattern selection mechanisms at finite distance from the bifurcation values [24].

2. Experimental technique

2.1. The gel strip reactor

The reactor is made of a narrow flat piece of polyacrylamide hydrogel. This transparent, relatively inert gel avoids any advective fluid motion inside the reaction cell. Two types of reactor geometry are currently used in the experimentation. One is a rectangular strip 20 mm long; the other (fig. 1) is an annular strip with 42 mm outer diameter. In both cases the strips have a width of 3 mm and a thickness of 1 mm. These porous strips are squeezed between a white "Ertalite" bottom plate and a glass or plexiglass cover which allows for observations from the top. Monitoring is provided by a CCD camera. Images are usually stored on video tapes and ultimately processed on a NUMELEC (Pericolor 2000) image analyzer. The long edges (or the rims) of the gel are set in contact with the content of two vigorously stirred tanks (fig. 1) where chemical compositions are kept constant by continuous flow of fresh solutions. The feed of the tanks is provided by peristaltic pumps which show less than 2% flow-rate drifts over 24 h. The thermoregulation of the system and the precooling of input flows are

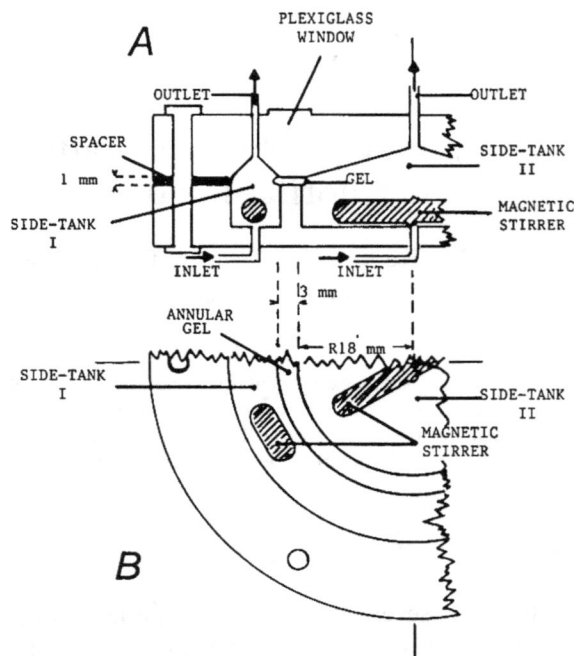

Fig. 1. Sketch of the annular gel strip reactor. (a) Cross-section (half part); (b) top view (quarter part).

performed by a waterjacket; the temperature is maintained at $7.0 \pm 0.4°C$.

2.2. The gel preparation

The hydrogel is prepared by dissolving in a 100 ml deionized water solution: 17.7 g acrylamide, 0.10 g N,N'-methylene-bisacrylamide, 0.7 g ammonium persulfate, 1.0 g triethanolamine, and 2.8 g Thiodène (a soluble starch from PROLABO used as triiodide (I_3^-) color indicator). The polymerizing solution is spread to form a thin film and left to react for 1 h. The whole preparation is performed at 0°C. The resulting sheet of polymer is then thoroughly washed and set to swell in purified water for at least 24 h before the reactor strip is cut out. During the swelling, the volume of the gel increases by a factor of three. The dry weight of the polymer is 5% of the wet gel. The average pore size, determined by the standard light scattering technique, is about 60 Å. Our hydrogel is thus essentially water in a loose grid matrix of polymer, so that the diffusion of small species should not differ much from that found in pure water.

2.3. The reaction

The chemical reaction selected for our study is the chlorite (ClO_2^-)–iodide (I^-)–malonic acid (MA) oscillatory reaction. It has played an essential role in the development of oscillating and stationary chemical front structures [6, 7] in "Couette reactors", which are quasi-one-dimensional open spatial reactors. This reaction is among the very few which can both exhibit transient oscillatory behavior in a batch stirred reactor [25, 26] and trigger waves, if solutions are poured in a thin layer onto a petri dish [25]. Kinetic studies by Lengyel et al. [26] have shown that the batch oscillatory behavior of this system is linked to that of the more recently discovered chlorine dioxide (ClO_2)–iodine (I_2)–malonic acid (MA) oscillatory reaction for which a skeleton mechanism has been proposed [26]. In this model, the strongly self-inhibited reaction between chlorite and iodide plays a fundamental part in controlling the switching between a slow iodide producing and a fast iodide consuming processes.

2.4. The experimental procedure

Fresh chemicals are stored in three separate flasks containing respectively a basic sodium chlorite solution, a neutral iodide solution, and a solution of malonic and sulphuric acids. They are pumped and mixed into the stirred tanks, on each side of the gel strip, as follows: chlorite and iodide solutions on one side (side I), malonic acid and iodide solutions on the other side (side II). Notice that, at large pH values, the reaction between chlorite and iodide is very slow [27]; in our conditions, no significant reaction is observed after 3 h. The residence time in each side-tank is fixed at 15 min. With this feeding mode, neither of the solutions in the side-tanks is significantly reactive on its own and uniform compositions are

easily achieved along the edges of the gel strip. In this paper, the chemical parameter values $[X]_0$ correspond to the concentration of species X in the stirred side-tanks. Experiments are usually started by flowing simultaneously premixed solutions of chemicals in each side-tank. Chemical gradients and patterns are allowed to settle spontaneously. Each composition is maintained for at least 6 h. A number of test experiments operated over 36 h show no significant difference with the pattern organization obtained after the first 6 h, save some fading of the color contrast probably due to the slow hydrolysis of the starch indicator.

3. Experimental results

Our feed conditions, where the chlorite (the oxidant) is fed only along one edge of the gel, lead to a monotonically decreasing oxidation capacity of the solution in the gel reactor from edge I to the malonic acid (a reducer) rich edge II. Iso-oxidation capacity lines then naturally build up parallel to the feed boundaries.

During the set of experiments reported here, the chlorite, the base, and the sulphuric acid concentrations were kept unchanged. For the fixed value of $[MA]_0 = 1.0 \times 10^{-2}$ M and increasing iodide feed concentrations, the sequence of stationary patterns shown in figs. 2a–2c was obtained. At low $[I^-]_0$ (fig. 2a), a single stable sharp color change was observed between the two edges. This trivial concentration pattern, which develops parallel to the feed boundaries, corresponds to a continuous increase in $[I_3^-]$ from edge I, where iodine species are mainly present as IO_3^- and HIO_2, to edge II, where the iodine species are mainly present as iodide, iodine, and triiodide. The local sharp change of color reveals the formation of a stable chemical front. This front is associated with the well known autocatalytic switching mechanism [28] of the chlorite–iodide reaction which, in particular, leads to a sharp decrease of the iodide concentration, by several orders of magnitude.

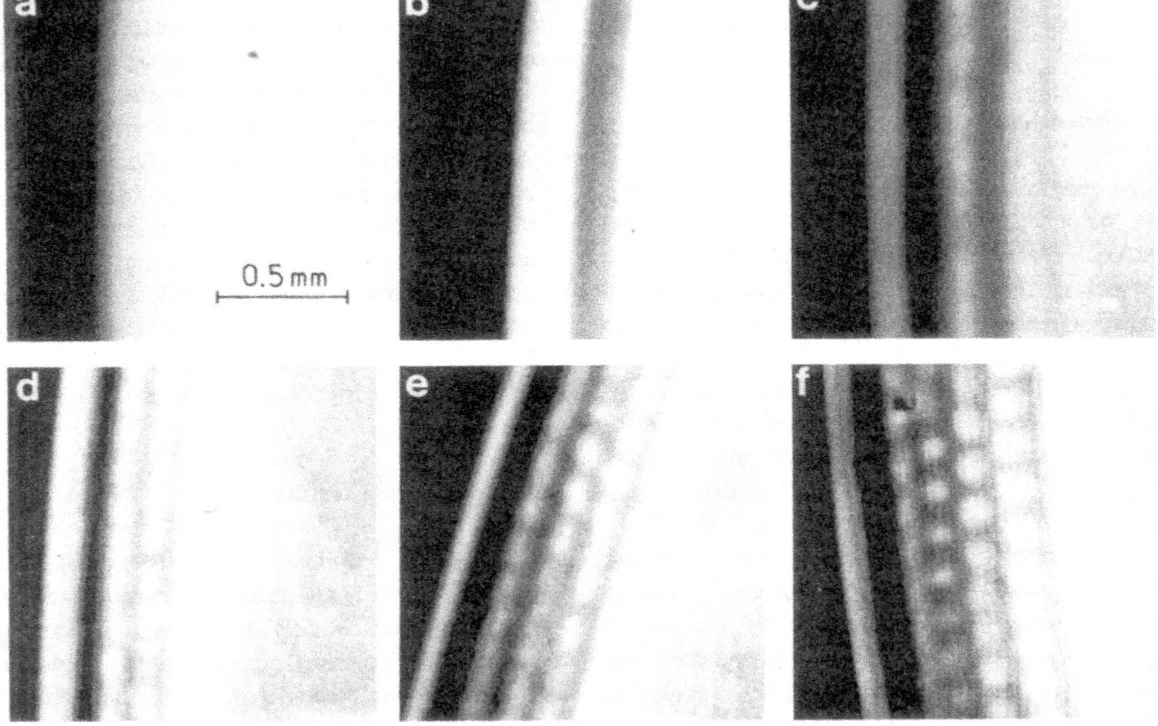

Fig. 2. Sustained chemical patterns in the annular gel strip reactor. Contrast enhanced images of the central portion of the reactor: Dark regions correspond to reduced iodine states colored in blue by the starch–triiodide complex, clear zones correspond to oxidized iodine states. The slight curvature of the first dark front (on the left of each image) is due to the circular geometry of the reactor. All other structures develop parallel to this front. All pictures are at the same scale and set so that edges I and II correspond respectively to the right- and left-hand sides of the pictures. Experimental conditions were as follows. Temperature: 4°C. Boundary feed compositions: fixed concentrations: $[H_2SO_4]_0^I = 1 \times 10^{-2}$ mol/ℓ, $[NaClO_2]_0^{II} = 2.4 \times 10^{-2}$ mol/ℓ, $[NaOH]_0^{I, II}$ = 3×10^{-3} mol/ℓ, $[Na_2SO_4]_0^{I, II} = 3 \times 10^{-3}$ mol/ℓ with $[CH_2(COOH)_2]_0^I$: (a,b,c) 3.3×10^{-3} mol/ℓ; (d,e,f) 8.3×10^{-3} mol/ℓ, and $[I^-]_0^{I, II}$: (a) Single stationary font; 1.0×10^{-3} mol/ℓ. (b) Stationary triple-front structure; 1.33×10^{-3} mol/ℓ. (c) Multiple stationary band pattern; 1.66×10^{-3} mol/ℓ. (d) Fuzzy modulated band pattern; 1.33×10^{-3} mol/ℓ. (e) Triple-line spot pattern (Turing structure); 1.66×10^{-3} mol/ℓ. (f) Quadruple-line spot pattern (Turing structure); 2.66×10^{-3} mol/ℓ.

On increasing $[I^-]_0$, a first nontrivial concentration pattern forms. It consists in the development, parallel to the previous sharp front, of a thin dark (reduced iodine) band in the clearer (oxidized iodine) region (fig. 2b). This stationary structure is similar to that previously observed in the Couette reactor with asymmetric boundary feed conditions [6, 7]. In this latter reactor the diffusive transport process is provided by small scale turbulent mixing. The effective diffusion coefficient, identical for all species in the solution, is then of the order of 0.1 cm^2 s^{-1} and the characteristic spacing between bands is typically 20 mm. In our present gel reactors, the typical distance between bands is 0.2 mm. Assuming that in the two experiments, the characteristic reaction times are of the same order of magnitude, we have the following approximation: $l_1/l_2 \sim (D_1/D_2)^{1/2}$, where the l_i's and D_i's are respectively the characteristic lengths and associated average diffusion coefficients of the two systems. This leads to an estimate of diffusion coefficients in the gel experiments of the order of 10^{-5} cm^2 s^{-1}, which is typically what is found for small ionic species in free aqueous solution. On further increasing $[I^-]_0$, up to three additional dimmer

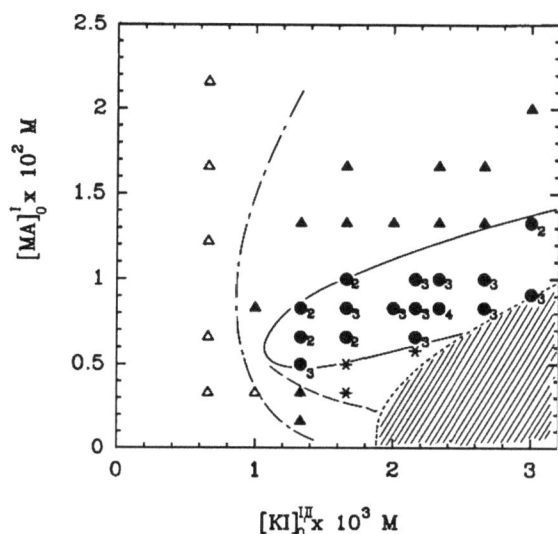

Fig. 3. Spatial structures phase diagram in the (iodide, malonic acid) constraint space. Fixed constraints as in fig. 2. ($\triangle$) Single stationary front structure (fig. 2a); ($\blacktriangle$) stationary triple-front structure (fig. 2b); ($*$) stationary multiple-band pattern (fig. 2c); ($\bullet$) Turing structures (figs. 2d, 2e, 2f). Hatched region correspond to a domain where iodine precipitates are observed.

dark bands develop parallel to the feed boundaries, in the oxidized region (fig. 2c).

At higher malonic acid concentration (e.g. $[MA]_0 = 2.0 \times 10^{-2}$ M), another completely new type of pattern is observed beyond a critical value of $[I^-]_0$ (figs. 2d–2f). Ranges of clear spots develop in the direction parallel to the feed boundaries. Indeed, all the previous bands – except the first clear one starting from reservoir II – lose their azimuthal invariance and break into rows of clear spots in a slightly darker background. This is a genuine *spontaneous symmetry breaking* phenomenon relative to the boundary feed geometry. As shown in the sequence of pictures (figs. 2d–2f) these structures become more easily discernible with increasing $[I^-]_0$. The spot patterns develop over large regions of the reactor, often extending over the full length of the reactor.

As shown in the phase diagram (fig. 3), the region where these symmetry breaking patterns are clearly observed is surrounded by regions where at least three stationary front structures

are obtained. Along the transition line, spot patterns often shear the space with regions of fuzzy band structures. As the iodide concentration is further increased, away from the transition line, the extension of the spot regions readily grows at the expense of the plain stripe structure. Inside the domain where the spot structure is observed, the number of visible rows of spots varies; the figures next to the symbols (fig. 3) stand for the number of clearly visible lines of spots. At high $[I^-]_0$ values, iodine crystals are observed to form along the first sharp stationary front along side I and obscures the neighbouring concentration patterns (hatched domain in fig. 3). Within our present experimental accuracy, the patterns appear and disappear reproducibly, without hysteresis, as a function of the control parameters. The sharpness of the spot patterns increases as the control parameter values move away from the transition line. An accurate detailed study of the bifurcation to the patterns is presently out of reach because of the weak color contrast due to the low concentration of colored species. Fig. 4 presents an example of the periodic light intensity profile observed parallel to the feed edges. Though there are some irregularities in the intensity modulations, the patterns have a well defined wavelength λ, in this case $\lambda \simeq 0.17$ mm. The wavelength of these longitudinal structures seems intrinsic. The development of these structures

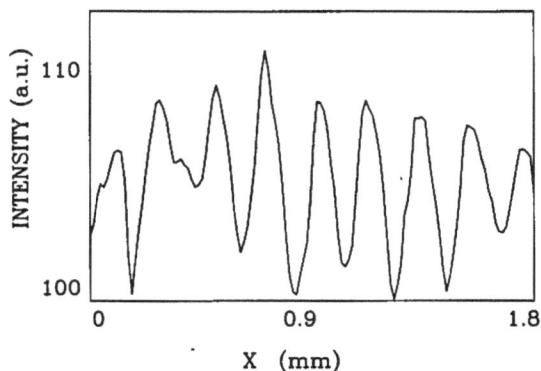

Fig. 4. Light intensity profile along the line of spot structures, parallel to the feed boundaries of the gel strip reactor. Experimental conditions as in fig. 2f.

does not sensitively depend on whether they are performed in the circular (periodic boundary conditions) or in a rectangular reactor (no flux boundary conditions). Their characteristic wavelength is not related to any of the geometric dimensions of the reactors including the thickness $\geq 5\lambda$. In the explored (iodide, malonic acid) section of the phase diagram, the wavelength of this transverse pattern does not change by more than 15%. Taking into account all the abovementioned properties, the structures formed of lines of stationary periodic concentration profiles, parallel to the feed edges, meet all the requirements of Turing patterns: stationarity, spontaneous symmetry breaking, intrinsic wavelength.

4. Discussion

The necessary conditions for Turing patterns are stringent. First, the kinetics should include at least one positive feedback loop (e.g. autocatalysis or substrate inhibition) and at least one negative feedback process (e.g., back inhibition). Oscillatory reactions necessarily fulfill this prerequisite. Though there is not yet a good detailed reaction scheme for the chlorite–iodide–malonic acid reaction, kinetic mechanisms of a number of subreaction systems are now well established [26–28].

Secondly, the major positive feedback species must diffuse significantly slower than the main negative feedback species. Of particular interest to us are the calculations by Lengyel and Epstein [29] using a two-variable skeleton model of the ClO_2–I_2–MA reaction. They show that, in this subsystem, a Turing instability would occur if I^- (the species controlling the positive feedback loop) diffuses slower than ClO_2^- (the species controlling the negative feedback process). It is still unclear if the natural small difference in the diffusion coefficients between these species (or other intermediate species of the reaction), in aqueous solution, is sufficient to trigger the onset

of a Turing instability. Some theoretical studies show that, under particular conditions, these differences may be arbitrarily small [30, 31]. Another possibility is an enhancement of the difference in the diffusion coefficients due to diffusion processes in the gel. Species that preferentially adsorb on specific sites of the polymer matrix would exhibit weaker apparent diffusion coefficients. Such trapping sites for I_3^- and I_2 could be provided by the immobilized starch. However, the Turing structures have recently been shown to develop even in the absence of starch [32]. Adsorption sites could also be provided by the amide functions on the gel; though the polymer only represents 5% by weight of the hydrogel, the concentration of amide function is as high as 0.7 M, which is much greater than any other species in the system. Diffusion and adsorption measurements of different species in the gel are presently in progress in order to better characterize the origin cf the observed Turing structures. It is also stimulating to notice that, the domain of Turing instability recently calculated by Lengyel and Epstein [29] on a model of the ClO_2–I_2–MA, qualitatively develops as in our phase diagram (fig. 3): At constant oxidant concentration, the region of Turing structures grows in the direction of increasing malonic acid and iodine species concentration [29]. Our latest experiments show that Turing structures can indeed be produced with the ClO_2–I_2–MA reaction [33].

One must also bear in mind that, in all cases, the reported observations are based on top views of the patterns. Thus, taking into account the characteristic dimensions of the different structures in comparison with the size of the gel strip, the representations on fig. 2 correspond, in fact, to two-dimensional projections of actually three-dimensional structures. In the region where the Turing-type patterns clearly develop over three or more lines (figs. 2e, 2f) the projections usually fit with a body-centered cubic pattern in agreement with the most stable three dimension Turing patterns predicted by Walgraef et al. [34]. A number of defects in the regular hexagonal arrangement

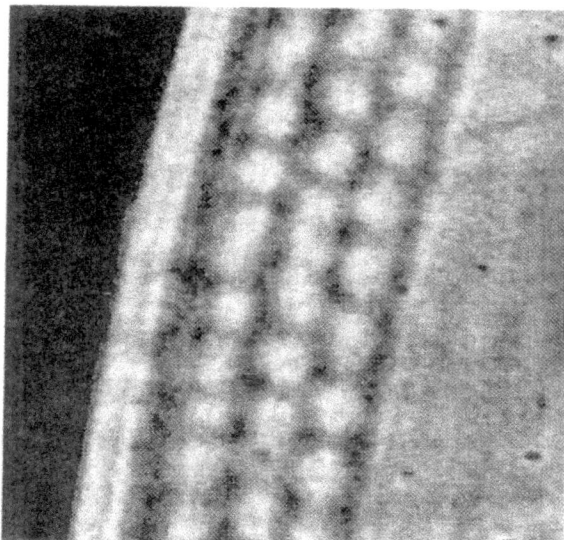

Fig. 5. Example of a fully developed irregular Turing structure. Notice the distortion of the regular hexagonal arrangement. Experimental conditions as in fig. 2 with $[CH_2(COOH)_2]_0^I = 9 \times 10^{-3}$ mol/ℓ; $[I^-]_0^{I, II} = 3 \times 10^{-3}$ mol/ℓ.

can also be seen from place to place along the region of spot patterns as exemplified in fig. 5. This could be due to some inherent small inhomogeneities of gels or to poorly controlled boundary conditions. Yet some preliminary calculations of Turing patterns in a two-dimensional reaction–diffusion system, with transverse concentration gradients introduced to mimic the experiments, show the same type of defects [35]. Taking into account that the side-tanks are thoroughly mixed and contain nonreactive solutions, and that the size of the gel inhomogeneities is several orders of magnitude smaller than the wavelength of the patterns, we think rather that the experimentally observed dislocations are also a direct consequence of the externally driven composition gradients.

Though the most spectacular result of the above experiments is the formation of the symmetry breaking patterns, the band patterns also merit some attention. The origin of these multiple band structures is still unclear. They could correspond to steep front structures induced by the imposed asymmetric boundary feed conditions, as discussed by Arneodo and Elezgaray [31, 36]. The formation of these stiff standing front structures does not require any differences in the diffusion coefficients of the species. They are not directly related to the Turing-type structures. However, stationary band patterns with more than three stationary fronts have never been observed in a Couette reactor where the effective diffusion coefficients of all species are forced to strict equality by the turbulent diffusion process [37]. This could mean that they also require some differences in the diffusion coefficients. Theoretical calculations by Dewel et al. [38] on patterning phenomena in strongly anisotropic media suggest that these band structures could still result from a Turing instability, which initially develops only in one direction. Experimentally, this case would be very difficult to distinguish from the previous one. One more possibility is that some of these band structures (beyond the third front starting from side II) are an illusion due to the stacking of different planes of spot structures which, for some region of parameter space, arrange themselves in a staggered way so that, on the average, they are only viewed as lines of different intensities.

5. Conclusion

As initially suggested by one of us [39] and further emphasized by Arneodo et al. [31], the approach to Turing structures in confined systems by introducing chemical gradients has been a successful method. Iodine–oxychlorine reaction systems [40] have now gained a rather unique status among isothermal oscillating and bistable reactions in liquid phase. Thus, to evaluate the full significance of this method, other reactions should be tested in the future. In this respect, the formidable expansion, during the last decade, in the design of homogeneous oscillating reactions should favor the discovery of Turing structures in other similar chemical systems.

Since biological systems are naturally gradient driven systems due to their feeding through membranes, attention should now be turned to the problem of pattern formation in nonuniform media instead of the more traditional space-uniform feed hypothesized in most theoretical calculations [15, 16, 41, 42].

Acknowledgements

We are indebted to A. Arneodo, P. Borckmans, G. Dewel and Q. Ouyang for stimulating discussions, and to I. Lengyel and I.R. Epstein for making their results available prior to publication. This work has been supported by the Venture Research Unit of British Petroleum.

References

[1] J. Boissonade, in: Dynamic and Stochastic Processes. Theory and Applications, eds. R. Lima, L. Streit and R. Vilela-Mendes, Lecture Notes in Physics, No. 355 (Springer, Berlin, 1990) p. 76.

[2] W.Y. Tam, W. Horsthemke, Z. Noszticzius and H.L. Swiney, J. Chem. Phys. 88 (1987) 3395;
G. Skinner and H.L. Swiney, Physica D 40 (1991) 1–16;
G. Kshirsagar, Z. Noszticzius, W.D. McCormick and H.L. Swiney, Physica D 49 (1991) 5–12, these Proceedings.

[3] Z. Noszticzius, W. Horsthemke, W.D. McCormick, H.L. Swiney and W.Y. Tam, Nature 329 (1987) 619;
N. Kreisberg, W.D. McCormick and H.L. Swiney, J. Chem. Phys. 91 (1989) 6532.

[4] E. Dulos, J. Boissonade and P. De Kepper, in: Nonlinear Wave Processes in Excitable Media, eds. A.V. Holden, M. Markus and H.G. Othmer (Plenum, New York, 1990) pp. 423–433.

[5] W.Y. Tam, J.A. Vastano, H.L. Swiney and W. Horsthemke, Phys. Rev. Lett. 61 (1988) 2163;
W.Y. Tam and H.L. Swiney, Physica D 46 (1990) 10.

[6] Q. Ouyang, J. Boissonade, J.C. Roux and P. De Kepper, Phys. Lett. A 134 (1989) 282:
P. De Kepper, Q. Ouyang, J. Boissonade and J.C. Roux, in: Dynamics of Exotic Phenomena in Chemistry, eds. M. Beck and E. Kòrös, Reac. Kinet. Catal. Lett. (Budapest) 42 (1990) 275–288;
J. Boissonade, Q. Ouyang, A. Arneodo, J. Elezgaray, J.C. Roux and P. De Kepper, in: Nonlinear Wave Processes in Excitable Media, eds. A.V. Holden, M. Markus and H.G. Othmer (Plenum, New York, 1990).

[7] Q. Ouyang, V. Castets, J. Boissonade, J.C. Roux, P. De Kepper and H.L. Swinney, J. Chem. Phys., submitted for publication.

[8] C. Vidal and A. Pacault, in: Evolution of Order and Chaos, ed. H. Haken (Springer, Berlin, 1982) p. 74.

[9] S.C. Müller, in: From Chemical to Biological Organization, eds. M. Markus, S.C. Müller and G. Nicolis (Springer, Berlin, 1988) p. 83.

[10] J.J. Tyson and J.P. Keener, Physica D 32 (1988) 327.

[11] V.I. Krinsky, in: Self-Organization of Autowaves and Structures Far from Equilibrium, ed. V.I. Krinsky (Springer, Berlin, 1984) p. 9.

[12] V.S. Zykov, Simulation of Wave Processes in Excitable Media (Manchester Univ. Press, Manchester, 1989).

[13] R.J. Field and M. Burger, eds., Oscillations and Traveling Waves in Chemical Systems (Wiley, New York, 1985).

[14] A. Pacault, Q. Ouyang and P. De Kepper, J. Stat. Phys. 48 (1987) 1005.

[15] G. Nicolis and I. Prigogine, Self-organization in Nonequilibrium Chemical Systems (Wiley, New York, 1977).

[16] H. Haken, Synergetics, an Introduction (Springer, Berlin, 1977).

[17] V. Castets, E. Dulos, J. Boissonade and P. De Kepper, Phys. Rev. Lett. 64 (1990) 2953.

[18] A.M. Turing, Phil. Trans. R. Soc. London B 327 (1952) 37.

[19] A.M. Zhabotinsky and A.N. Zaikin, J. Theor. Biol. 40 (1973) 45;
M. Orbàn, J. Am. Chem. Soc. 102 (1980) 4311;
K. Sholwalter, J. Chem. Phys. 73 (1980) 3735.

[20] P. Möckel, Naturwissenschaften 64 (1977) 224;
M. Gimenez and J.C. Micheau, Naturwissenschaften 70 (1983) 90;
D. Avnir and M. Kagan, Nature 307 (1984) 717.

[21] J.C. Micheau, M. Gimenez, P. Borckmans and G. Dewel, Nature 305 (1983) 43.

[22] I. Nagypál, G. Basza and I.R. Epstein, J. Am. Chem. Soc. 108 (1986) 3635.

[23] S. Chandrasekhar, Hydrodynamic and Hydromagnetic Stability (Oxford Univ. Press, 1961);
H.L. Swinney and J.P. Gollub, eds., Hydrodynamic Instabilities and the Transition to Turbulence (Springer, Berlin, 1981).

[24] J.D. Murray, Mathematical Biology (Springer, Berlin, 1989)

[25] P. De Kepper, I.R. Epstein, K. Kustin and M. Orbán, J. Phys. Chem. 86 (1982) 170.

[26] I. Lengyel, G. Rabai and I.R. Epstein, J. Am. Chem. Soc. 112 (1990) 4606.

[27] A.J. Indelli, J. Phys. Chem. 68 (1967) 3027;
D.M. Kern and C.-H. Kim, J. Am. Chem. Soc. 87 (1965) 5309;
J. De Meeus and J. Sigalla, J. Chim. Phys. Chim. Biol. 63 (1965) 453.

[28] O. Citri and I.R. Epstein, J. Phys. Chem. 91 (1987) 6034.

[29] I. Lengyel and I.R. Epstein, Science, submitted for publication.

[30] J.E. Pearson and W. Horsthemke, J. Chem. Phys. 90 (1989) 1588.

[31] A. Arneodo, J. Elezgaray, J. Pearson and T. Russo, Physica D 49 (1991) 141–160, these Proceedings.

[32] Q. Ouyang and H.L. Swinney, private communication (June 1990).

[33] I. Nagypál, E. Dulos, V. Castets, J. Boissonade and P. De Kepper, in preparation.

[34] D. Walgraef, G. Dewel and P. Borckmans, Adv. Chem. Phys. 49 (1982) 311.

[35] J. Boissonade, V. Castets, E. Dulos and P. De Kepper, in: Bifurcation and Chaos: Analysis, Algorithms, Applications, eds. T. Küpper, F.W. Schneider, R. Seydel and H. Troger, (Birkhäuser, Basel, 1991), p. 64.

[36] J. Elezgaray and A. Arneodo, in: New Trends in Nonlinear Dynamics and Pattern Forming Phenomena: The Geometry of Nonequilibrium, eds. P. Coullet and P. Huerre (Plenum, New York, 1990) p. 21;

A. Arneodo and J. Elezgaray, Phys. Lett. A 143 (1990) 25.

[37] W.Y. Tam and H.L. Swinney, Phys. Rev. A 36 (1989) 1211.

[38] G. Dewel, D. Walgraef and P. Borckmans, J. Chim. Phys. (Paris) 84 (1987) 1335;
G. Dewel and P. Borckmans, Phys. Lett. A 84 (1987) 1335;
G. Dewel and P. Borckmans, in: Patterns Defect and Materials Instabilities, eds. D. Walgraef and N. Ghoniem, NATO Advanced Studies (Kluwer, Dordrecht), in press.

[39] J. Boissonade, J. Phys. (Paris) 49 (1988) 541.

[40] P. De Kepper, J. Boissonade and I.R. Epstein, J. Phys. Chem. 94 (1990) 6525.

[41] H. Meinhardt, Models of Biological Patterns Formation (Academic Press, New York, 1982).

[42] A. Babloyantz, Molecules, Dynamics and Life (Wiley, New York, 1986).

Physica D 49 (1991) 170–176
North-Holland

Dynamics of fronts, nuclei and patterns in 2D random media

W. Ebeling, A. Engel, L. Schimansky-Geier and Ch. Zülicke

Sektion Physik/04, Humboldt-Universität zu Berlin, Invalidenstrasse 42, O-Berlin 1040, Germany

In this paper we consider the influence of static and dynamic random perturbations of the properties of reactive media on the formation of structures and propagation of fronts. First we study the front and nuclei dynamics in uniform media. For systems with two order parameters the existence of stable nuclei is shown. Then we describe several effects on propagating fronts and nucleation dynamics resulting from random perturbations.

1. Introduction

We consider reaction–diffusion systems with scalar or vector order parameters X governed by partial differential equations of the type

$$\partial_t X(r, t) = f(X) + D \Delta X. \qquad (1.1)$$

As is well known, eq. (1.1) applies to a wide class of physical, chemical and biological systems [1]. We are interested mainly in thermodynamically open systems that have moving interfaces separating spatial regions in different states of the reaction. Examples are the phenomena observed in combustion [2], catalysis [3], and ionization [4].

The main aim of this work is the study of the influence of static or dynamic perturbations of the properties of the active media. This will be modelled by addition of a random part to eq. (1.1)

$$\partial_t X(r, t) = f(X) + D \Delta X + \sqrt{2\varepsilon}\, g(X)\, \xi(r, t). \qquad (1.2)$$

For this stochastic source term we assume generally Gaussian statistics given by

$$\langle \xi(r, t) \rangle = 0,$$

$$\langle \xi(r', t')\, \xi(r, t) \rangle = S(|t' - t|, |r' - r|). \qquad (1.3)$$

S is the correlation function of the applied temporal noise and spatial disorder, which has to be specified for concrete situations. Investigations in this direction are performed mainly on two topics: (i) the generation of new structures induced by noise [5], and (ii) the effect of noise on existing stable spatially nonuniform structures [6, 7]. We will be concerned with stable trigger fronts and will investigate the influence of noise on the velocity of the front motion.

2. Basic dynamical effects in uniform media

Let us begin with the simplest case: $X = n(r, t)$ is a scalar order parameter and the reaction function has the potential U, i.e.,

$$\partial_t n = -\partial_n U + D \Delta n. \qquad (2.1)$$

Defining the Ginzburg–Landau functional

$$\Phi[n(r, t)] = \int dr \left[U(n) + \tfrac{1}{2} D(\nabla n)^2 \right], \qquad (2.2)$$

we may write

$$\partial_t n = \frac{\delta \Phi[n]}{\delta n} \qquad (2.3)$$

The system tries to minimize the functional and will move along a pathway of steepest descent in the space of density distributions. Let us consider the case that $U(n)$ possesses the minima $n_1, n_2, \ldots, n_s$ with arbitrary $s = 1, 2, 3, \ldots$. Then we can distinguish three relaxational processes:

(i) For very short times the potential part of $\Phi[n]$ is locally minimized, i.e., the next minimum is reached. This leads to 2D subregions, which are divided by interfaces of different shape.

(ii) By minimization of the kinetic part of $\Phi[n]$, smooth interfaces ranging over the diffusion length l_0 are formed.

(iii) The slowest relaxational process is the optimization of the different contributions of $\Phi[n]$. Consider an interface between the stable states n_l and n_m. The interface dynamics is driven by the difference of the potential on both sides of the interface $U(n_l) - U(n_m)$ and by the curvature of the interface, which is defined by

$$c(r) = \frac{\delta O(n)}{\delta V(n,r)} = \operatorname{div} e. \qquad (2.4)$$

Here $V(n)$ and $O(n)$ denote the volume and the surface of the part of space with a density larger than n, and e is the normal unity vector. Introducing the front velocity by

$$v = \frac{\partial_t n}{|\nabla n|} \qquad (2.5)$$

we get from eqs. (2.1) and (2.4)

$$v = D(c_{cr} - c) \qquad (2.6)$$

with the critical curvature

$$c_{cr} = \frac{\partial_n U}{D|\nabla n|} - \frac{\nabla n \cdot \nabla |\nabla n|}{|\nabla n|^2} \qquad (2.7)$$

The second contribution in (2.7) disappears for inflection lines (i.e., the lines with maximal gradients). Further, introducing a normal (stationary) profile n_0 we get approximately

$$c_{cr} = \frac{U(n_l) - U(n_m)}{D \int_l^m de \, (\partial n_0/\partial e)^2}. \qquad (2.8)$$

One sees from (2.6) that $v_0 = Dc_{cr}$ is the velocity of a plane interface. From the existence of a critical curvature it follows that any part of the interface with a smaller curvature moves forward and any part with a larger curvature moves backward. Circles with a radius smaller than c_{cr}^{-1} will shrink and circles with a radius larger than c_{cr}^{-1} will expand.

Note that the interface description makes sense only in the case of steep and well separated minima and describes only the slowest relaxation process (iii). In a special coordinate system $r = [x, z]$, where the z-direction is perpendicular to the front, we get the description [8]

$$n(x,z,t) = n_0 \left(\frac{z - \phi(x,t)}{[1 + (\nabla\phi)^2]^{1/2}} \right), \qquad (2.9)$$

$$\frac{\partial_t \phi}{[1 + (\nabla\phi)^2]^{1/2}} = v_0 + D \nabla \cdot \frac{\nabla\phi}{[1 + (\nabla\phi)^2]^{1/2}}, \qquad (2.10)$$

where $\phi(x,t)$ denotes the local front position. These are the basic dynamic phenomena for the case of scalar order parameters. A static structure requires either plane fronts with $U(n_l) = U(n_m)$ or curved fronts having exactly the critical radius; both situations are either structurally or dynamically unstable.

We are extending now our considerations to two-component systems like

$$\partial_t n = f(n,T) + D \Delta n,$$
$$\partial_t T = W(n,T) + \chi \Delta T. \qquad (2.11)$$

We shall show that here even stable nuclei may exist. To be definite we consider the following thermokinetic model. We assume that the scalar order parameter n is bistable. The thin interphase at $|r| = R(t)$ separates an inner heat producing phase ($|r| < R$) from a surrounding passive phase ($|r| > R$). The corresponding heat balance equation in terms of renormalized quantities reads

$$\partial_t T = \bar{T} - T + wH(R - |r|) + \Delta T. \qquad (2.12a)$$

Here $\bar{T}$ is the bath temperature, w the heating rate and H the Heaviside step function. The basic assumption of this approach is that the interface of the n variable is relatively narrow and shows a slow dynamics, which is slaving the dynamics of the T variable. The motion of the interface itself is determined by the local temperature $\bar{T}$ due to

$$d_t R = G(T_0 - \bar{T}), \quad G \ll 1, \quad \bar{T} = T(R, t). \qquad (2.12b)$$

This temperature dependence is chosen in such a way, that at temperatures lower than a certain equilibrium temperature T_0 the hot phase grows ("reversed temperature dependence") [9]. This **model corresponds to the nullcline picture represented** in fig. 1. The scalar factor G in eq. (2.12b) arises from different time scales of the n and T processes ($G \sim \tau_T/\tau_n$). The boundary conditions for a domain centered at the coordinate origin are given by

$$\partial_t T(0, t) = 0, \quad T(\infty, t) = \bar{T}. \qquad (2.12c)$$

The one-dimensional case was studied in an earlier work [9]. We have shown with $T_0 = 1$ the existence of stable stationary solutions of the size

$$R_0 = \tfrac{1}{2} \ln\left(\frac{\tfrac{1}{2}w}{\bar{T} + \tfrac{1}{2}w - 1} \right). \qquad (2.13)$$

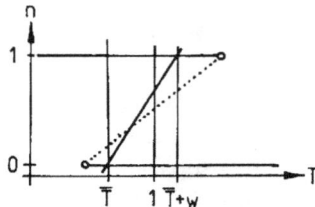

Fig. 1. Nullclines of the thermokinetic model (2.12).

Here we shall study the two-dimensional case. For this problem curvature effects are essential. The corresponding modification of the equilibrium temperature is known from thermodynamics as the Gibbs–Thomson relation

$$T_0 = 1 - \Sigma c, \qquad (2.14)$$

where Σ is the surface tension. Introducing this expression into eq. (2.12b) we find the velocity of the n variable corresponding to eq. (2.6). It reads for radial-symmetric nuclei ($c = 1/R$)

$$d_t R = G\left(1 - \bar{T} - \frac{\Sigma}{R}\right). \qquad (2.15)$$

The quasistationary interface temperature $\bar{T}$ is found from eq. (2.12a) with

$$\Delta = \frac{1}{r}\partial_r + \partial_r^2, \quad G \to 0. \qquad (2.16)$$

It yields the radius dynamics (2.15) [10]

$$d_t R = G\left(1 - \bar{T} - \frac{\Sigma}{R} - wRK_0(R) I_1(R)\right), \qquad (2.17)$$

where K_0 and I_1 are modified Bessel functions [11]. Certain approximations and a numerical solution of eq. (2.17) lead to the following behaviour (cf. figs. 2a and 2b): In the temperature interval $1 - w < \bar{T} < 1 - \tfrac{1}{2}w$ the situation is similar to the case of one scalar order parameter. There exists a critical unstable nucleus – hot nuclei shrink if they have undercritical size; overcritical nuclei will grow.

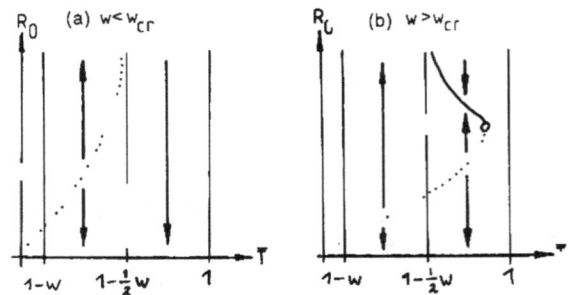

Fig. 2. Dependence of the radius of 2D stationary nucleus on the bath temperature $\bar{T}$.

The behaviour in the temperature interval $1 - \frac{1}{2}w < \overline{T} < 1$ depends strongly on the heating rate w. If it is subcritical,

$$w < w_{cr} \approx 4.2\Sigma, \qquad (2.18)$$

the uniformly cold phase is the only stable solution – the formation of stable nuclei is suppressed. Otherwise, for supercritical heating rates $w > w_{cr}$ the system exhibits under the condition

$$1 - \frac{1}{2}w < \overline{T} < 1 + 0.026 - 1.1(w\Sigma)^{1/2} \qquad (2.19)$$

a large stable and a smaller unstable nucleus. This means that if a overcritical hot nucleus occurs, it will relax into the stable configuration. We emphasize that, contrary to the 1D case, the necessary heating rate for 2D nuclei is determined by the value of the surface tension Σ (eq. (2.18)). Also the 2D stable nuclei are always smaller than the 1D domains.

Thus the existence of a second reaction component leads to certain modifications of the picture for the one-component case. Let us point out that the picture derived here for a thermokinetic model remains valid for other two-component systems if the front of one reactive component is narrow and the other reactive variable is slaved. Stable droplets require however a special picture of the isoclines. It seems to be possible that the isocline picture shown in fig. 1 may be realized, e.g., for the light-sensitive Ru-catalyzed Belousov–Zhabotinsky reaction [12] or the Ce-version [13].

3. Interaction of interfaces with quenched static disorder

So far we have only discussed systems with idealized homogeneity in space. Due to various reasons in all real physical systems there are influences of disorder of different types. We first deal with static randomness represented, for example, by impurities in solids or inhomogeneities in chemically reacting systems. There is a wealth of interesting phenomena related to equilibrium properties of interface configurations in random fields [14]. In this section we focus on the interaction of moving interfaces with static disorder. Including the Gaussian random field term in the reaction–diffusion equation of the bulk order parameter we find a modified equation of motion for the interface coordinate $\phi(x, t)$, restricted to the case $|\nabla\phi| \ll 1$, the equation reads

$$\partial_t\phi = v + D\,\partial_x^2\phi + (2\varepsilon)^{1/2}h(x, \phi) \qquad (3.1)$$

where the Gaussian random field $h(x, \phi)$ is the projection of the bulk disorder on the Goldstone mode. Eq. (3.1) can be solved by interaction in the random field strength $\varepsilon^{1/2}$. There are at least two physically interesting effects of the disorder. First the interface will move with locally differing velocity giving rise to a roughening of the front which, counterbalanced by the surface tension, will show up as a statistical width

$$\Delta\phi = \left[\overline{(\phi - \overline{\phi})^2}\right]^{1/2}.$$

Second the interface propagation is hindered by disorder; hence there will be a correction $\Delta v = v - \overline{\partial_t\phi}$ to the interface velocity (fig. 3a). From eq. (3.1) we find to first order in $\varepsilon^{1/2}$ [8]

$$\Delta\phi \sim \left(\frac{\varepsilon}{Dv}L\right)^{1/2} \qquad (3.2)$$

where L denotes the length of the interface. Since $\nabla\phi \sim \Delta\phi/L$ the use of the approximate equation (3.1) is justified for sufficiently large interface velocity v.

The first contribution to the velocity correction Δv is of second order in the random field strength and reads [8, 15]

$$\Delta v \sim \frac{\varepsilon}{(Dv)^{1/2}\Delta^{3/2}}, \qquad (3.3)$$

where Δ denotes the correlation length of the random field h. Eq. (3.3) is valid for $\Delta v \ll v$ only,

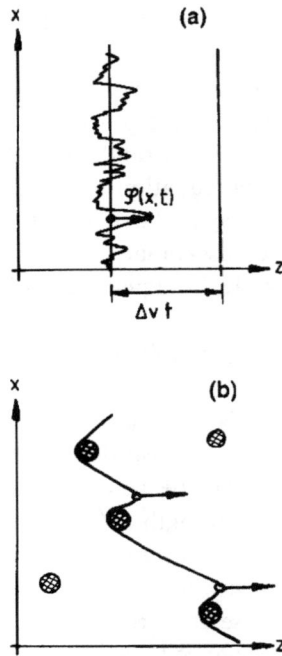

Fig. 3. Front propagation in quenched disorder. (a) Static disorder roughens a moving interphase and slows down the motion. (b) The front propagation is hindered by isolated static obstacles.

i.e., we must have

$$v \gg v_{\text{pin}} \sim \frac{1}{D^{1/3}\Delta} \qquad (3.4)$$

where v_{pin} gives the order of magnitude of the pinning threshold. For $v < v_{\text{pin}}$ the interface is unable to move in the presence of static disorder of the type discussed. Considering v_{pin} as a free parameter with the estimate (3.4), we obtain a rough approximation of the effective front velocity:

$$v_{\text{eff}} = v - \frac{v_{\text{pin}}^{3/2}}{v^{1/2}}. \qquad (3.5)$$

In particular in chemical systems there is sometimes another type of static disorder of interest. Consider a two-dimensional system with randomly distributed spheres with radius r and with different chemical properties. Inside the sphere the plane front velocity will hence take the value

$\tilde{v}$ different from v. If $\tilde{v} < v$ the spheres will hinder the front propagation, i.e., they will act as static, randomly distributed obstacles. This is the usual effect of impurities or inhomogeneities in chemical reacting systems. However, in special catalytic systems one may find also the opposite case $\tilde{v} > v$. Now the modified chemical properties inside the spheres give rise to an enhanced front velocity inside the spheres and therefore to a local acceleration of the moving interface.

For the first case one can calculate the time t^* necessary to overcome a single obstacle and one finds [16]

$$t^* \sim \frac{R^2}{D}\left(\frac{\tilde{v}}{v}\right)^2. \qquad (3.6)$$

Moreover the obstacles can stop the front only if both the distance a between them is smaller than twice the critical radius and the strength $|\tilde{v}|$ is sufficiently large. This means the average distance a has to fulfill the two inequalities

$$a < 2R_k = 2\frac{D}{v}, \qquad (3.7a)$$

$$a < (Dt^*)^{1/2} = r\frac{|\tilde{v}|}{v}. \qquad (3.7b)$$

If inequality (3.7a) is violated, the front breaks through between neighbouring obstacles and proceeds in its motion; if inequality (3.7b) is violated, elastic forces arising from the deformation of the interface are able to pull the front over the obstacles [16].

In the case of accelerating regions $\tilde{v} > v$ we find an average increase δv of the front velocity given by $\delta v = c(\tilde{v} - v)$, where c denotes the concentration of spheres with different chemical properties. Approximating the deviations of the local front velocity from this average by a Gaussian random field we get from eq. (3.3) a negative correction to this naive estimate of order

$$\Delta v \sim \frac{(\tilde{v} - v)^2 c^{5/4}(1 - c)}{[D(v + c(\tilde{v} - v))]^{1/2}}. \qquad (3.8)$$

For $c \ll 1$ we have therefore $\Delta v \sim c^{5/4}$ and the negative correction Δv is smaller than the average increase $\delta v \sim c$ so that the interface moves faster than without the catalytic regions, as expected.

4. Interfaces and nuclei in noisy media

Quite new questions arise if the disorder becomes time dependent. In the simplest case this means that the disorder is correlated only for a certain time τ; after that every correlation is neglible. Such a situation can be modeled by the application of noise on nonlinear systems in extended media.

Again we project the dynamics on the Goldstone mode and find a stochastic Langevin equation for the front position (first we restrict to the 1D case)

$$d_t \phi = \frac{(2\varepsilon)^{1/2}}{d} \int d\rho \, e^{v\rho/D} \, d_\rho n_0(\rho)$$

$$\times g(n_0(\rho + \phi)) \xi(\rho + vt, t)$$

$$d = \int d\rho \, e^{v\rho/D} [d_\rho n_0(\rho)]^2; \quad \rho = z - vt. \quad (4.1)$$

Several situations can be investigated. Here we assume additive colored noise, i.e., $g(n) \equiv 1$, and the correlation function

$$S(r, t) = \frac{1}{\pi \Delta \tau} \exp(-r^2/\Delta^2) \exp(-t^2/\tau^2). \quad (4.2)$$

In this case the effect of noise leads to a diffusional motion of the front position, comparable with the roughening obtained above [6]. (We never find pinning due to the finite correlation time of the noise.)

$$\langle [\phi(t) - \phi(0)]^2 \rangle = \mathscr{D}t, \quad (4.3)$$

where the value of the effective diffusion coefficient $\mathscr{D}$ depends strongly on the ratio of the interface thickness l_0 to the effective correlation length $\lambda = (\Delta^2 + v^2\tau^2)^{1/2}$. During the time τ the correlation is transported by the front with velocity v, which leads to the new correlation length $v\tau$. The effective diffusion coefficient will be maximal if λ/l_0 becomes vanishingly small. In this case the interface that is the most unstable region of the front is very flat and the noise is quasi-white. $\mathscr{D}$ becomes small for narrow fronts and strong correlations [7, 17].

Let us briefly discuss the 2D case with circular symmetry. In section 2 we pointed out the existence of critical radii. Obviously the diffusional motion of the interface makes possible the escape over the unfavourable critical configuration. Assuming white noise with intensity ε we find the Langevin dynamics for the radius R (the stochastic analog to eq. (2.6))

$$d_t R = D \left(\frac{1}{R_{cr}} - \frac{1}{R} \right) + \eta_R(t), \quad (4.4)$$

where

$$\langle \eta_R(t) \rangle = 0,$$

$$\langle \eta_R(t') \eta_R(t) \rangle = \frac{1}{2\pi R} \frac{\delta(t' - t)}{d}.$$

Introducing the Schlögl potential as an approximation of eq. (2.2)

$$\Delta F(R) = -\pi R^2 \int_{n_1}^{n_3} dn f(n) + 2\pi R D d, \quad (4.5)$$

where n_1 and n_3 are the stable stationary states of the deterministic dynamics, we get the mean time a radius needs for growing stochastically to reach the critical value [18]

$$T(R_0 \to R_{cr}) \sim d \int_{R_0}^{R_{cr}} dR \int_0^R dR'$$

$$\times 2\pi R \exp \left(-\frac{\Delta F(R') - \Delta F(R)}{\varepsilon} \right). \quad (4.6)$$

The situation considered above is quite similar to nucleation processes in equilibrium phase transitions, where T is a measure for the inverse nucleation rate.

5. Conclusions

We have studied in this paper the dynamics of stable states in multistable reaction–diffusion systems under the influence of static or dynamic random fields. The consideration is limited to the 2D case and to sufficiently strong nonlinear contributions to the rate function, which guarantees that the interface regions are sufficiently narrow. For one-component systems the local front velocity in uniform systems is proportional to the difference between the critical curvature and the actual curvature. This behaviour is modified for the two-component case where even stable droplets may exist. If static obstacles are embedded in the active medium, the front velocity will depend on such parameters as the strength, distance, diameter and other parameters of the obstacles and their distribution. A typical phenomenon discussed in this context is the pinning of the front at the obstacles. The most complicated case refers to space- and time-dependent fluctuations. For small correlation times it yields a diffusional motion of the interface. Hence unfavourable situations may be overcome after some time with a certain probability. We have shown this for the nucleation of an undercritical nucleus to become overcritical.

References

[1] Y.A. Vasilev, Yu.M. Romanovsky, D.S. Chernavsky and V.G. Yakhno, Autowave Processes (Deutscher Verlag der Wissenschaften, Berlin, 1987).
[2] Ya.B. Zeldovich, G.I. Barenblatt, V.B. Librovich and G.M. Makhviladze, The Mathematical Theory of Combustions and Explosions (Nauka, Moscow, 1980) [English transl.: Consultants Bureau, New York, London, 1985].
[3] V.I. Krinsky, ed., Autowave and Structures Far From Equilibrium (Springer, Berlin, 1984).
[4] W. Ebeling, A. Förster, D. Kremp and M. Schlanges, Physica A 159 (1989) 285.
[5] A.S. Mikhailov, Phys. Rep. 184 (1989) 307.
[6] A.S. Mikhailov, L. Schimansky-Geier and W. Ebeling, Phys. Lett. A 96 (1983) 453.
[7] H. Malchow and L. Schimansky-Geier, Noise and Diffusion in Nonequilibrium Bistable Systems (Teubner, Leipzig, 1985).
[8] A. Engel, W. Ebeling, R. Feistel and L. Schimansky-Geier, in: Selforganization by Nonlinear Irreversible Processes, eds. W. Ebeling and H. Ulbricht (Springer, Berlin, 1986) p. 110.
[9] Ch. Zülicke, A.S. Mikhailov and L. Schimansky-Geier, Physica A 163 (1990) 559.
[10] Ch. Zülicke, Ph.D. Thesis, Humboldt-Univ. Berlin (1990).
[11] M. Abramowitz and A. Stegun, (eds.), Pocketbook of Mathematical Functions (Harri Deutsch, Thun, Frankfurt, 1984).
[12] W. Ebeling, J. Stat. Phys. 45, 5/6 (1986) 891.
[13] J.J. Tyson, in: Oscillations and Traveling Waves in Chemical Systems, eds. R.J. Field and M. Burger (Wiley, New York, 1985) p. 93.
[14] T. Nattermann and J. Villain, Phase Transitions 11 (1988) 5.
[15] M.V. Feigelman, Zh. Eksp. Teor. Fiz. 85 (1983) 1851 [in Russian].
[16] A. Engel and W. Ebeling, Phys. Lett. 122 A (1987) 20.
[17] L. Schimansky-Geier, A.S. Mikhailov and W. Ebeling, Ann. Phys. (Leipzig) 40 (1983) 277.
[18] L. Schimansky-Geier and W. Ebeling, Ann. Phys. (Leipzig) 40 (1983) 10.

Physica D 49 (1991) 177–181
North-Holland

Structure multistability in spatially modulated reaction–diffusion systems

Yu.D. Kalafati and Yu.A. Rzhanov

Institute of Radio Engineering and Electronics of USSR Academy of Sciences, Marx Avenue 18, 103907 Moscow, USSR

We describe spatially inhomogeneous structures that occur in bistable reaction–diffusion systems which can be described by one or two rate equations for the kinetic variables. Attention is devoted mainly to the conditions for determining the occurrence of spatially inhomogeneous structures.

1. Introduction

In this paper we consider the spatially inhomogeneous structures (SIS) which occur in bistable reaction–diffusion system (RDS) with spatially modulated parameters. Periodic modulation can be caused, for example, by periodically arranged inhomogeneities. In the simplest case the effect of bistability in RDS can be described by only the equation for the kinetic variable (usually the temperature T or the concentration of chemical reagent n) which characterize the system. The normalized equation reads

$$\tau_n \frac{dn}{dt} = F_n(n, S) + l_n^2 \nabla^2 n, \qquad (1)$$

where $F_n(n, S)$ is a nonlinear function with three roots that correspond to the homogeneous states of the system (two of them, n_1 and n_3, are stable; n_2 is unstable), τ_n is the relaxation time, and l_n is the diffusion length. $F_n(n, S)$ describes the source of reagent n homogeneously distributed in space. It contains the term that determines the generation rate of the reagent and recombination term that is responsible for the relaxation. The generation rate depends on the bifurcation parameter S.

Consider a one-dimensional bistable RDS described by eq. (1). If the system is homogeneous and the nonequilibrium reagent is generated homogeneously, the only stable SIS is the switching

front. The switching front is a transition region distinguishing in space two stable homogeneous states of the system. This solution is unique. The front velocity V is determined by the expression [1]

$$V = \frac{\int_{n_1}^{n_3} F(n, S)\, dn}{\int_{-\infty}^{+\infty} (dn/dz)^2 \, dz} \qquad (2)$$

V is positive if the region occupied by the state n_3 grows in time and negative otherwise. It follows from (2) that V becomes zero if

$$\int_{n_1}^{n_3} F(n, S)\, dn = 0.$$

This condition is not rough (in a mathematical sense) and is violated under a small variation of the bifurcation parameter S.

Real bistable spatially distributed systems always contain inhomogeneities. The presence of inhomogeneities makes the condition for the stability of a motionless front rough, and $V = 0$ holds for a finite region of S. This effect was observed for the first time to the best of our knowledge in ref. [2], and for random distribution of inhomogeneities was considered analytically in ref. [3].

More complicated phenomena occur in the bistable system if the inhomogeneities are ar-

0167-2789/91/$03.50 © 1991 – Elsevier Science Publishers B.V. (North-Holland)

ranged periodically with a period comparable to the correlation length of the corresponding homogeneous system. In this case the effect of switching front multistability has been demonstrated [4, 5]. To obtain the analytical results we approximated in refs. [4, 5] the nonlinear function F_n with a piecewise linear one:

$$F(n, S) = n_1 - n + (n_3 - n_1)\Theta(n - n_2(S)), \quad (3)$$

where $\Theta(n)$ is the Heaviside function, n_1 and n_3 are the carrier concentrations corresponding to the stable states, and n_2 is the critical concentration corresponding to the unstable state ($n_1 < n_2 < n_3$). In our model a change of the bifurcation parameter influences only the value of n_2 (n_1 and n_3 do not depend on S). It is useful to note that if $n_2 = n_{2M} \equiv (n_1 + n_3)/2$, the velocity of the front in the model equals zero. We consider the inhomogeneities as a set of interchanging homogeneous domains of two different types. The domains of type "+" have width (in one-dimensional case) L_+, and $n_2 = n_{2+} < n_{2M}$. The domains of type "−" have width L_-, and $n_2 = n_{2-} > n_{2M}$. Under the conditions corresponding to the "+" ("−") domains the front velocity of the homogeneous system is positive (negative). The normalized equation reads:

$$\tau_n \frac{\partial n}{\partial t} - l_n^2 \frac{\partial^2 n}{\partial z^2} = n_1 - n + (n_3 - n_1)\Theta(n - n_2(z)).$$
$$(4)$$

Carrying out the stability analysis of the stationary solution of eq. (4), we have shown [4] that the solution is stable if each domain is switched on as a whole ($n_\pm(z) > n_{2\pm}$ for all z inside the domain) or switched off as a whole ($n_\pm(z) < n_{2\pm}$), where $n_\pm(z)$ is the distribution of concentration inside the domain. If at some point $z^*, n_\pm(z^*)$ equals $n_{2\pm}$, then the solution loses its stability. In refs. [4, 5] we have found and classified the set of heteroclinic structures Q_N ($n \to n_1$ for $z \to +\infty$, $n \to n_3$ for $z \to -\infty$) – the switching fronts – with N local maxima in the transition region (the case

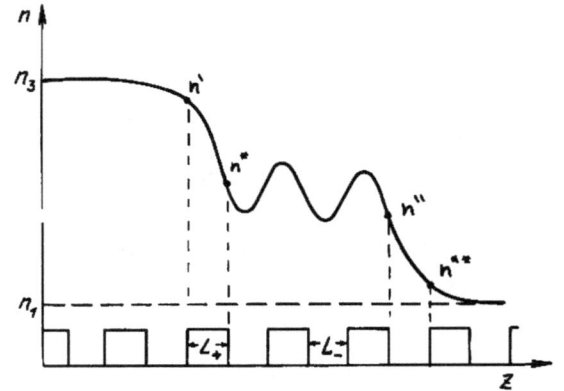

Fig. 1. The spatial distribution of concentration n for the structure Q_2. The arrangement of the domains L_+ and L_- is shown on the z-axis.

with $N = 2$ is shown in fig. 1). Fig. 2 shows the regions corresponding to the stable motionless structures Q_N in the plane (n_{2+}, n_{2-}). The value of N is given in the upper right corner of the corresponding rectangle region of stability. The analytic expressions for the critical values $n_N^*, n_N^{**}, n_N', n_N''$ (see figs. 1 and 2) can be found elsewhere [5].

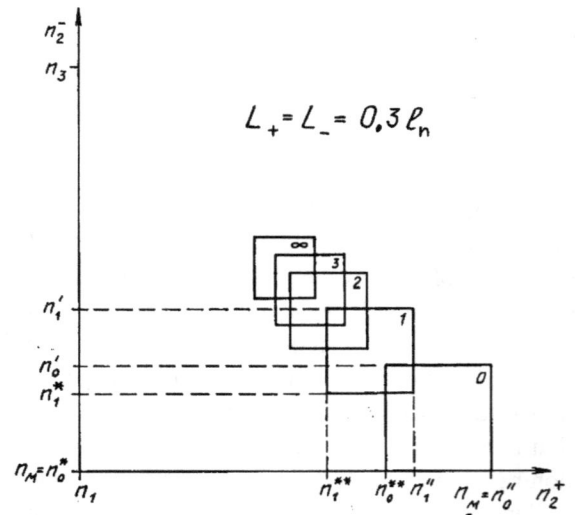

Fig. 2. The regions of stability of the structures Q_N with N maxima in the parameter plane (n_{2+}, n_{2-}). See explanation in the text.

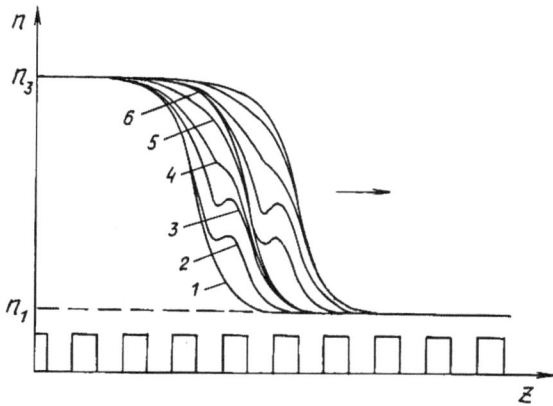

Fig. 3. The profiles of concentration during switching wave propagation when the structures Q_0 and Q_1 appear on the front one after the other. Different numbers denote different times.

The region in fig. 2 satisfying the following conditions

$$n_1^{**} < n_{2+} < n_0^{**} \qquad n_0^* < n_{2-} < n_1^*$$

corresponds to the moving front with more complicated dynamics than that of the homogeneous case. The structures Q_0 and Q_1 appear on the front one after the other (fig. 3).

In refs. [4, 5] we reached the conclusion that all of the complicated fronts between homogenous

states n_1 and n_3 appear due to the existence of the stable periodic structure n_p with the same period as the period of domains (inhomogeneities). In the present paper we consider the switching fronts between this periodic structure and the homogeneous states n_1 and n_3 (fig. 4).

2. Analytical consideration

Following the model described in section 1, we consider the motionless front between the homogeneous state n_1 ($z < 0$) and the periodic structure n_p ($z > 0$) (see fig. 4). Inside the domain "+" or "−", eq. (4) is linear and its solution can be written in the form:

for $z > 0$

$$n_{k+} = A_{k+} \exp(-z/l_n)$$
$$\qquad + B_{k+} \exp(z/l_n) + n_3,$$
$$n_{k-} = A_{k-} \exp(-z/l_n)$$
$$\qquad + B_{k-} \exp(z/l_n) + n_1,$$

for $z < 0$

$$n = A \exp(z/l_n) + n_1,$$

where k is the domain number ($k = 0, 1, 2, \ldots$)

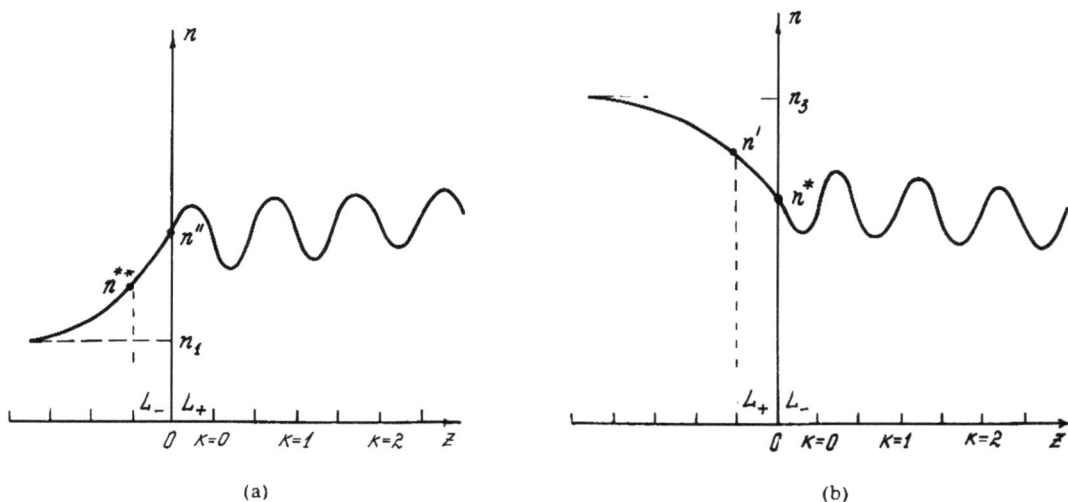

(a)

(b)

Fig. 4. The distribution of concentration in the front (a) between n_1 and n_p and (b) between n_p and n_3.

and A_k, B_k, A are constant coefficients. Using matching and boundary (the solution must be bounded for all k) conditions, one can obtain

$$A_{k+} = \tfrac{1}{2}(n_3 - n_1)$$
$$\times \left[1 - R_- + (1 - R_- + R_+ R_-) R_+^k R_-^k \right] /$$
$$(1 - R_+ R_-),$$

$$B_{k+} = \tfrac{1}{2}(n_3 - n_1) R_+ (1 - R_-)/(1 - R_+ R_-),$$

$$A = \tfrac{1}{2}(n_3 - n_1)(1 - R_+)/(1 - R_+ R_-), \qquad (5)$$

where $R_+ = \exp(-L_\pm / l_n)$.

The expressions (5) allow us to calculate n^{**} and n'', which determine the stability of the front (fig. 4):

$$n^{**} = n_1 + \tfrac{1}{2}(n_3 - n_1) R_- (1 - R_+)/(1 - R_+ R_-),$$

$$n'' = n_1 + \tfrac{1}{2}(n_3 - n_1)(1 - R_+)/(1 - R_+ R_-).$$

For the case of the motionless front these values must satisfy the following inequalities:

$$n^{**} < n_{2+} < n''. \qquad (6)$$

Similarly, we consider the motionless switching front between the periodic structure and the homogeneous state n_3, and we find

$$n^* = n_3 - \tfrac{1}{2}(n_3 - n_1)(1 - R_-)/(1 - R_+ R_-),$$

$$n' = n_3 - \tfrac{1}{2}(n_3 - n_1) R_+ (1 - R_-)/(1 - R_+ R_-).$$

The inequalities that determine the front stability in the last case read

$$n' < n_{2-} < n^*. \qquad (7)$$

The intersection of stability regions (6) and (7) of two motionless fronts $n_1 - n_p$ and $n_p - n_3$ coincides with the rectangle denoted by sign "∞" in fig. 3, which means that all three states n_1, n_3,

and n_p exist in equilibrium with each other in this region of parameters.

3. Conclusions

The periodic stable stationary inhomogeneous state (some kind of induced superlattice), discussed above, is not unique. In some region of parameters the periodic states with periods $(iL_+ + jL_-)$, where i and j are odd, are possible [4, 6]. It should be pointed out that chaotic structures possibly exist, as was discussed in ref. [6] and also in ref. [7] in connection with the problem of optical information storage. Homoclinic structures (satisfying the same boundary conditions for $z \to \infty$ and $z \to -\infty$) can also be investigated in RDS by the method described above. In two-component RDS (described by two kinetic variables) the spatial modulation of the parameters can lead to the stabilization of structures which are unstable in RDS with homogeneous parameters. This effect is of extreme importance in the vicinity of the bifurcation that describes the appearance of a moving structure from the motionless one [8, 9].

The spatially inhomogeneous structure can lose its stability and make a transition to another structure if the bifurcation parameter S is changed. This kind of instability can lead to interesting applications in the solid state devices with N- or S-type voltage–current characteristics, if their parameters are spatially modulated. For example, in a film with phase transition semiconductor–metal or in a distributed Josephson junction, a multivalued voltage–current characteristic is expected.

References

[1] A.C. Scott, Active and Nonlinear Wave Propagation in Electronics (Wiley, New York, 1970).
[2] V.F. Pastushenko and Yu.A. Chizmadjev, Proceedings of the 2nd All-Union Symposium on Oscillation Processes in

Biological and Chemical Systems, Vol. 2 (1971) 305 p. [in Russian].

[3] A.V. Gurevich and R.G. Mintz, Usp. Fiz. Nauk. 142 (1984) 61 [in Russian].

[4] Yu.A. Rzhanov and Yu.D. Kalafati, Opt. Commun. 70 (1989) 1618.

[5] Yu.D. Kalafati and Yu.A. Rzhanov, Izv. VUZ Radiofizika 32 (1989) 966 [in Russian].

[6] A.A. Nepomnyashchy, in: Nonlinear Waves I. Dynamics and Evolution, eds. A.V. Gaponov-Grekhov, M.I.

Rabinovich and J. Engelbrecht, Research Reports in Physics (Springer, Berlin, 1989) p. 103.

[7] W.J. Firth, Phys. Lett. A 125 (1987) 375.

[8] Yu.D. Kalafati and Yu.A. Rzhanov, Izv. VUZ Radiofizika 32 (1989) 569 [in Russian].

[9] Yu.D. Kalafati and Yu.A. Rzhanov, in: Nonlinear Waves I. Dynamics and Evolution, eds. A.V. Gaponov-Grekhov, M.I. Rabinovich and J. Engelbrecht, Research Reports in Physics (Springer, Berlin, 1989) p. 159.

Physica D 49 (1991) 182–197
North-Holland

Chapter 4. Waves and patterns in biological systems

Spiral waves in normal isolated ventricular muscle

Jorge M. Davidenko, Paul Kent and José Jalife

Departments of Pharmacology and Physiology, SUNY Health Science Center at Syracuse,
766 Irving Avenue, Syracuse, NY 13210, USA

Rotating waves of electrical activity may be observed in small ($20 \times 20 \times < 1$ mm) pieces of normal ventricular epicardial muscle from the sheep heart. Application of an appropriately timed voltage gradient perpendicular to the wake of a quasiplanar wavefront gives rise to sustained vortex-like reentry which may last indefinitely, unless another stimulus of appropriate characteristics and timing is applied to terminate the arrhythmia. Once established, the vortex pivots at a frequency of 5 to 7 Hz around a small elongated core (~ 2 by 4 mm) which only develops small electrotonically mediated depolarizations. Imaging of the entire course of local activation and recovery may be accomplished with high temporal and spatial resolution using the voltage-sensitive dye di-4-ANEPPS and a 10×10 photodiode array system focused onto the preparation. Two- and three-dimensional reconstructions of the optical images generated by the local fluorescence response to the circulating waves provide accurate maps of the distribution of voltage across most of the surface of the tissue and gives insight into the dynamics and mechanisms of initiation, maintenance and termination of the arrhythmia. The overall results suggest that neither dispersion of action potential duration (APD) nor tissue anisotropy are essential for the induction or maintenance of reentry. A transient nonuniformity in refractoriness created by the appropriately timed voltage gradient is sufficient to establish the circulating activity. Anisotropy and dispersion of APD serve only to produce a nonuniform topographical distribution of conduction velocity and excitable gap (i.e., that interval during which part of the circuit is excitable). Annihilation of the reentry by an appropriately timed stimulus is the result of collision of the reentrant wavefront with the stimulus-induced excitation wave propagating in the opposite direction. These results may be useful in the understanding of the mechanisms of ventricular tachycardia and may have important implications in the development of new and more specific antiarrhythmic therapies.

1. Introduction

Spiral waves have been recognized as a behavior that is common to a wide variety of excitable media [1–4]. Several terms have been used to describe this particular type of self-sustaining wave propagation, including "rotors", "reentrant vortices", two-dimensional "spiral waves" and three-dimensional "scroll waves". In addition, the dynamics and topological properties of this type of activity have been extensively analyzed in the physical, chemical, and biological literatures (see refs. [1, 2] for review). Recent studies in the heart [5–7] have focused on certain predictions which have emerged as a result of theoretical studies on the dynamics of generic excitable media. Since rotating vortices of electrical activity are demon-

strable in normal myocardial tissues [5–10], it might be possible to use analytical tools derived from studies in other excitable media to provide insight into the dynamics of reentrant activity in the heart.

To date, most studies on functionally determined reentrant ventricular arrhythmias have relied upon the assumption that initiation and maintenance of reentry requires at least some degree of nonuniformity in one or more of the electrophysiological characteristics of the cardiac tissue. For example, since the studies of Han and Moe [11], it has been surmised that nonuniform dispersion of refractoriness (or action potential duration) in neighboring areas of the myocardium is a requirement for the establishment of unidirectional block and reentry. Accordingly, a premature electrical impulse should be blocked in that area in which refractoriness is greatest, and reentry should occur as a result of slow conduction proceeding retrogradely through an alternate path. Another possibility is that reentry, slow conduction and block are the result of nonuniform anisotropic propagation [10, 12, 13]. Indeed, recent studies support the participation of anisotropy in the development of reentrant tachycardia [10, 12], and attempt to explain the arrhythmia in terms of differences in safety factor for propagation in transverse versus longitudinal fiber direction and of extremely slow conduction across lines of apparent block.

Another possibility is that neither nonuniform dispersion of refractoriness nor anisotropy are crucial for the induction or maintenance of reentry. In fact, it may be possible that only a transient local inhomogeneity of refractoriness may be sufficient to initiate sustained vortex-like reentry in otherwise normal myocardium. In the so-called "pinwheel experiment" of Winfree [2], a premature voltage pulse is delivered to the center of an idealized homogeneous excitable medium on the wake of a perfectly planar wave initiated sometime earlier at one end of the medium. The intersection ("critical point") of a critical circular isogradient line created by the voltage pulse with a critical recovery line that follows the planar wave, results in the formation of a rotor. In this particular case, there would be two critical points, each giving rise to a counter-rotating vortex. On the basis of such a theoretical experiment, Winfree made the prediction that spiral waves may arise in two-dimensional areas of normal myocardium, whereas scroll waves would be demonstrable in normal ventricular myocardium in which the three-dimensional wall thickness was preserved. In either case, rotating excitation waves should occur even if nonuniform dispersion of refractoriness and anisotropy were not included.

Experimental evidence supporting these predictions has been obtained in the multiple electrode mapping experiments of Shibata et al. [5] and Frazier et al. [6]. However, although those studies did demonstrate that spiral waves of electrical depolarization resulting in periodic activity of the heart may be induced by an epicardial shock of appropriate strength and timing, they only provided information on the wave front and did not allow any measurements of the recovery characteristics of the tissue during the reentrant process. Thus, direct comparison of the experimental results with the theoretical predictions have not yet been done.

We have developed an in vitro model of cardiac excitation. The model enables the study of two-dimensional propagation and vortex-like reentry in small pieces of normal ventricular epicardial muscle by means of high resolution optical mapping and voltage-sensitive dyes [7]. As a prerequisite for using optical mapping in the study of cellular electrophysiologic mechanisms underlying reentrant arrhythmias, we have established the following criteria: (1) the preparation should be completely immobile and thin enough to allow homogeneous viability during long periods of superfusion; (2) all cellular electrophysiological parameters should be within the normal range as demonstrated by microelectrode recordings; and (3) an appropriate set of stimulation parameters should result in reproducible *sustained* reentrant tachycardia. The results demonstrate that self-sustaining two-dimensional rotating waves may be initiated in these preparations by single premature electrical stimulation. Direct mapping of such waves by means of high resolution optical methods allows the accurate analysis of the entire activation–recovery process and gives insight into the detailed mechanisms of reentrant ventricular tachycardias.

2. Methods

2.1. Dissection of the preparation

Experiments were conducted in isolated sheep epicardial muscle preparations. Young sheep (10–20 kg) were anesthetized with sodium pentobarbital (36.5 mg/kg i.v.). The heart was rapidly removed and placed in warm, oxygenated Tyrode solution. After clearly identifying the fiber orientation by gross anatomical inspection, square pieces of epicardium (about 2 cm by 2 cm by 0.5 mm) were cut with a dermatome. Extreme care was exerted throughout the dissection procedure to prevent damage to the epicardial surface of the tissue. Suitable preparations were immediately transferred to a Plexiglas chamber and continuously superfused with oxygenated Tyrode solution. The composition of the solution was (in mM): NaCl, 130; KCl, 4; $NaHCO_3$, 24; NaH_2PO_4, 1.2; $MgCl_2$, 1; $CaCl_2$, 1.8; and glucose, 5.6. Solutions were saturated with a gas mixture of 95% O_2 and 5% CO_2, temperature was maintained at $37 \pm 0.5°C$, and pH was 7.4.

2.2. Recording and stimulation techniques

To avoid interference with the optical signals, stimulating and recording electrodes were embedded in the wax bottom of the chamber (see fig. 1). The preparation was firmly pinned to the floor of the chamber to ensure that its bottom surface made good contact with all electrodes. The chamber was mounted on a vibration isolation table. Transmembrane potentials were recorded using glass microelectrodes filled with 3 M KCl and connected to a WPI dual microprobe system. Stimulation of the preparation was performed through a square array for Ag–AgCl bipolar electrodes (black bars in fig. 1). Each pole of the electrodes was long enough (20 mm) to make contact with one side of the tissue. An electrode combination consisting of opposite pairs was used to create premature voltage gradients perpendicular to a propagating wave initiated by a basic stimulus. A pair of Pulsar 6i stimulators (Frederick Haer Co.) was used as the stimulus source.

2.3. High-resolution optical mapping

The use of voltage-sensitive dyes and optical recording techniques enabled us to produce high-resolution isopotential maps as well as two- and three-dimensional fluorescence graphics, by

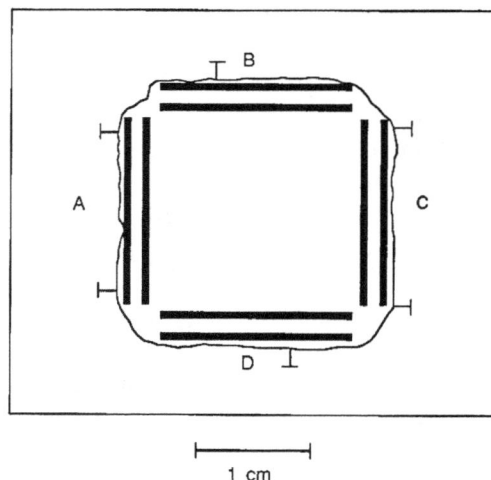

Fig. 1. Schematic of the isolated tissue preparation. Black bars represent Ag/AgCl electrodes.

simultaneously recording from 100 fine-grained points in a matrix covering a tissue surface which ranged between 14.06 and 225 mm^2, depending on lens amplification. The preparations were stained with di-4-ANEPPS (2.5 µg/ml). No obvious deterioration of the preparation or the fluorescence occurred with this dye and the signal-to-noise ratio (> 30-to-1) was better than that observed with other dyes (WW781 or RH414). Application of the dye was carried out 1 h after tissue equilibration (see above), by continuous superfusion of a recirculating volume (250 ml) of the dye-containing Tyrode. The dye was allowed to bind for 20–30 min. To prevent mechanical interference, the electromechanical uncoupler diacetyl monoxime (DAM, 10 to 20 mM) was added to the superfusate prior to the beginning of the recordings. Under these conditions, the stained preparation remained stable and immobile for long periods of time (typically 6 h or longer) while its electrophysiological properties were maintained within the normal range.

2.4. Optics and electronics

We have designed our own optics for imaging cardiac muscle rather than using a commercially available fluorescence microscope. As shown by

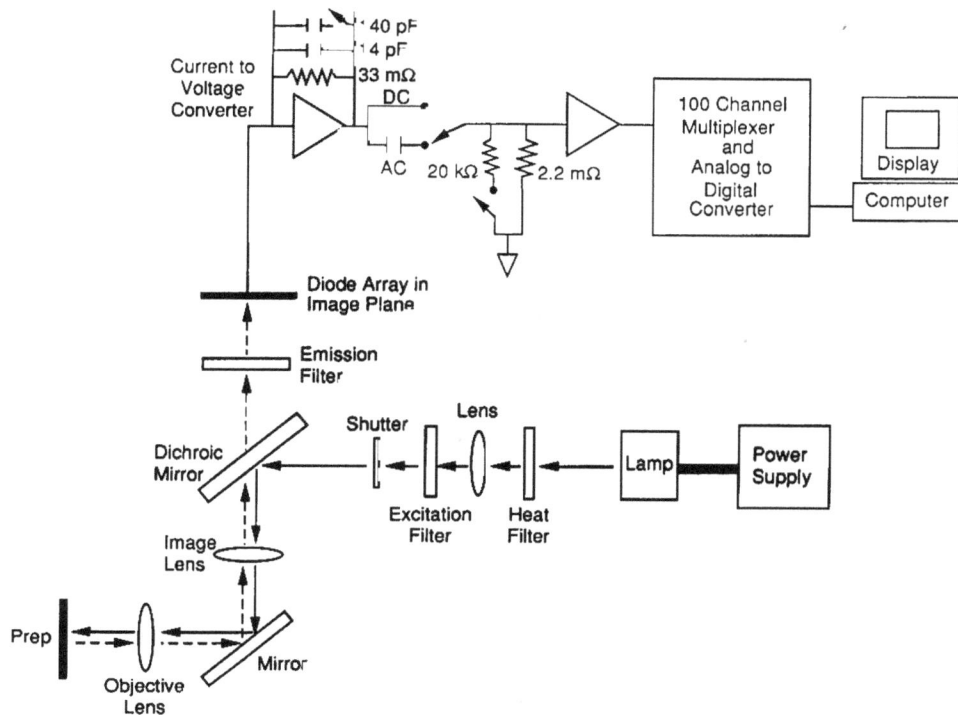

Fig.   Schematic of the optical mapping setup.

the schematic diagram of fig. 2, a real fluorescent image of the object stained is projected onto a 10 × 10 photodiode array (Centronics LTD), utilizing epi-illumination. The light from a tungsten–halogen lamp is collimated and made quasimonochromatic using an interference filter together with a KG-3 Schott heat filter. The light is then reflected 90° (45° incidence) from a dichroic mirror and focused onto the preparation. With this system a uniform illumination of the preparation (8–9 mm) is achieved. The emitted fluorescent wavelengths from the preparation are transmitted through the emission filter. An image is formed at the focal plane of a lens placed before the dichroic mirror. The ratio of the focal length of this lens and the objective lens determines the optical magnification. With this configuration magnifications from 1 to 7× have been utilized. For instance, with a magnification of 4× and with a photodiode array measuring 1.5 cm on each side, a 3.75 mm × 3.75 mm area of the object is imaged on the array. Furthermore, with

every photodiode measuring 1.4 mm on each side, each detector receives light from a 350 μm × 350 μm area of the object. In order to position and focus the tissue onto the array, a laser beam (not shown) is employed such that its reflection off the tissue, which is focused on the photodiode array by the optics, is brought to its minimal diameter. By always shining the beam onto the same point and bringing its image onto the same position at the photodiode array, the preparation is consistently positioned.

The output of each photodiode is connected to a current–voltage converter (feedback resistance = 33 Mohms). The signal is filtered with a low-pass filter and may be ac or dc. With ac recordings, time constants of either 100 ms or 10.3 s may be selected and these signals are again amplified by a factor of 1000–5000. For the dc coupled signal the second amplifier has unity gain. The outputs of the amplifiers are connected to seven 16-channel A/D boards (RC Electronics) to achieve 100 channels. The A/D boards are run in a frame mode, triggered by an external clock. In this manner, upon receiving a pulse from the external clock, the 100 channels are sequentially digitized (12 bits) in about 100 μs (approximately 1 μs per channel). The digitized data are then transferred to memory with a 22-bit DMA board. With 4 Mbyte of memory, 2 million samples (each being 12 bits) can be collected. After a trial, the data residing in high memory are then transferred to low memory, piece by piece, appropriately un-shuffled and then transferred to a 40 Mbyte hard disk. After processing the data the signals are visualized on a Tektronix storage oscilloscope with the X–Y display driven by the output of a D/A converter. A hard copy of the data can be obtained by replacing the storage scope with a Hewlett Packard X–Y plotter. Alternatively, files can be transferred to a SUN 4/110 workstation for further processing and graphics display.

2.5. Two- and three-dimensional fluorescence plots and isopotential maps

To map the electrical activity, activation at each recording site was arbitrarily defined as the time at which the detected fluorescence reaches 50% of its maximal amplitude. Activation was considered to have failed if the maximum fluorescence at a given recording site was less than 30% of the maximal amplitude detected for any cell of the array. In addition, correlation with intracellular recordings of transmembrane potential was used in some instances to verify the criteria for determining failure of activation. Two- and three-dimensional fluorescence maps were constructed at different points in time. In the three-dimensional plots (see fig. 8 below) the z axis represents the intensity of fluorescence change (equivalent to potential change). Voltage distribution at different instants was also studied with isopotential maps. To improve visualization of the activation process, isopotential lines only included the higher 70% of the fluorescence values. Values from 0 to 30% were not differentiated from the background.

2.6. Histological studies

To clearly define the anatomical characteristics of the tissue (thickness of the specimen, fiber orientation, presence of bands of connective tissue, and presence of abnormal or damaged tissue) three preparations were analyzed at the end of the experiments through light microscopy. Specimens were fixed in formalin, embedded in paraffin and stained with hematoxylin-eosin. Slices of the preparation allowed a detailed analysis of the morphology of the tissue from the epicardial surface to the cut surface.

2.7. Experimental protocols

2.7.1. Induction of reentry
The preparation was driven with trains of ten basic stimuli (S_1) at a constant basic cycle length (BCL) of 300–500 ms, delivered through electrode pairs A, B, C or D (see fig. 1). The duration of S_1 was 2–4 ms and the intensity was twice the diastolic threshold. The trains were separated by a 1000 ms pause. After verification of the presence of a stable quasi-planar S_1 activation pattern, a premature stimulus (S_2) (2–4 ms), was delivered perpendicularly to S_1 by using the poles of electrodes positioned at opposite sides of the preparation. For instance, if S_1 was delivered through electrode A, then S_2 was delivered through electrodes B($+$) and D($-$). S_2 intensity and S_1–S_2 intervals were used as input parameters for the induction of reentrant activity. This protocol allowed the determination of the "vulnerable domain" of the preparation by plotting the incidence of single responses and multiple or sustained reentry as a function of both S_1–S_2 interval and S_2 intensity (see fig. 5 below). Stimulation was immediately stopped when sustained vortex-like reentry was established.

2.7.2. Annihilation and reinitiation of reentry
In some cases, self-sustaining activity was terminated by electrical stimulation. Moreover, after determining the range of parameters which re-

sulted in the induction of sustained tachycardia, the same procedures were repeated using a different configuration of S_1 and S_2 stimulation, in an effort to determine the role of fiber orientation in the establishment of the arrhythmia.

3. Results

3.1. Isopotential map of a quasi-planar wave

In previous attempts to map electrical activity in heart tissue with optical monitors [14–17], investigators have been unable to overcome some serious limitations of the technique, including a poor signal-to-noise ratio and a large mechanical artifact, as well as an uncertain correlation between the amplitude of the fluorescence signal and the actual voltage. Most of these problems have been recently overcome in our laboratory by the use of the highly sensitive dye di-4-ANEPPS [7], and by constantly superfusing the tissue with diacetyl monoxime (DAM). DAM has been shown to completely suppress contractibility while producing negligible effects on the electrical activity [18]. In addition, close inspection of the fluorescence signals generated by the voltage-sensitive dye reveals a remarkable similarity with transmembrane potentials recorded with microelectrodes (fig. 3A). One should consider, however,

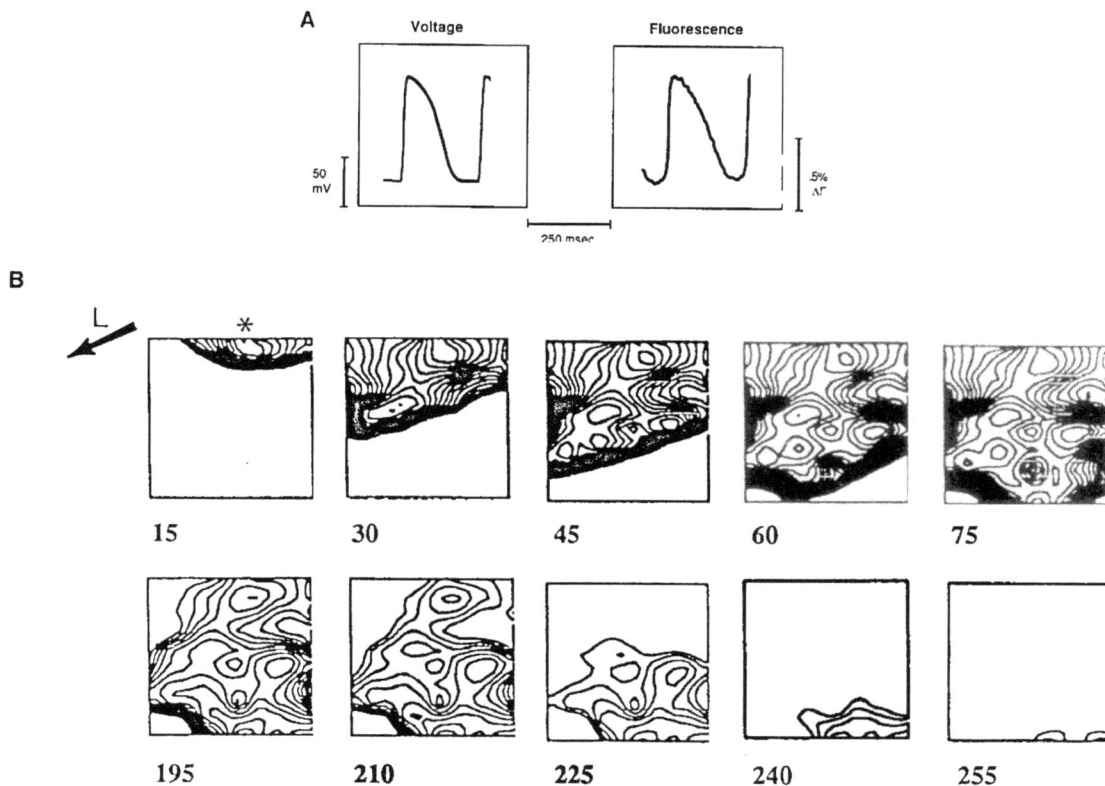

Fig. 3. (A) Transmembrane activity recorded with an intracellular microelectrode (left) and fluorescence changes detected by one photodiode (right). Both recordings were obtained from the same region of the preparation. ΔF is the fluorescence change from background fluorescence. (B) Isofluorescence (equivalent to isopotential) maps of activation (upper frames) and repolarization (lower frames) during one basic response. Numbers at the bottom of each frame represent time in milliseconds. To improve visualization, isopotential lines were intentionally displayed at small intervals. Only the higher 70% of the potential values were differentiated from the background (resting) fluorescence.

that each photodiode receives fluorescence from an area of 0.12–1.96 mm², so that the signal generated represents the activity of approximately 20–150 cells. Under these conditions, the fluorescent "action potential" is expected to be somewhat distorted. Indeed, as shown in fig. 3A, a comparison of the duration of the optical signal obtained with a typical magnification of 4× (corresponding to an area of 0.12 mm² monitored by a single photodiode) during normal propagation, with that recorded from the same region of the preparation with an intracellular microelectrode, demonstrates that the "optical action potential" may be 10–25% longer than the action potential obtained with microelectrodes and may have a lower upstroke velocity.

Thus, the use of voltage-sensitive dyes and optical mapping with a photodiode array provides excellent temporal resolution and gives direct access to the analysis of the electrical events occurring in the tissue throughout the entire excitation–recovery cycle. Moreover, the spatial resolution is also very high as illustrated in fig. 3B, which shows a series of isopotential maps constructed to determine the course of a quasi-planar propagating wave. The impulse was initiated by a single extracellular electrical stimulus (asterisk) applied to a thinly sliced piece (21 × 17 × 0.5 mm) of sheep epicardial muscle. The upper frames show the excitation and progression of the wavefront toward the bottom margin of the tissue. Conduction is faster in the direction parallel to the longitudinal axis of the fibers (arrow labelled L). After excitation of the upper two thirds of the tissue, the quasiplanar wavefront proceeds more or less diagonally toward the bottom and excites the entire preparation within 75 ms. The lower frames show the recovery process. Clearly, although repolarization follows the activation front, its time course is nonuniform and lasts 255 ms. This pattern of relatively uniform activation followed by nonuniform repolarization was observed in all eight preparations studied and could not be attributed to inhomogeneities in terms of morphology or fiber orientation. In three

of such preparations histological inspection revealed normal undamaged myocardial tissue within 500 μm from the epicardial surface.

3.2. Theoretical bases of reentry initiation

The idea to create a model of spiral wave reentry in small pieces of normal cardiac tissue was based on theoretical concepts and computer simulations [2–4, 19], as well as on a recently reported in vivo model of reentrant tachycardia in normal hearts [5, 6]. Fig. 4 shows the theoretical basis of the model. A planar wave of excitation is intersected by a voltage gradient which directly excites part of the tissue. Actual propagation of action potential starts at the region directly affected by the diagonal voltage gradient from tissue that is distal to the S_2 electrode. This new excitatory wave can only propagate from the directly excited region (which is totally refractory)

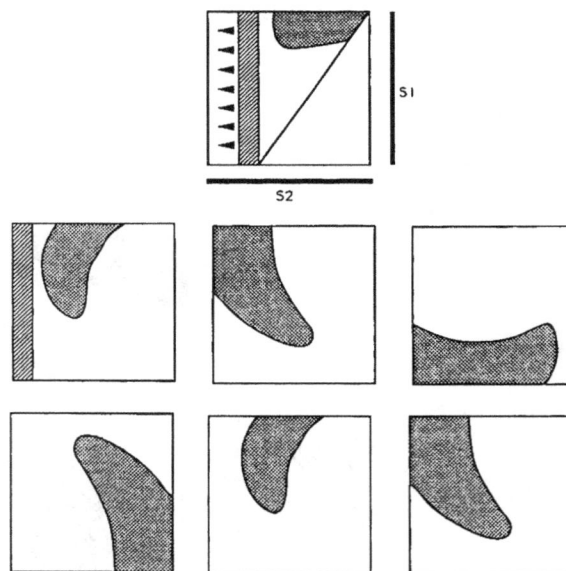

Fig. 4. Schematic illustrating the hypothesis for the induction of spiral waves in isolated cardiac tissue. Arrows represent the direction of a planar wave front (diagonally shaded region). Crosshatched region represent the rotating process (i.e., central part of a spiral wave). Black bars represent the electrodes utilized for basic (S_1) and premature (S_2) stimulation. The diagonal line represent the border of the region which was directly excited by S_2 simulation.

towards the tail of the previous basic wave. The presence of tissue in different recovery states results in curling of the excitatory wave and the generation of a self-sustaining reentrant activity.

Thus, the theory predicts that an appropriate stimulus gradient should result in the establishment of a spiral wave in normal myocardium. In addition, successful initiation of a vortex would occur only when the voltage gradient created by S_2 falls within certain window of vulnerability of the cardiac cycle which in fact corresponds to a "vulnerable domain" with low and high time limits as well as low and high limits of stimulus intensity [2]. Hence, a critical voltage gradient of proper timing must exist which always results in sustained vortex-like reentry. Such a critical should be bounded by regions of lesser or higher gradients which give rise to non-sustained reentry or no reentry at all.

3.3. The vulnerable domain

Some support for the critical voltage gradient hypothesis has been obtained in multiple-electrode mapping experiments in the open-chested dog. After elaborate measurements and modelling derived from the mapping experiments in the whole heart, Frazier et al. [6] estimated the critical voltage gradient for sustained reentry to be about 5 V/cm for S_2 stimuli occurring approximately at the end of the effective refractory period, which corresponds closely to theoretical prediction [2]. We have not yet undertaken a quantitative study of the critical voltage gradient for sustained vortex-like reentry in our isolated preparations. However, we have roughly estimated the vulnerable domain in some preparations by utilizing an S_1–S_2 protocol and simply measuring the S_1–S_2 interval and S_2 voltage range which results in nonsustained or sustained reentry. A representative experiment is illustrated in fig. 5. The preparation was driven with trains of ten basic stimuli (S_1) at a BCL of 430 ms, applied through electrode pair A (see fig. 1). S_2 stimuli were delivered after every 10th S_1 response

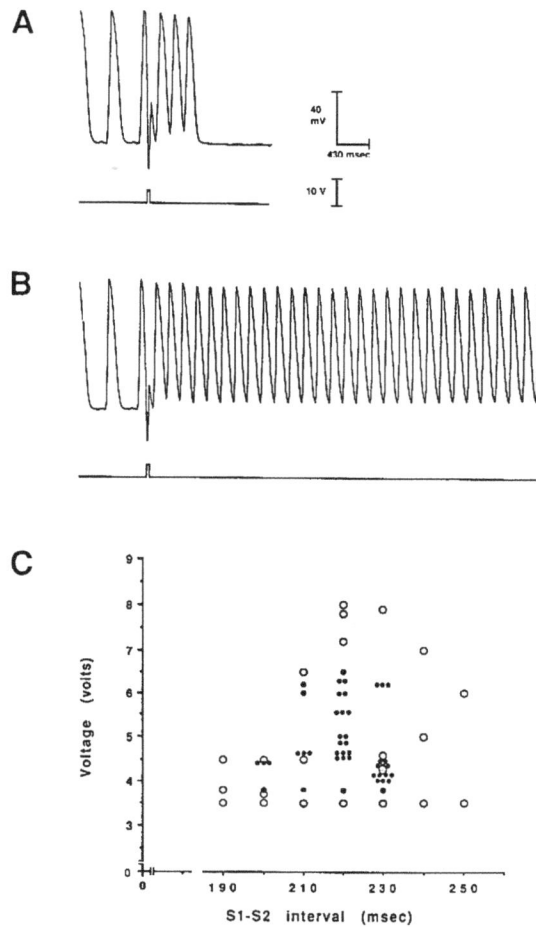

Fig. 5. Determination of the vulnerable domain. (A, B) Intracellular potential (top trace) and current amplitude for premature stimulation (bottom trace) recorded during the induction of nonsustained (A) and sustained reentry (B). In both cases, premature stimulation was delivered at an S_1–S_2 interval of 230 ms. The intensity of S_2 was 4.5 V in (A) and 4.3 V in (B). (C) Plot showing the incidence of repetitive activity at different combinations of S_2 intensities and S_1–S_2 intervals. Open circles represent the absence of reentrant activity. Dots represent nonsustained reentrant events. Stippled circle represents sustained reentry.

through electrodes B(+) and D(−) which were perpendicular to the S_1 electrode pair. The S_1–S_2 interval and S_2 intensity were changed systematically over wide ranges to determine the vulnerable domain. Panels A and B show typical microelectrode recordings (top tracings) and S_2 stimulus intensities (bottom tracings) which re-

sulted in nonsustained (panel A) and sustained
(panel B) repetitive activity at a period of 177 ms
(frequency = 5.6 Hz). In panel C, we have plotted
the incidence of single premature responses (open
circles), nonsustained bursts of reentry (black
dots) and sustained reentrant activity (stippled
circle) as a map of voltage versus S_1–S_2 interval.
Clearly, in this experiment there was a single
S_1–S_2 interval at 230 ms in which an S_2 stimulus
of 2 ms and 4.25 V resulted in sustained repetitive activity that lasted indefinitely. At higher and
lower currents, nonsustained repetitive dis-
charges occur but only within a narrow window of
S_1–S_2 intervals and voltages. Such a window is

surrounded by early and late S_1–S_2 intervals and
high as well as low voltages which do not result in
reentry. These results are similar to those of the
experiments of Frazier et al. [6] in the whole
heart.

3.4. Mapping of sustained reentry

The complete activation–recovery process associated with an already established sustained
reentry is readily visualized in the isopotential
maps presented in fig. 6. The vortex-like activity
was induced using the S_1–S_2 protocol depicted in
fig. 4. The numbers below each frame are in ms

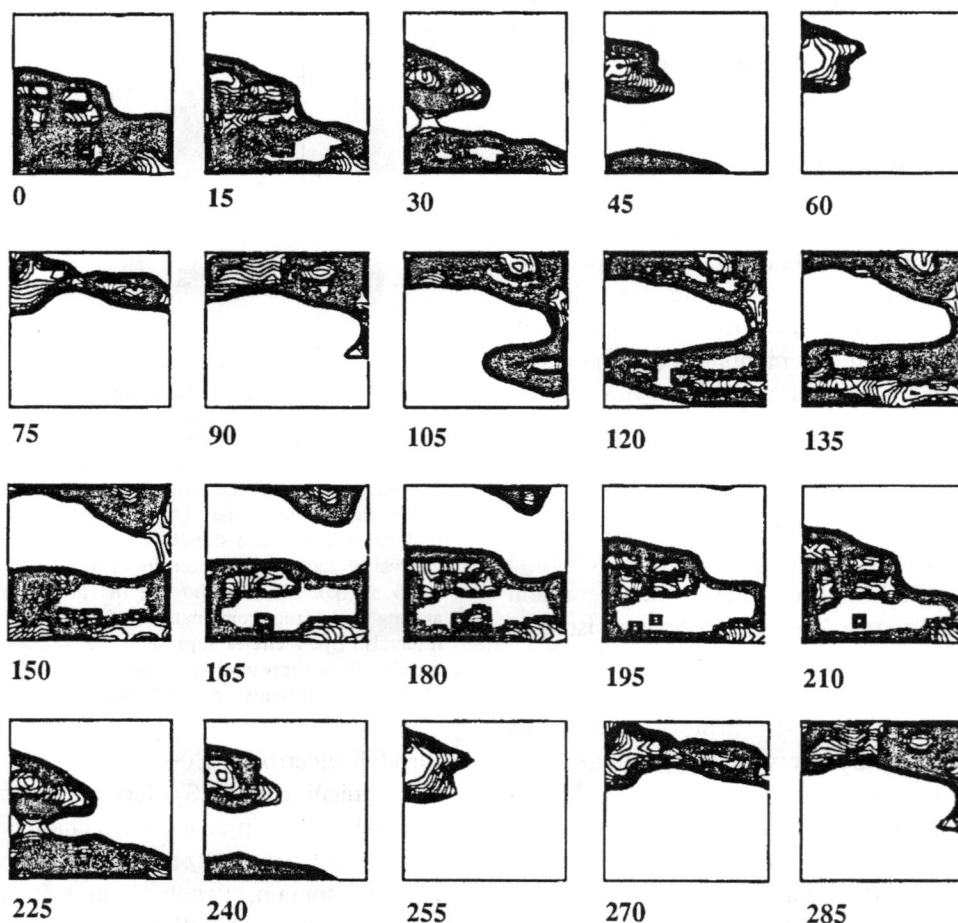

Fig. 6. Isopotential maps obtained during clockwise self-sustaining reentrant activity.

and indicate the time from the first frame (zero time is arbitrary). The excitation wave rotated in a clockwise direction for an indefinite number of cycles around an elongated central core of low voltage with apparent subthreshold activity (< 30 mV). The main axis of this region is parallel to the fiber orientation. Snapshots were taken at equally spaced intervals of 15 ms and they represent approximately one and one quarter turns of a rotor with a cycle length of about 205 ms. Note that although the activation sequence is remarkably constant from one cycle to the next, within a given cycle, the speed of propagation is highly nonuniform, being faster in the longitudinal (horizontal) than in the transverse direction of the fibers. Moreover, a rough estimate of ~ 80 ms for the local excitable gap (i.e., the time interval during which a given part of the tissue is excitable) may be obtained by measuring the time to activation of the upper right corner during the first 7 snapshots. The results show quite clearly vortex-like activity in this small piece of isolated anisotropic myocardium. Moreover, as the electrical wave rotates over and over through the same tissue, it gives the impression that if the preparation were perhaps twice as large, one could observe the activity folding into itself to form a spiral wave.

3.5. Termination of sustained reentry

Sustained vortex-like reentry may be annihilated by a single properly timed electrical stimulus. Fig. 7 shows an example in which a rotating activity is interrupted by electrical stimulation. The first 8 frames show a complete revolution of the vortex (revolution time 210 ms). An electrical stimulus was applied at time 276 ms (frame No. 10). The impulse was delivered to the entire left border of the preparation. However, only a small region at the bottom of this border (excitable gap) was excited. As a result of this, a new excitatory wave was created, which propagated in a counterclockwise manner. A few milliseconds after the stimulus was delivered (time 300 ms)

both waves collided, which terminated the reentrant activity.

3.6. Clockwise and counterclockwise vortices

Uniform and nonuniform anisotropy may certainly play a role in the initiation, location and size of the rotor, as well as in the revolution time and direction of rotation. However, all these aspects may also depend on the arrangement or configuration of the stimulating electrodes. In fact, we were able to modify both the rotation of the vortex and the location of the center of the rotor (core), by changing the position of the stimulating electrodes. The three-dimensional fluorescence maps presented in fig. 8 show the sequence of activation during two episodes of reentrant excitation generated in the same preparation. In these plots, the X and Y axes represent the position of the wavefront and the Z axis represents the change in fluorescence from rest. In both cases, basic stimulation applied from electrode S_1 generated a leftward moving wavefront (not shown), which was perpendicular to the main axis of the fiber (arrow labelled "L"). In fig. 8a, a premature pulse (S_2), applied to the bottom of the preparation, resulted in the formation of a vortex with counterclockwise rotation. After interrupting this circulation, it was possible to initiate a different arrhythmia by now applying S_2 to the top portion of the tissue (fig. 8b). In this case, the rotation of the vortex was clockwise and the region of low voltage activity was larger. Although conduction velocity was nonuniform in either case, the spatial distribution of conduction velocities was different in both. The total revolution time increased from 170 ms in panel a to 190 ms in panel b. The differences between both episodes are also appreciated in the rotation and main axis of the orthogonal vectogram displayed in panel c. These results demonstrate that, although tissue anisotropy modifies the characteristics of the reentrant process, it is not essential for the initiation or maintenance of the rotating excitation.

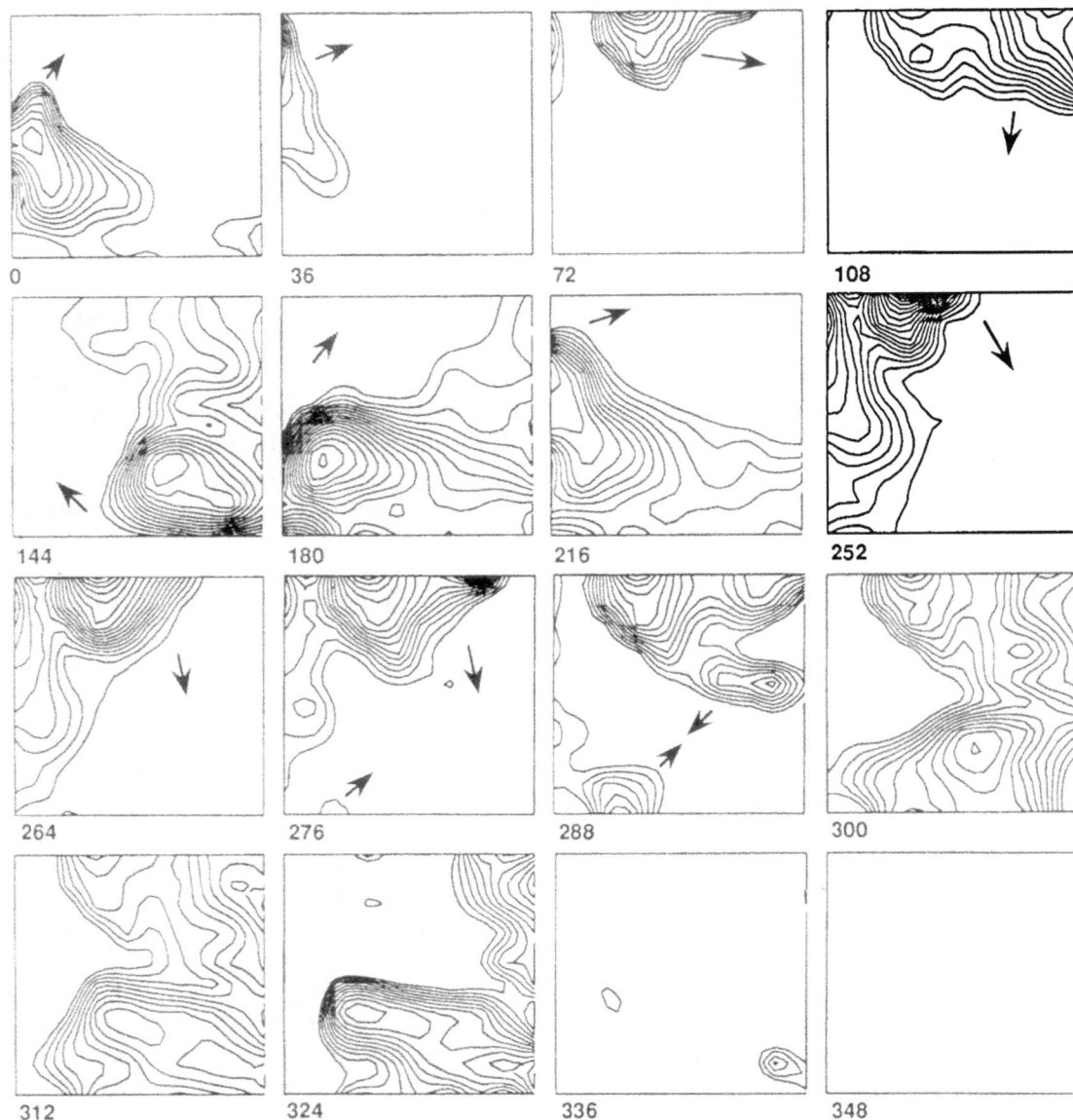

Fig. 7. Isopotential maps obtained during annihilation of the rotating activity. An electrical impulse applied at time 276 ms creates a counterrotating wave which collides with the reentrant activity, thus terminating the arrhythmia.

4. Discussion

4.1. The optical setup

Most experimental studies on two-dimensional reentry have been based on multiple extracellular recording techniques to analyze the initiation of the arrhythmia, as well as the sequence of wavefronts and the dynamical changes of the region of block or slow conduction [6, 9, 12]. Extracellular electrodes are useful for detecting with limited accuracy the activation time of the area underneath the electrodes. Yet, in spite of the large spatial resolution offered by these techniques, they only provide information on the "head" of the activation front. Intracellular microelectrode techniques, on the other hand, provide more detailed and direct information but, unfortunately, only a few simultaneous recordings may be obtained [18]. The use of monophasic action potential recordings is also limited by the damage and electrophysiological changes induced by the suction electrodes.

Recently, high-resolution optical methods using voltage-sensitive dyes have been introduced for simultaneous recording of membrane potentials from multiple sites in a variety of excitable tissues, including the heart [14, 20]. For example, Morad and Salama [14] showed that the amplitude of an optical signal can be linearly related to membrane potential and that such a signal is equivalent in time course to an intracellular recording. Only two studies have focused on the

optical mapping of reentrant activity [17, 21]. In both cases, reentry was obtained in frog or guinea pig atrial tissue. One of the major technical difficulties for optical recording from the heart is the mechanical artefact. Investigators have attempted to overcome such a limitation, by sandwiching the heart between two glass windows and pressing Lucite pads against the two other sides of the heart [15]. In other cases the preparations were superfused with calcium-free solutions [17]. We believe that the introduction of diacetyl monoxime to the optical mapping technique represents a major advancement in terms of suppression of motion artefact and maintenance of normal electrophysiological properties. Clearly, in our experiments fluorescence changes associated with local activity give an excellent representation of the voltage defection associated with that activity (fig. 3A). Thus, mapping of the electrical activ-

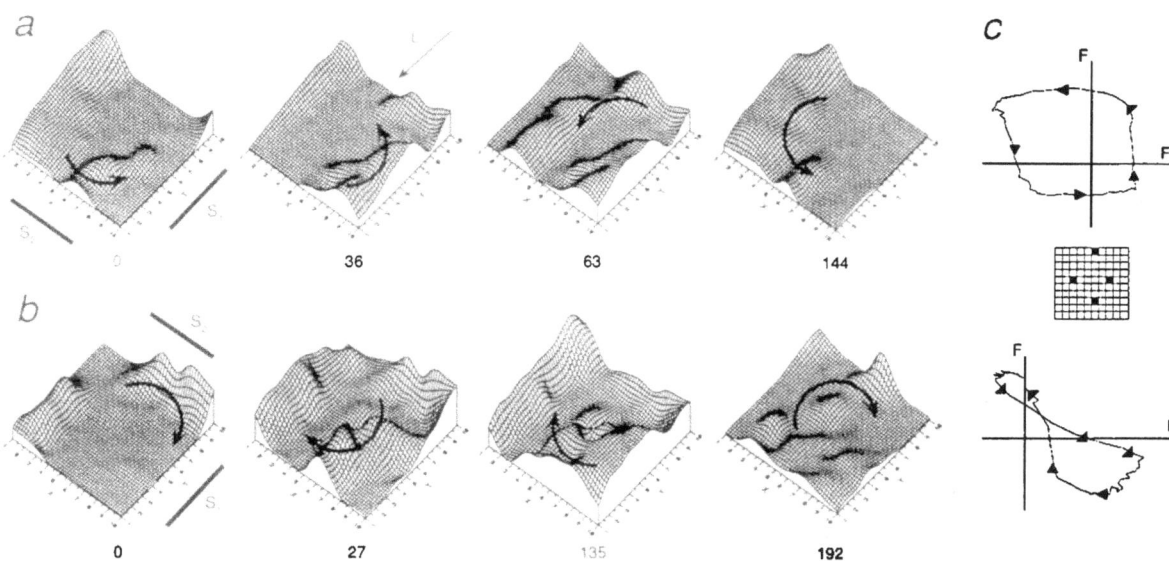

Fig. 8. Two different types of reentrant activity generated in the same preparation. (a) Anticlockwise rotating reentry: the arrow labelled L on the top indicates the longitudinal axis of the fibers. Numbers on the bottom of each frame indicate time in ms from the first frame. The activity was initiated by applying basic stimulation (S_1) to the right border and a single premature stimulus (S_2) to the bottom border. An elongated region of low-voltage activity is apparent from X 7–8, Y 9 to X 5–6, Y 5. Total rotation time 170 ms. (b) Clockwise rotating reentry initiated by applying S_1 to the right border and a single S_2 to the top border. The region of low-voltage activity extends over a larger area (X 4–7, Y 9 to X 4–6, Y 5). Total rotation time 195 ms. (c) Orthogonal fluorescence vectograms obtained by plotting the difference through one reentrant cycle between two points on the X axis versus that between two points on the Y axis, as indicated in the schematic of the diode array (center). The rotation and main axis of the loops dramatically changed from one episode to the other.

ity is achieved with high temporal and spatial resolution from 100 points representing the optical field detected by the 10×10 photodiode array.

4.2. The mechanism of reentry

Reentrant excitation in cardiac muscle has been the subject of intense investigation for more than 75 years. Ever since the classical studies of Mines [22], in which reentry occurs around an anatomical obstacle, through the recent demonstration [9] that reentry may be maintained even in the absence of anatomical obstacle, to the more recently postulated anisotropic reentry [12], most mechanisms proposed by experimental electrophysiologists have postulated that preexisting functional or anatomical inhomogeneities in the myocardium are essential for the establishment of reentrant rhythm disturbances. The present study focuses on voltage-like reentry in normal myocardium generated in the absence of anatomical obstacle.

Several mechanisms have been proposed to explain vortex-like reentry in two-dimensions. In 1977, Allessie et al. [9] demonstrated that the vortex of electrical activity which may be induced in a small piece of rabbit atrial tissue consists of normally excitable fibers whose resting potential is decreased during the arrhythmia, but show perfectly normal action potentials during stimulation at low frequency.

Traditionally, it has been assumed that unidirectional block and reentry require the presence of nonuniform dispersion of refractoriness [11, 23, 24] resulting from intrinsic differences in refractory periods of neighboring cells. Accordingly, a premature impulse should be blocked in that area in which refractoriness is greatest, and reentry should occur if tissue proximal to the region of block has recovered sufficiently to allow reactivation. Although the premature impulse may enhance the inhomogeneity, dispersion of recovery has usually been assumed to be an intrinsic property of the tissue involved. This hypothesis has

been supported by a large body of circumstantial experimental evidence. For example, ventricular fibrillation threshold decreases when dispersion of refractoriness is increased by premature stimulation, adrenaline, sympathetic stimulation, ischemia or drugs [11, 23–25].

Alternatively, slow conduction and block have been explained in terms of nonuniform anisotropic propagation [12, 26] or a combination of anisotropic propagation and nonuniform dispersion of repolarization [13]. A premature beat may be blocked in the direction parallel to the long axis of the fibers, which is supposed to have a low safety factor for propagation [13], and conduct slowly in the direction perpendicular to the long axis of the fibers. Recently, however, these concepts have been challenged by studies demonstrating that premature beats are more prone to block in the transverse direction [27, 28]. Nevertheless, there are a number of studies that support the participation of anisotropy in the development of reentrant tachycardia. Preeminent among those are the experiments of Wit et al. [10], who have shown that the line of block during the first beat of a reentrant tachycardia occurs perpendicular to the fiber orientation (block of longitudinal propagation). On the other hand, once the arrhythmia has been established, the line of block is oriented parallel to the long axis of the fibers. As yet, no mechanistic explanation of these interesting but puzzling results has been put forth.

Yet another possibility is that reentry is analogous to spiral waves in other excitable media. Vortex-like activity has been known to occur in excitable media for more than 20 years by chemists, applied mathematicians and theoretical biologists. The present study focuses on some testable predictions originated by the analogy of the normal myocardium to generic excitable media in which spiral wave formation has been extensively studied. On the other hand, since the ultimate goal of most previous electrophysiological studies was to determine the mechanisms of reentrant arrhythmias occurring in diseased hearts, it makes perfect sense that they were based on experimental models in which nonuniformity of the electrical properties of the tissue was enhanced either by ischemia or other means. However, as demonstrated by our experiments, vortices of electrical activity analogous to spiral waves in homogeneous excitable media may also be present in normal quasi-homogeneous ventricular muscle. Thus, it would seem that neither nonuniform dispersion of refractoriness nor anisotropy should be crucial for the induction or maintenance of reentry. In the case of our model, the premature stimulus itself may be sufficient to induce transient local inhomogeneity of refractoriness and unidirectional block and lead to the initiating of reentry. This idea is supported by the observation that both the direction of the rotation and location of the core were different in a given preparation when the parameters of S_1-S_2 stimulation were changed (see fig. 8).

4.3. Spiral waves in isolated cardiac tissue

Rotating excitation could be initiated in several ways in two-dimensional computer models of excitable media in which nonuniform dispersion of refractoriness or anisotropy were not included [4, 19]. More recently, realistic membrane models, in which several ionic currents are incorporated to describe the electrical activity of the cardiac cell, have been shown to support spiral waves at a period of about 180 ms [29]. It is interesting that such period is within the range (140–210 ms) observed in our experimental preparations. In this regard, the myocardium may be considered to be a typical excitable medium, in which vortex-like reentry may be initiated by the impact of the geometrically graded recovery of the medium to a previous excitation (i.e., the "tail" of the wavefront) with a geometrically graded transverse premature stimulus. On the basis of his pinwheel experiment, Winfree [2] made some testable predictions derived from this model: (i) Reentrant activity may arise in normal homogeneous tissue. (ii) The location of the center of the vortex (core) depends on the intensity of the premature impulse as well as its timing. (iii) The vulnerable period is in fact a vulnerable domain with low and high time limits as well as with low and high limits of stimulus intensity. Stimuli of exceedingly high intensity may produce the intersection of the critical voltage gradient with the critical recovery at a point close or beyond the border of the tissue. On the other hand, stimuli of exceedingly low intensity may be insufficient to generate the critical gradient. (iv) Appropriate stimulation conditions may result in a pair of mirror-image rotors which are similar to the figure of 8 reentry described in whole heart experiments [30]. In this theory, dispersion of stimulus strength is equally as important as dispersion of recovery in the creation of reentry by suprathreshold critical stimulation [2].

· Some of these concepts have been supported by an in vivo model of electrically induced sustained ventricular tachycardia in which multiple extracellular recordings were used [5]. Moreover, Frazier et al. [6] observed that the lines of block in their experimental model of reentry are not related to fiber orientation or to transmural rotation of fibers from endocardium to epicardium. However, adequately detailed comparison of the theoretical models with intracellular recordings from actual cardiac tissue still remained a technical challenge. The results presented here represent the first step toward achieving that goal.

In our opinion, the following observations strongly argue in favor of the hypothesis that reentrant excitation in isolated cardiac tissue is analogous to spiral waves in other excitable media: (i) Sustained reentry could be consistently generated in small two-dimensional pieces of normal ventricular myocardium. The characteristics of the stimulus parameters used to initiate the activity were derived from principles which govern spiral waves formation. (ii) Reentrant activity seemed to revolve around an elongated organizing center (i.e., a rotor) which seems to be smaller than previously thought and is consistent with the concept of "phase singularity" [1, 2]. (iii) The periodicity, direction of rotation and location of the core during vortex-like activity in the isolated tissue could be modified at will by changing the arrangement of stimulating electrodes responsible for the initiation of the self-sustaining activity. (iv) Reentry occurred only under the conditions in which the S_2 stimulus fell within a certain vulnerable domain (not just an interval). Such a domain is consistent with theoretical predictions and experimental results in the whole heart. (v) Finally, analogous to what has been observed for mutually annihilating waves in chemical excitable media [1], application of an appropriately timed stimulus during the vortex-like activity produced a counterrotating wave which collided head-on and annihilated the reentrant wavefront.

However, before interpreting our results, one should consider that, unlike other excitable media, normal cardiac tissue is by no means homogeneous. It is indeed possible that some degree of dispersion of refractoriness might have been responsible, or at least contributed to the initiation of the reentrant activity. The latter possibility will be very difficult to rule out in the light of the experiment demonstrating that, even in the presence of a planar wavefront, the repolarization process which follows that front is nonuniform (see fig. 3B). Thus, although our experiments do support the idea that spiral wave activity may occur in cardiac tissue, some of the concepts derived from more traditional electrophysiological studies for the understanding of reentrant cardiac activity have not been ruled out and should be taken into consideration for the analysis of the arrhythmia.

Acknowledgements

We acknowledge the invaluable collaboration of Dr. Donald C. Michaels and Dr. Dante Chialvo. We thank Wanda Coombs and Joanne Getchonis for their skillful technical assistance and La Verne Gilbert for secretarial and administrative assistance. This work was supported in part by grants HL29439, HL40991 and HL39707 from N.I.H.

Glossary

Action potential: Transmembrane voltage change occurring in an excitable cell in response to a stimulus greater than a given threshold.

Arrhythmia: Absence of normal heart rhythm. The term arrhythmia is generally used for high or low rate rhythm whether regular or irregular.

Cardiac anisotropy: Specific orientation of cardiac cells which results from their rod-like elongated shape. Cardiac anisotropy is the basis for directional differences in transverse versus longitudinal propagation velocity of the electrical impulse.

Diastolic threshold: Minimum current intensity necessary to generate an action potential.

Electromechanical uncoupler: Agent that suppresses contractility without affecting the electrical activity of the cells.

Epicardial: Refers to the cardiac muscle comprising the external surface of the heart.

Excitable gap: Time interval during which a given portion of a reentrant circuit remains excitable. It is calculated as the difference between the rotation period and the refractory period.

Lines of apparent block: Elongated epicardial regions of very slow conduction velocity recorded during reentrant activity by means of extracellular electrodes.

Premature stimulus: Electrical stimulus applied at an interstimulus interval which is briefer than that of previous stimuli.

Reentry: According to Winfree's definition [2] "reentry means reactivation of a patch of tissue by an action potential that returns without ever dying out."

Refractoriness: Time interval following excitation during which the cell remains inexcitable.

Superfusion: Exposure of the outer surface of a biological preparation to a running solution.

Ventricular tachycardia: Particular form of high rate arrhythmia originated in the ventricles. It may be quite periodic.

Unidirectional block: Situation in which an area of tissue allows propagation of the electrical impulse in one direction only.

References

[1] A.T. Winfree, When Time Breaks Down, The Three-Dimensional Dynamics of Electrochemical Waves and Cardiac Arrhythmias (Princeton Univ. Press, Princeton, NJ, 1987).

[2] A.T. Winfree, J. Theor. Biol. 138 (1989) 353.

[3] A.M. Pertsov and V.G. Fast, Kardiologia N 5 (1987) 75.

[4] V.S. Zykov, in: Simulation of Wave Processes in Excitable Media, ed. A.T. Winfree (Manchester Univ. Press, Manchester, New York, 1987) p. 7.

[5] N. Shibata, P. Chen, E.G. Dixon, P.D. Wolf, N.D. Danieley, W.M. Smith and R.E. Ideker. Am. J. Physiol. 255 (1988) H891.

[6] D.W. Frazier, P.D. Wolf, J.M. Wharton, A.S.L. Tabg, W.M. Smith and R.E. Ideker, J. Clin. Invest. 83 (1989) 1039.

[7] J.M. Davidenko, P.F. Kent, D.R. Chialvo, D.C. Michaels and J. Jalife, Proc. Natl. Acad. Sci. USA 87 (1990) 8785.

[8] M.A. Allessie, F.I.M. Bonke and F.J.C. Schopman, Circ. Res. 39 (1976) 168.

[9] M.A. Allessie, F.I.M. Bonke and F.J.G. Schopman, Circ. Res. 41 (1977) 9.

[10] A.L. Wit, S.M. Dillon, J. Coromilas, E.A. Saltman and B. Waldecker, in: Mathematical Approaches to Cardiac Arrhythmias, ed. J. Jalife. (Ann. NY Acad. Sci., New York, 1990), in press.

[11] J. Han and G.K. Moe, Circ. Res. 14 (1964) 44.

[12] S.M. Dillon, M.A. Allessie, P.C. Ursell and A.L. Wit, Circ. Res. 63 (1988) 182.

[13] M.S. Spach, P.C. Dolber and F.J. Heidlage, Circ. Res. 65 (1989) 1612.

[14] M. Morad and G. Salama, J. Physiol. (London) 292 (1971) 167.

[15] G. Salama, R. Lombardy and J. Elson, Am. J. Physiol. 252 (1987) H384.

[16] G. Salama and M. Morad, Science 191 (1976) 485.

[17] B.C. Hill and K.R. Courtney, Ann. Biomed. Engin. 15 (1987) 567.

[18] T. Li, N. Sperelakis, R.E. Teneick and J.R. Solaro, J. Pharmacol. Exp. Theor. 232 (1985) 688.

[19] F.J.L. van Capelle and D. Durrer, Circ. Res. 47 (1980) 454.

[20] L.B. Cohen, and B.M. Salzberg, Rev. Physiol. Biochem. Pharmacol. 85 (1978) 33.

[21] T. Sawanoboti, Y. Hirano, A. Hirota and S. Fujii, Am. J. Physiol. 247 (1984) H185.

[22] G.R. Mines, J. Physiol. 46 (1913) 349.

[23] E.N. Moore, J.F. Spear and L.N. Horowitz, Am. J. Cardiol. 32 (1973) 814.

[24] C.S. Kuo, K. Munakata, C.P. Reddy and B. Surawicz, Circulation 67 (1983) 1356.

[25] B.F. Hoffman, P.F. Siebens, P.F. Cranefield and C.M. Brooks, Circ. Res. 3 (1955) 140.

[26] M.J. Schalij, M.A. Allessie, W.J. Lammers and F.V. Kaam, Circulation (Abstr) 76 (1987) IV-113.

[27] C. Delgado, B. Steinhouse, M. Delmar, D. Chialvo and J. Jalife, Circ. Res. 67 (1990) 97.

[28] C.W. Balke, M.D. Lesh, J.F. Spear, A. Kadish, J.H. Levine and E.N. Moore, Circ. Res. 63 (1988) 879.

[29] M. Coutermanche and A.T. Winfree, in: Science at the John von Newmann National Supercomputer Center, Vol. 3, ed. G. Cook (Consortium for Scientific Computing, Princeton, NJ, 1990) p. 79.

[30] N. El-Sherif, R.A. Smith and K. Evans, Circ. Res. 49 (1981) 255.

Physica D 49 (1991) 198–213
North-Holland

Spatial long-range interactions in squid giant axons

Yoshiro Hanyu[1] and Gen Matsumoto[2]

Electrotechnical Laboratory, Molecular and Cellular Neuroscience Section, Tsukuba City, Ibaraki 305, Japan

Membrane states and their bifurcation characteristics are studied for the squid giant axon as a function of external Ca^{2+} concentration, temperature and externally applied current step. It is well known that the membrane states are divided into two states, resting (R) and spontaneous oscillation (O), according to the Ca^{2+} concentration contained in the solution surrounding the axon. The present experiments further clarified that each of these states was further subdivided into higher (H) and lower (L) temperature phases, according to temperature. The spatially unclamped axon in the higher-temperature phase, either in the R or O state, can bifurcate to produce limit-cycle oscillations of action potentials. The bifurcation parameters are external Ca^{2+} concentrations and externally applied current for the axon in the O and R states, respectively. Both the axon in the lower-temperature phase and the spatially clamped axon in the higher-temperature phase can bifurcate to produce intermittent oscillations of action potentials. The bifurcation characteristics at or between the higher- and lower-temperature phases are closely related to the spatial properties of the preoscillatory fluctuations along the axon, suggesting that a particular spatial interaction is responsible for the periodically oscillatory dynamics and the bifurcation to it. The molecular origin of the spatial interaction possibly originates from the specific distribution of Na channels, which may be regulated by subaxolemmal cytoskeletons. Electron microscopic experiments and other evidence to support this idea are described.

1. Introduction

Nonlinear and nonequilibrium systems show a rich diversity of spatiotemporal structures. As a system moves far away from equilibrium, it loses stability and transits to an oscillating state at a critical bifurcation parameter. Several kinds of bifurcation types, including Hopf bifurcation, are observed in physical, chemical and biological systems such as fluid systems [1, 2], laser systems [3], chemical reaction systems [4–7] and biological systems [8–13]. However, characteristic properties in the subcritical transition region of the Hopf bifurcation have not been well studied, except in electrical systems, because of its poor sensitivity and resolution. Temporal properties of the fluctuation of electrical circuits were studied and analyzed in detail for their advantages in terms of electrical phenomena [14, 15].

We have shown that the squid giant axon is an excellent system for the study of the Hopf bifurcation because of its advantage in that the activity and its associated fluctuations are electric with orders of 10^{-5}–10^{-1} V in amplitude and of 10^{-4}–10^{-2} s in time [16, 17]. Thus far, the squid giant axon is the only system in which we can study the spatiotemporal fluctuations in the vicinity of and associated with the Hopf bifurcation. From these studies, we have demonstrated that nerve excitation can be grasped as a transition, assisted by externally applied current, from a state with an asymmetrically stable fixed point to a limit-cycle state [17]. This model was proposed on the basis of the finding that the spatial long-range interaction was crucial for generation of action potentials, in sharp contrast with the present view that nerve excitation takes place as a result of the voltage-dependent opening and closing of totally independent Na and K channels embedded in membranes [18].

The present experiments aim at obtaining further insights into the spatial long-range interac-

[1]Present address: Department of Physics, Faculty of Science and Technology, Keio University, 3-14-1 Hiyoshi, Kohoku-ku, Yokohama City, Kanagawa 223, Japan.
[2]To whom correspondence should be addressed.

tion among Na channels and at providing some reasonable explanation for the discrepancy between our previous model of nerve excitation and the conventional view.

2. Materials and methods

2.1. Materials

Giant axons of squid (*Doryteuthis bleekeri*) were used. The squid were collected in Sagami Bay, transported to the Electrotechnical Laboratory in Tsukuba City and maintained in a small, circular, closed-system aquarium until the experiments were performed. The giant axon, 400–600 μm in diameter and 30–180 mm in length, was carefully isolated under a dissecting microscope. The axon thus prepared was placed in a chamber filled with artificial seawater (ASW or 40 Ca–ASW) consisting of 460 mM NaCl, 10 mM KCl and 40 mM CaCl$_2$, buffered with 30 mM Tris.–HCl, pH and osmolarity being adjusted to 8.2 at 10°C and 980 mOsm at 37°C.

2.2. Experimental procedures

Membrane potentials were simultaneously measured at 2 to 4 locations along the axon by inserting internal glass-pipette electrodes of the Ag–AgCl type filled with 0.6 M KCl aqueous solution into both cut-open ends [9]. The same type of potential electrode was immersed in the external solution to measure the reference potential. The external solution was grounded through a platinized platinum plate immersed in the solution. Stimulating current was delivered to the axon through two types of internal platinized platinum wire electrodes (50 μm in diameter) classified by their conductive length. The length of one type (point-type electrode) was 0.4 mm to avoid spatial clamping of the axon, and the other type (spatial-clamp type) was over 3 mm, depending on the length required to clamp the axon spatially.

In order to characterize membrane states and the bifurcation from one state to another, we varied the bifurcation parameters: external Ca^{2+} concentration, temperature and externally applied current. External Ca^{2+} concentrations contained in ASW were changed while other components and factors in the solution, such as ionic species and their concentrations, osmolarity and pH, were kept constant. The osmolarity of 980 mOsm was maintained by adjusting the amount of Tris. The temperature of the external solution surrounding the axon was kept constant within ±0.1°C at any temperature from 1.0 to 25.0°C [19]. The step or pulse current was delivered through one of the two types of internal current electrodes electrically connected to a pulse generator (type SEN 7103, Nihon-Koden Co. Ltd.) through a 470 kΩ resistor. For experiments on spatially clamped (spatially homogeneous in the electric sense) axons, we used the spatial-clamp-type electrode. The length of the conductive portion of the platinized platinum wire was determined on the basis of how long the axon was spatially clamped. In some experiments, we inserted a platinized platinum wire over the axon to clamp the axon spatially, even when the current was not necessarily applied. For experiments on spatially unclamped axons, we used the point-type electrode for current delivery to approximately stimulate under the spatially unclamped condition.

2.3. Electron microscopic observation

For the electron microscopic observation of the subaxolemmal cytoskeleton, the axon was chemically prefixed by intracellularly perfusing it first with the standard internal solution 400 KF and then with the chemical fixative containing 3.0% glutaraldehyde and 0.1 M sodium cacodylate buffer (pH 7.4), the osmolarity being adjusted to 980 mOsm by sucrose. During this perfusion, the temperature surrounding the axon was kept constant, higher or lower than 10°C, according to the experimental purpose. The solution of 400 KF consisted of 400 mM KF and 20 mM Tris.–HCl buffer (pH 7.25), the osmolarity being adjusted by sucrose to 980 mOsm. The fixative was further perfused for 5 min after the axonal excitability was lost. The axon thus prefixed was then bathed in the fixative solution containing 3% glutaraldehyde, 0.5% tannic acid, 4 mM $CaCl_2$, 205 mM NaCl and 0.1 M sodium cacodylate buffer (pH 7.4) at a temperature above or below 10°C for 15 min. Fixation was further continued for 2 h at 4°C after the axon was cut into small pieces. After being rinsed with an aqueous solution of 205 mM NaCl, 4 mM $CaCl_2$ and 0.1 M cacodylate buffer, the specimen was postfixed with 1% OsO_4 containing 0.1 M phosphate buffer (pH 6.2), dehydrated in graded concentrations of ethanol, stained en bloc with 1% aqueous uranyl acetate for 2 hours at room temperature and embedded in Epon 812. Thin sections were prepared by cutting with a diamond knife, doubly stained with uranyl acetate and lead citrate, and examined under an electron microscope (type LEM 2000, Akashi Corp.) at an accelerating voltage of 100 kV.

3. Results

3.1. A diagram exhibiting membrane states as a function of external Ca^{2+} concentration and temperature

It is well known that the squid giant axon retains two stable states according to its environmental conditions such as external Ca^{2+} concentration [17, 20–23], internal pH [24] and externally applied steady current [25]; one stable state is the resting state with a stable focal point and the other one is the spontaneously oscillating state. In the present study, we have carefully characterized membrane states in greater detail as a function of external Ca^{2+} concentration ($[Ca^{2+}]_e$), temperature (T) and externally applied steady current. The result is summarized in fig. 1, where ASW and Ca^{2+}-reduced ASW were used

for external solutions, and the externally applied current was given through an internal platinized platinum electrode (the point-type electrode) with a conductive portion of 0.4 mm in length. It is confirmed in the present experiment that the resting (R) and spontaneously oscillating (O) states are located in higher and lower Ca^{2+} concentration regions, respectively. At the same time, it is seen in the figure that the transition boundary smoothly changes as a function of $[Ca^{2+}]_e$ and T; more Ca^{2+} ions are needed to realize the resting state at lower temperatures. The transition can be represented on the state diagram by the thick solid line T_1. Moreover, we have found that the two states, R and O, can be further classified into two states; the classification is determined by the response to the externally applied steady current in the case of the resting state and by the oscillating behavior in the case of the spontaneously oscillating state.

In the resting state, the spatially unclamped axon can generate only one or two action potentials followed by subthreshold potential oscillation at lower temperatures or stable long-lasting oscillations of action potentials at higher temperatures, when it is stimulated by a superthreshold outwardly flowing step current (fig. 2). The axon in the former state (R_L) made a transition to the latter one (R_{II}) as a function of $[Ca^{2+}]_e$ and T (fig. 1); that is, for the axon immersed in ASW containing 40 mM and 32 mM Ca^{2+}, it took place at 10 and 7°C, respectively, represented by solid circles on the straight line T_2 in the state diagram (fig. 1). It was observed that the transition took place in a very narrow region of the two parameters, $[Ca^{2+}]_e$ and T. The oscillation state R_{II} emerged after accompanying critical oscillation (record 2 in fig. 2B). The current-induced oscillation was stable; the time intervals between neighboring action potentials continued regularly (or in a periodic fashion): the amplitude of the action potential was unchanged from one to another; the oscillation lasted as long as the current was externally driven. It was experimentally proven that the oscillation lasted stably for the 5 min

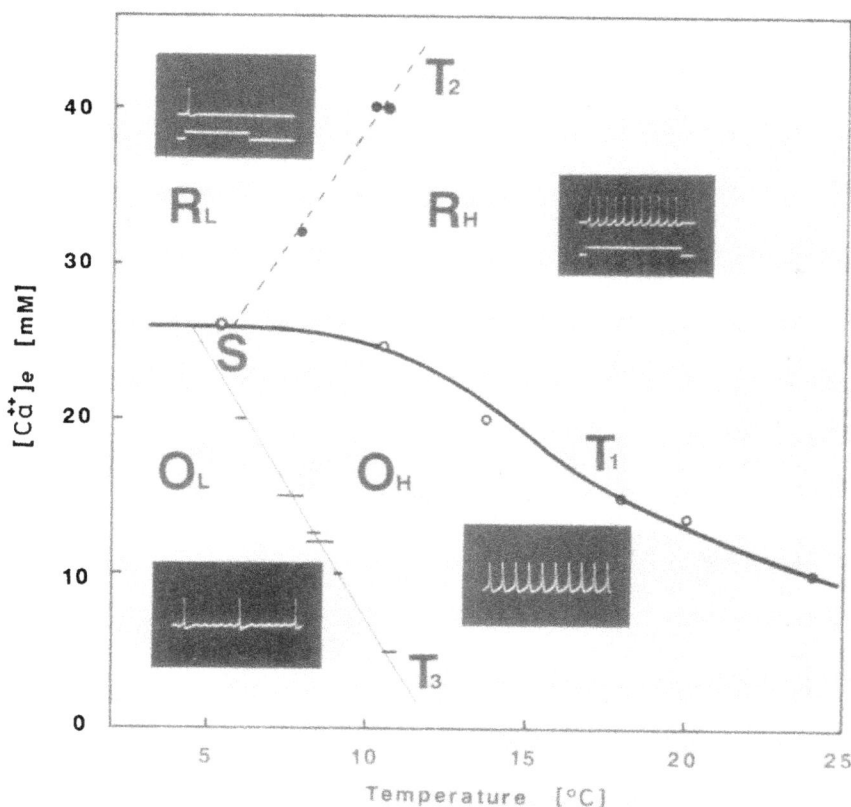

Fig. 1. Diagram illustrating two membrane states, the resting and spontaneous oscillation states (R and O, respectively), as a function of external Ca^{2+} concentration $[Ca^{2+}]_e$ and temperature. The transition boundary between these two states is shown as a thick solid curve (T_1). At the lower-temperature side, curve T_1 has a tendency to shift to higher $[Ca^{2+}]_e$, corresponding to the fact that the critical potential to induce the action potential becomes lower as the temperature drops. The two states, R and O, can be subdivided into higher- and lower-temperature phases. The resting substates, R_H and R_L, are classified according to the differences in their response characteristics to the external current stimulation. The spontaneous oscillation state is subdivided into O_H and O_L on the basis of characteristic oscillation behaviors: periodic in O_H and intermittent in O_L. The boundaries between R_H and R_L, and between O_H and O_L are illustrated as thin solid lines, T_2 and T_3, respectively. The lines of T_2 and T_3 almost intersect on curve T_1, forming a triple point S. The axons used in the present experiments were spatially unclamped (see text).

current step. These characteristics are seen in fig. 2B.

In the spontaneously oscillating state, the spatially unclamped axon generates periodically oscillating (limit-cycle) action potentials at higher temperatures, but its oscillation becomes intermittent at lower temperatures (fig. 1). State O, thus, can be divided into two substates: the higher-temperature phase O_H corresponds to the limit-cycle state and the lower-temperature phase O_L corresponds to the bursting or intermittently oscillating state. The transition is a function of $[Ca^{2+}]_e$ and T, represented by a straight line in

the state diagram, as illustrated by the thin solid line T_3 in fig. 1. It should be noted that the line T_3 approximately agrees with T_2 on the curve T_1, forming the triple point S of the transitions of the state diagram (fig. 1). The transition behavior was observed by varying the temperature at a fixed Ca^{2+} concentration. A typical experiment for the axon immersed in 10 Ca–ASW is illustrated in fig. 3, where the temperature was lowered from 8.9 to 8.2°C by 0.1 or 0.2°C steps. The axon exhibited a limit-cycle oscillation at 8.9°C but some irregularity (or aperiodicity) below 8.8°C, showing that the transition region is quite narrow on the state

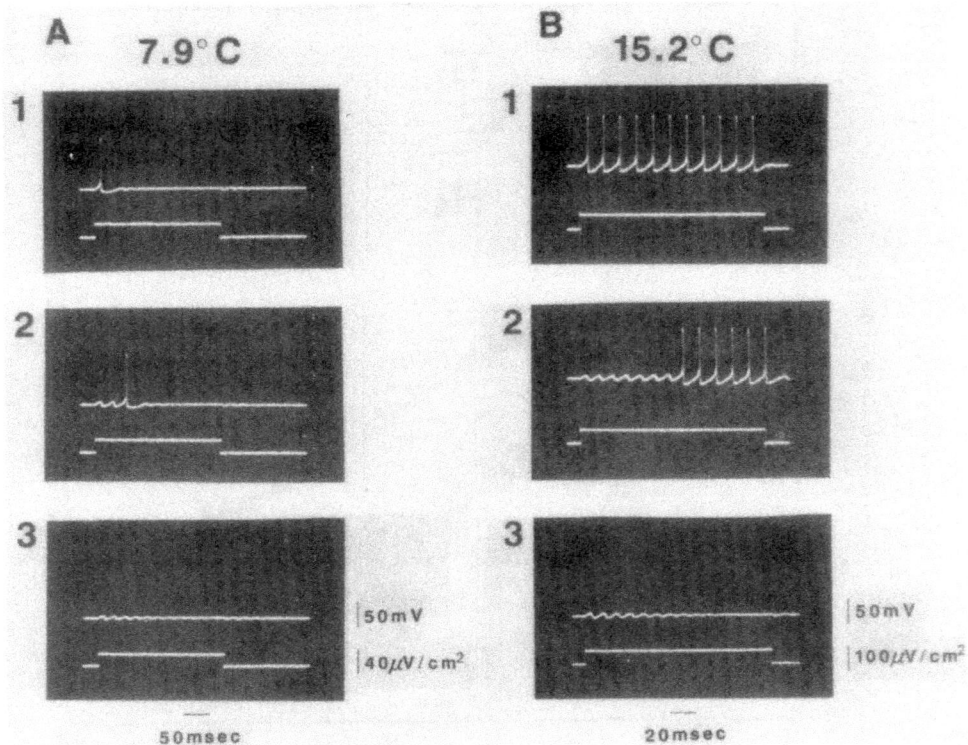

Fig. 2. Characteristic responses of the axon in the R_I (A) and R_{II} (B) states to a stimulating current step. Records 1, 2 and 3 represent typical potential responses to the suprathreshold, immediately above the threshold, and subthreshold currents, as well as their step current records, respectively. The axon was spatially unclamped (see text).

diagram. The irregularity appeared as an irregular mixture of action potential generation and subthreshold oscillations. The intermittent period during which action potentials are not generated increased as the temperature was lowered, as shown in the records of fig. 3. As the temperature was increased, on the other hand, from the bursting state O_I to the limit cycle state O_{II}, the stable limit-cycle oscillation did not appear throughout the unstable limit cycle such as seen at 8.8 and 8.7°C in fig. 3 (data not shown); that is, the bifurcation between O_I and O_{II} is not equivalent for temperature increase and decrease.

3.2. Bifurcation characteristics from the R_{II} to O_{II} state

Both spatial and temporal coherence become more and more enhanced for the spatially un-

clamped axon as the state R_{II} approaches O_{II}; that is, among fluctuating components of potentials around the resting (steady) potential, only the component with a specific frequency increases in its amplitude after the external Ca^{2+} concentration is reduced. The specific frequency agrees well with the macroscopic firing frequency at the O_{II} state [16, 17, 20–26]. At the same time, the spatial coherence of fluctuating potentials measured at two points 20–40 mm apart along the axon increases with their temporal coherence [17, 27, 28]. A typical experiment is illustrated in fig. 4, where amplitudes of the power spectral density (PSD) at the specific frequency (thin solid curve) and the resting potential (thick solid curve) are plotted against time measured after switching the external solution from 40 Ca–ASW (normal ASW) to 15 Ca–ASW at 16°C. A rather long time of 17.4 min was needed to obtain spontaneous

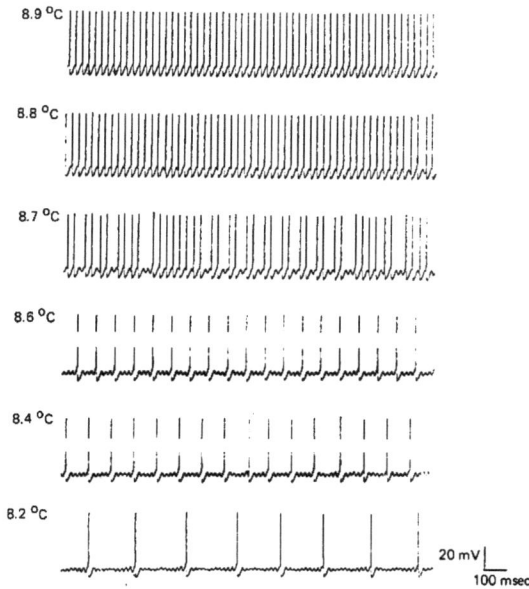

Fig. 3. Typical current records obtained for the axon to bifurcate from the O_H (uppermost record at 8.9°C) to O_L (other five records obtained at 8.8–8.2°C) states by lowering the temperature. Spontaneous oscillation of action potentials at 8.9°C is periodic (uppermost record), but some irregularity is observed at 8.8°C (second record from the top). When the temperature falls below 8.7°C, intermittently oscillating periods during which action potential generation is absent become more frequent and longer. The axon was nominally spatially unclamped since no wire electrode was inserted into the axon.

repetitive firing of action potentials after switching the external solution, suggesting that intracellular molecular reorganization or modification may contribute to inducement of the bifurcation, in addition to the extra-cellular Ca effect on screening of fixed charges of the external membrane surface [26]. After switching, the resting potential was first hyperpolarized and then depolarized. From 5 min after switching, the time when the resting potential became depolarized, the fluctuating potential component around 125.0 Hz increased rapidly and homogeneously over the axon with time. This is clearly shown by the PSD measurements of fig. 4, where the PSDs around 125.0 Hz measured at two different points 11 mm apart grow equally with time (see inset records). The fluctuating potentials are almost the same over the axon in their amplitude and time course,

but different only in their phase relation, as shown below. The phase difference at two distinct points along the axon was obtained by calculating the cross-power spectrum function $P_{nm}(f)$ between the fluctuating potentials $V_n(t)$ and $V_m(t)$, where f and t stand for frequency and time, respectively. The phase difference between the frequency components of V_n and V_m at f is obtained as the inverse tangent of the ratio of the real to the imaginary part of the cross-power spectrum $P_{nm}(f)$. The phase difference thus calculated for the fluctuating potential component at 125.0 Hz continuously changed with time until the spontaneous firing took place, shown as a broken curve in fig. 4. This suggests that even in state R_H, the 125.0 Hz component of the potential with an amplitude of 10–100 μV is generated at the specific location of the axon and propagates along the axon. In order to ascertain this idea, we measured the phase differences between V_1 and V_2, between V_2 and V_3, and between V_2 and V_4 (see inset of fig. 5), where V_i's ($i = 1, 2, 3, 4$) are fluctuating potentials measured at four different locations along the axon bathed in 15 Ca–ASW, each separated by 11 mm. The phase differences were obtained as a function of time after switching of the external solution from 40 Ca–ASW to 15 Ca–ASW. Typical results of the measurements at 11 and 15.5 min are shown as upper and lower records of fig. 5, respectively. In both records, the phase difference between V_1 and V_2 is opposite in sign to those between V_2 and V_3 and between V_2 and V_4. Further, the absolute value of the phase difference between V_2 and V_4 is twice that between V_1 and V_2 or between V_2 and V_3. The results indicate that the 125.0 Hz component is generated near location 2 and propagates to directions 1, 3 and 4. The specific location in the axon in the O_H state is known to be the pacemaker which generates action potentials spontaneously [17, 28]. The pacemaker can be formed since Na channels for the axon in the O_{II} state can interact in a long-range manner over the whole axon of 3–10 cm in length [17]. The present experiment suggests that this

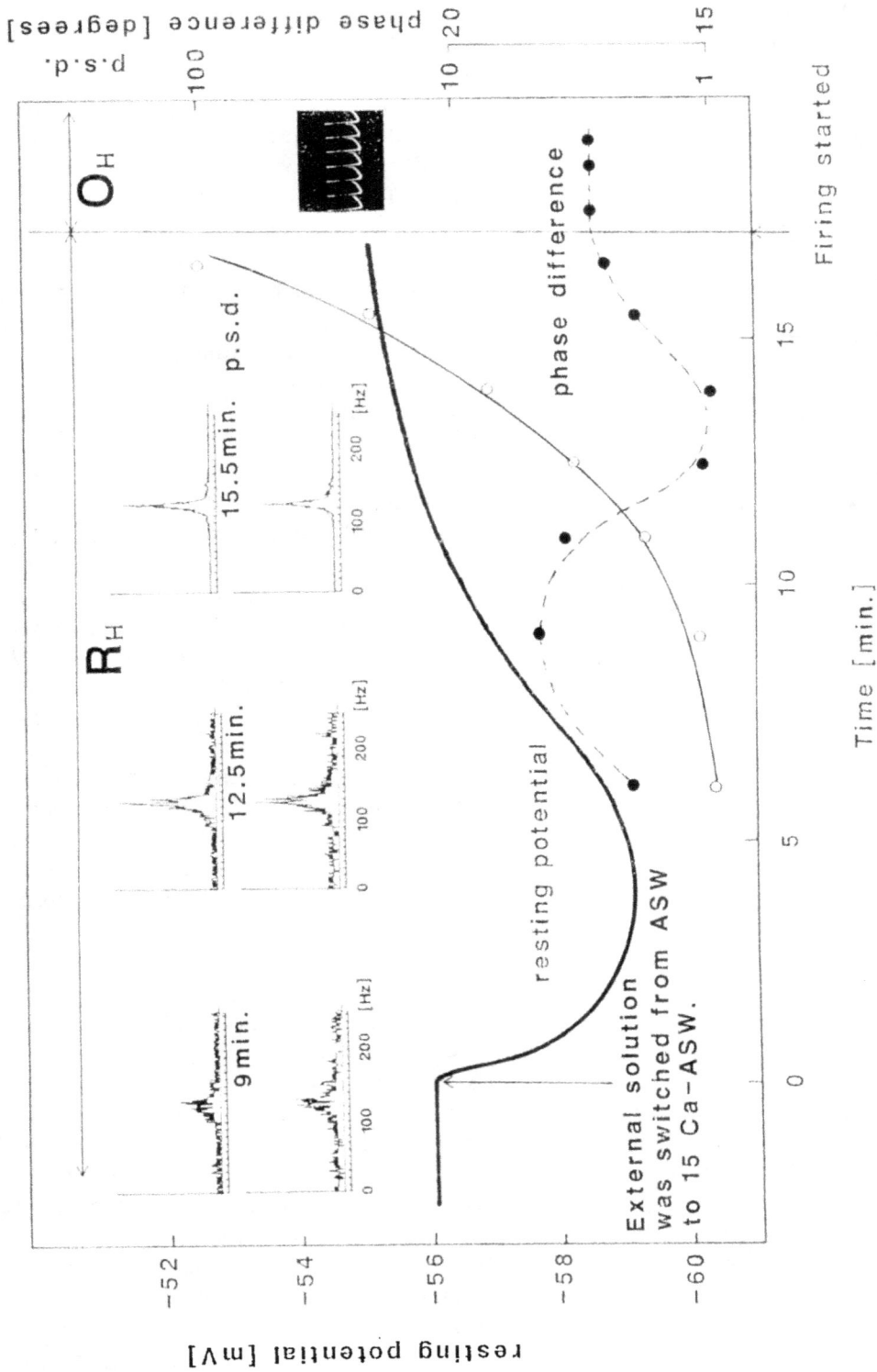

Fig. 4. Variation with time of the resting potential (thick solid curve), the peak amplitudes of power spectrum densities for a specific component of fluctuating potentials (thin solid curve, PSD) and phase difference between the specific components of fluctuating potentials measured at two points 11 mm apart from each other along the axon (broken curve). The peak amplitudes of PSD are normalized by that obtained at 6 min. The time is measured just after switching the external solution from 40 Ca-ASW to 15 Ca-ASW. The axon was in the R_H and O_H states before and after 17.4 min, respectively, since the spontaneous firing took place at 17.4 min. *Inset:* Power spectrum densities simultaneously measured at 9, 12.5 and 15.5 min at two points 11 mm apart, showing that temporal fluctuation increases homogeneously over the axon with time. The axon was spatially unclamped at 16°C since no wire electrode was inserted into the axon.

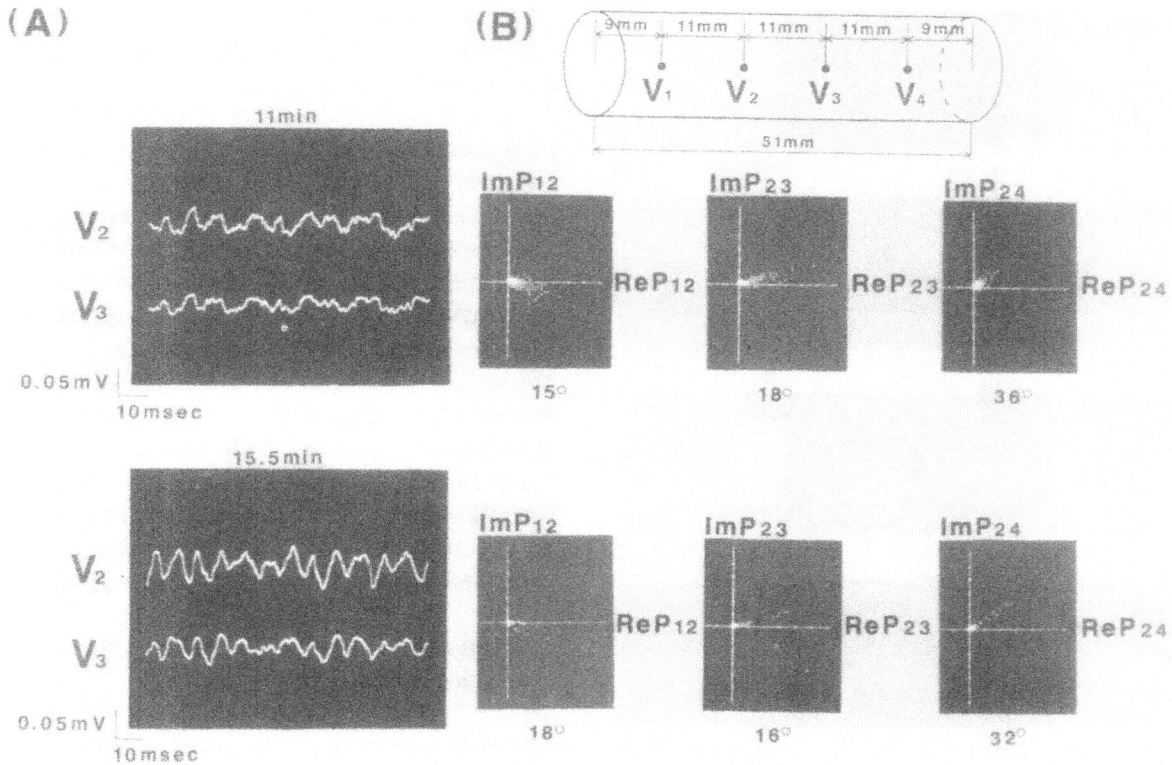

Fig. 5. Fluctuating potentials simultaneously measured at four different points, each 11 mm apart along the axon. (A) Typical real records of fluctuating potentials simultaneously measured at two points, V_2 and V_3 (see inset of schematic diagram of the axon and electrode configuration). Upper and lower records correspond to those obtained at 11.0 and 15.5 min for the axon in fig. 4, respectively. (B) Ratio of imaginary to real parts of the cross-power spectrum for fluctuating potentials simultaneously measured at two points between V_1 and V_2, V_2 and V_3, and between V_2 and V_4, respectively. Upper and lower records correspond to those obtained at 11.0 and 15.5 min for the axon in fig. 4, respectively. The axon was at 16°C and spatially unclamped since no wire electrode was inserted into the axon.

long-range spatial interaction has peculiar characteristics even for an axon in the R_{II} state, at least in that the axon depolarizes the resting potential after the external solution is switched from 40 Ca–ASW to Ca-reduced ASW [29]. This indicates that the long-range spatial interaction among Na channels is essential in inducing the bifurcation from the R_{II} to the O_{II} state.

3.3. Bifurcation characteristics from the R_I to the O_I state

Bifurcation characteristics from the R_I to the O_I state were also studied for the spatially un-

clamped axon in detail, and can be summarized as follows. (1) The transition was extremely rapid. A typical experiment is illustrated in fig. 6, where the external solution surrounding the axon at 5.4°C was switched from 40 Ca–ASW to 15 Ca–ASW at time 0. It is seen in the figure that the intermittent oscillation appeared 29 s after the switching. (2) A specific fluctuating component of potentials rapidly increased homogeneously over the axon. The phase difference of the fluctuating components observed at two points 15 mm apart was 0 or 1.5° at maximum until the firing finally started. Even if the phase difference exists, it is fluctuating, both in size and in sign.

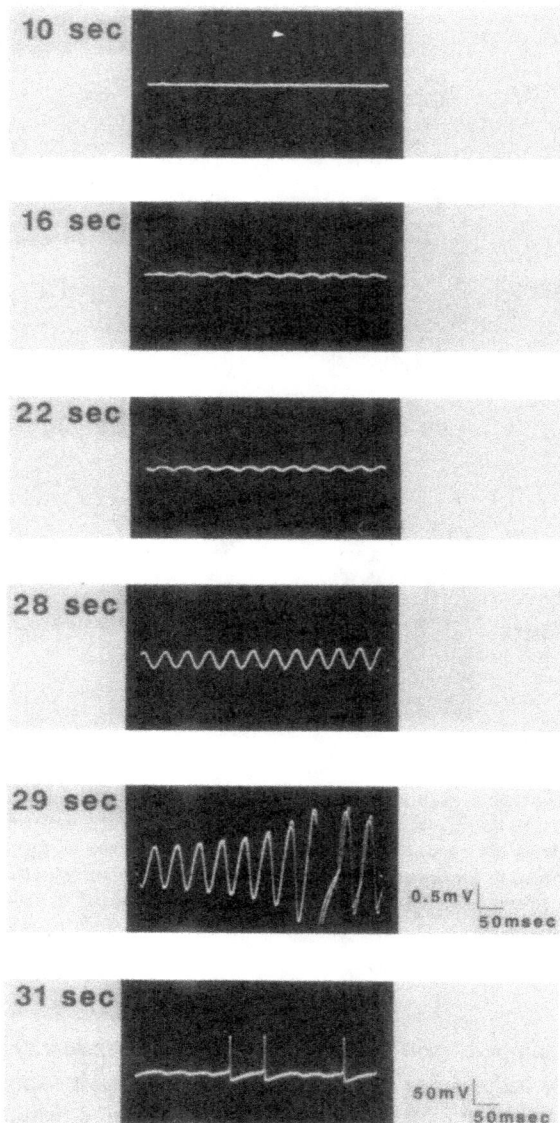

Fig. 6. Two superimposed records of fluctuating potentials (upper five records) and spontaneously induced action potentials (lowest record) for the axon at 5.4°C simultaneously taken at two points 15 mm apart along the axon after switching the external solution from 40 Ca–ASW to 15 Ca–ASW, as a function of time. It is seen that the R_L state (upper four records and up to 200–300 ms before the end of the fifth record of 29 s) becomes rapidly destabilized to bifurcate to the O_L state (lowermost record and the last 200–300 ms record of the fifth one). It is noted that the phase difference between the two intermittently oscillating action potentials is fluctuating from time to time. The time resolution for the record taken at 31 s is not enough to resolve the phase difference clearly.

(3) The resting potential was monotonically shifted to become hyperpolarized by 1.5–1.0 mV to initiate the firing.

All of these results indicate that the bifurcation characteristics from the R_L to the O_L state are different from those from the R_H to the O_H state. The cause of the bifurcation should be on the extracellular surface of the axon since the transition is quite rapid after switching of the external solution. The fluctuating potentials were found to accord, in terms of their spatiotemporal behaviors on growth, over every part of the axon at any time, suggesting that the oscillation is reinforced by external force homogeneously applied to the whole axon. In this respect, the transition behavior is close to that from the R_H to O_H state for the spatially clamped axon, as described below.

3.4. Bifurcation characteristics under spatially clamped conditions

To learn more about the property of the spatial interaction, the axon was spatially clamped by inserting a platinized platinum wire through the whole axon. Under the condition of this spatial clamp, the axon immersed in Ca^{2+}-reduced ASW was able to bifurcate to the spontaneously oscillating state even at higher temperatures. However, its bifurcation behavior differs from the bifurcation of the R_H to the O_H state, as shown in fig. 7. (1) External Ca^{2+} concentration had to be further reduced to induce the spontaneously oscillating state, indicating that the spatial clamp makes it difficult for the axon to bifurcate to the oscillating state. At 16°C, the axon should be immersed in 9 or less Ca–ASW to obtain the oscillating state, which should be compared with the case in which, under the spatially unclamped condition, the axon immersed in 17 Ca–ASW exhibited the oscillation spontaneously (see fig. 1). (2) The spontaneous oscillation induced by reducing external Ca^{2+} concentration even at higher temperatures fires intermittently under the spatially clamped condition, consisting of several suc-

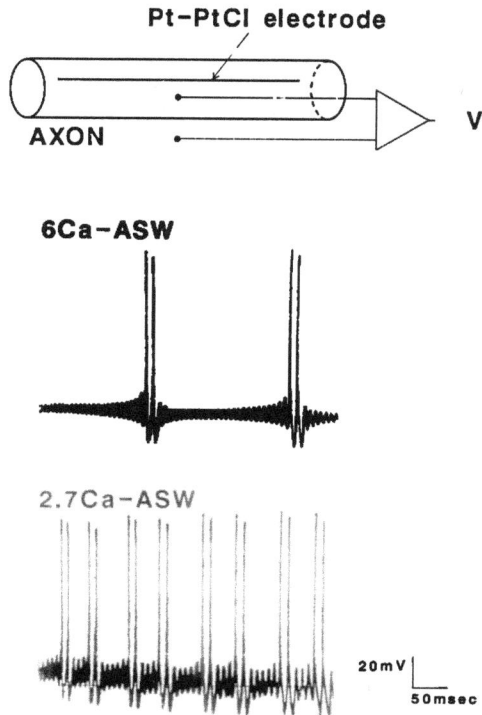

Fig. 7. Effect of the spatial clamp on the axon at the state in which spontaneous firing occurs in a periodic fashion at 16°C. The periodically fired axon immersed in 6 Ca–ASW (upper record) and in 2.7 Ca–ASW (lower record) was each spatially clamped by inserting a platinized platinum wire electrode over the whole region of the axon (see inset), and intermittent oscillations immediately followed. The intermittent period became shorter as external Ca^{2+} concentrations became more reduced.

cessive trains of action potentials followed by subthreshold oscillations. Typical patterns of the firing at 16°C are shown in fig. 7; the more reduced the Ca^{2+} concentration was, the shorter the intermittent period of time became. (3) Fluctuating potentials increased more and more in amplitude with time homogeneously over the axon after the external solution was switched to Ca^{2+}-reduced ASW.

The following phenomena of the spatially unclamped axon were, however, commonly observed: (a) the time to reach the oscillation state after switching to the reduced Ca–ASW was 12–45 min for the four axons tested, unchanged as for the unclamped case; (b) the resting (steady)

potential once hyperpolarized and then depolarized with time. During the depolarized period of time, the fluctuating component of potentials increased with time to yield the oscillation (bursting) state. These experiments indicate that the spatial clamp deprives at least one of the spatial freedoms for electrical activities along the axon that are essential for both the limit-cycle generation of action potentials and the phase difference of fluctuating potentials.

The spatially unclamped axon at the R_H state exhibited periodically repetitive action potentials to a suprathreshold step stimulation (figs. 1 and 2). However, it is well known in the physiology of squid giant axons [26, 30, 31] that this is not the case for the spatially clamped axon; that is, one or a few (two or three) successive trains of action potentials are followed by membrane depolarization which blocks action potentials from being generated in succession thereafter. The membrane depolarization caused by outwardly flowing steady current is believed to be due to accumulation of K^+ ions in the extracellularly periaxonal space surrounding the squid giant axon [26, 32]. However, in our present experiments, inhibition of the oscillation in succession by spatially clamping the axon at the O_H state does not originate from the K^+ ion accumulation. This is due to the fact that we did not give any externally flowing current through the platinized platinum wire for the spatial clamp and the axon did not appreciably exhibit membrane depolarization. Further, it is noted that the failure of the successive oscillation of action potentials for the spatially unclamped axon in the R_L state differs in their mechanism from the blockage of the successive oscillation for the spatially clamped R_H axon, both induced by external application of a current step. In the latter, the successive oscillation is blocked by the spatial clamping while, in the former, it is not realized in the in vivo axon. Therefore, we can say that the specific interaction, which is present in the R_H (O_H) axon but absent in the R_L (O_L) axon, is crucial for generation of the periodic oscillation, and that the spe-

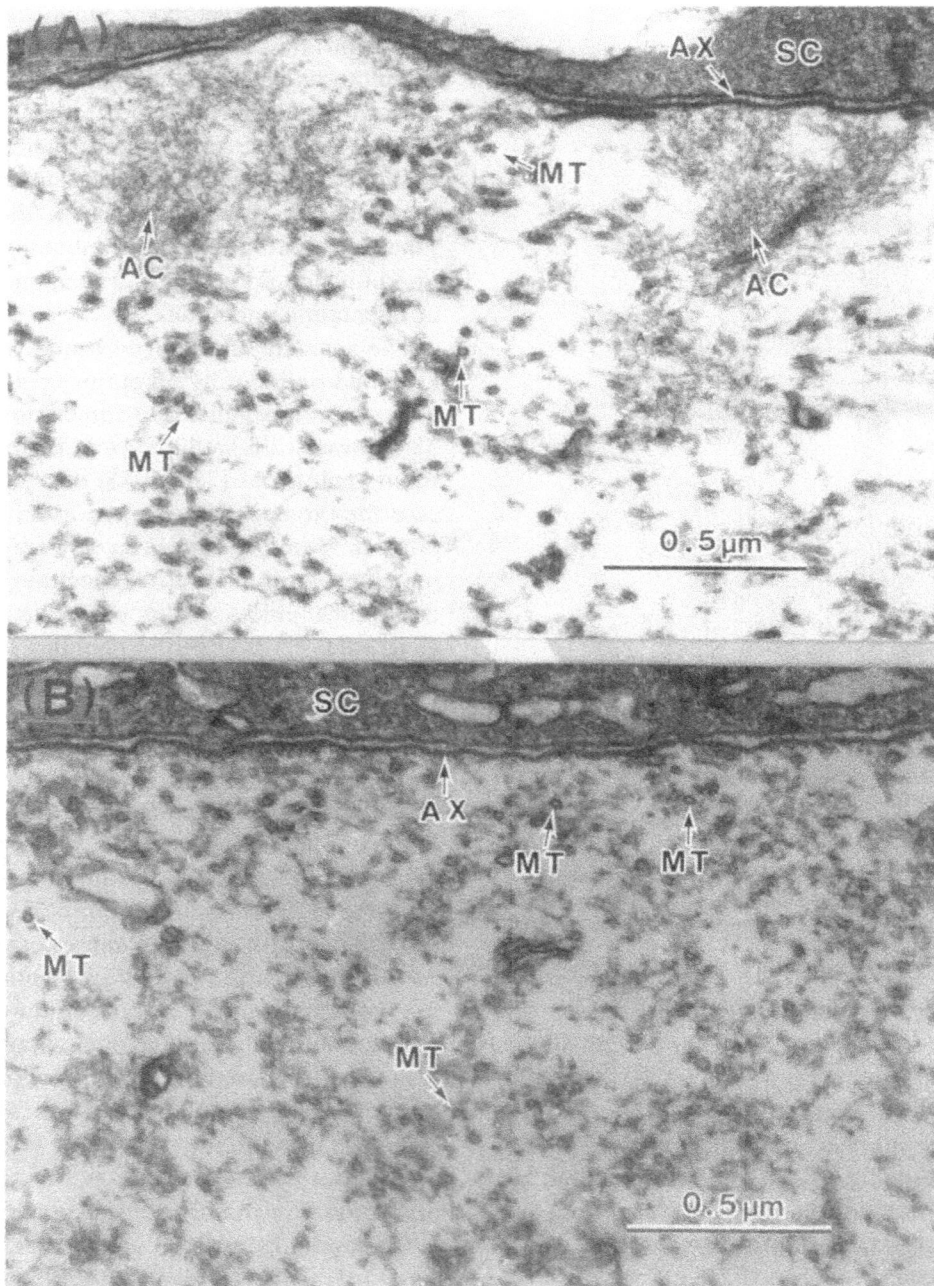

Fig. 8. Cross-sectional view of electron micrographs of the subaxolemmal cytoskeleton at 16°C (A) and 8°C (B). AX: axolemma, SC: Schwann cells, AC: actin filaments, MT: microtubules. (A) Actin filaments form spotlike clusters, sticking to the axolemma. In the figure, the two clusters are seen 1.5 µm apart from center to center. Microtubules running parallel to the longitudinal direction of the axon locate between them, some close to the axolemma. (B) Microtubules distribute homogeneously in the subaxolemmal region while actin filaments can barely be seen as clusters.

cific interaction can be blocked by the spatial clamping.

3.5. Towards elucidation of molecular mechanisms of the spatial long-range interaction

We found that electrical activities of the squid giant axon bifurcated with temperature through the strength of the spatial interaction. For the bifurcation, the transition took place at 10°C for the axon immersed in 40 Ca–ASW (see fig. 1). Corresponding to this transition, we found that the subaxolemmal cytoskeleton for the axon immersed in 40 Ca–ASW made a structural transition at 10°C; the subaxolemmal cytoskeleton is mainly constituted of microtubules and actin filaments. At higher temperatures, actin filaments are associated with the axolemma through a specialized two-dimensional meshwork, forming clusters just beneath the axolemma. Microtubules run parallel to the axolemma and are embedded in a fine meshwork. Some microtubules run along the axolemma among clusters of actin filaments and are associated laterally with the axolemma through slender connections. The axolemma at temperatures higher than 10°C, therefore, can be divided into microfilament-associated and microtubule-associated regions [33–36], which probably represent two important functional domains (fig. 8A). Further, we observed that, at temperatures lower than 10°C, both microfilament-associated and microtubule-associated regions disappeared and these filaments appeared to evenly distribute over the axon (fig. 8B). Thus, the structural changes in the subaxolemmal cytoskeleton correlate well with the electrical state changes, observed both above and below 10°C.

The other specialization consists of patches of an electron-dense subaxolemmal undercoating, where a high density of Na channels is most probably gathered in either of these regions of the squid axolemma [37]. On the assumption that Na channels can associate with either microfilament- or microtubule-associated regions, the axon immersed in 40 Ca–ASW retains local patches at

temperatures higher than 10°C where Na channels gather at high density, while Na channels distribute evenly over the axolemma below 10°C. This rearrangement of Na channel organization may be attributed to the molecular mechanism of the spatial long-range interaction, as will be discussed in section 4.

4. Discussion

It has been well known that the squid giant axon retains two stable membrane states, the resting (R) and spontaneous oscillation (O) states, according to the Ca^{2+} concentration contained in the solution surrounding the axon [20, 26]. The transition from the R to O state takes place by increase of temporal coherence [16, 17, 21–23]; that is, a component with a specific frequency among fluctuating potentials around the resting potential increases in amplitude with time after switching of the external solution to one with more reduced Ca^{2+} concentrations until finally the spontaneous firing appears stably with the same specific repetitive frequency (see figs. 4 and 6).

The present experiments have revealed that both the resting and the spontaneous oscillating states can be further subdivided into two phases according to temperature: that is, higher- and lower-temperature phases (fig. 1). This subdivision is determined by the temperature-dependent stability and instability of limit-cycle oscillations of action potentials, irrespective of whether they are spontaneously induced in the O state or induced by externally applied current in the R state. At higher temperatures, the limit-cycle oscillation is stable while, at lower temperatures, it becomes unstable. The transition temperatures, T_2 for the R state and T_3 for the O state, between the higher- and lower-temperature phases change as a function of external Ca concentrations. The T_2 accords with T_3 on the transition concentration of Ca for the bifurcation between the R and O states, to make a triple point S (see fig. 1). This indicates that the R state at higher-temperature phase (R_H) bifurcates to the higher-temperature phase (O_H) but not to the O state at the lower-temperature phase (O_L), and that the R state at the lower-temperature phase (R_L) always bifurcates to O_L but not to O_H. In other words, the temperature-dependent bifurcation takes place between R_H and O_H or between R_L and O_L but not between R_H and O_L or between R_L and O_H, suggesting that the bifurcation is governed by some temperature-dependent mechanism common to both the R and O states.

The question has therefore arisen as to what common mechanism regulates the temperature-dependent bifurcation. The bifurcation can phenomenologically be characterized by phase relations among the specific frequency component of fluctuating potentials along the axon as follows: (1) The R_H state bifurcates to the O_H state by lowering external Ca concentrations at higher temperatures. This bifurcation is characterized by the growth of spatiotemporal coherence for the specific frequency component of fluctuating potentials along the axon [17, 25, 28] (see also fig. 5). In particular, it is experimentally found (see fig. 5) that the specific frequency component of fluctuating potentials is generated at a given portion of the axon in the R_H state and is propagated to other portions along the axon, just as macroscopic action potentials are generated and propagated in the O_H state, suggesting that the presence of nonlinear local oscillators at mesoscopic levels with specific characteristic frequencies and strong coupling in the long range is important to induce the bifurcation. The phase difference of the local oscillation per distance is unchanged before and after the bifurcation [17] (see also fig. 4), showing that the propagation velocity is unchanged before and after the bifurcation. (2) Spatial clamping blocks the periodic oscillation of action potentials in the R_H or O_H state. The effect of spatial clamping on the O_H state is illustrated in fig. 7, where external Ca concentrations should be reduced more to induce

spontaneous oscillations of action potentials, and even the oscillation thus induced is not a limit cycle but a burst. In other words, the behavior of the spatially clamped axon in the O_H state becomes quite similar to the one in the O_L state. The response of the R_H axon under spatial clamping to the current step stimulation resembles that of the R_L axon, in the sense that the externally applied current step cannot generate limit-cycle oscillations of action potentials. All these experiments show that spatial clamping blocks the generation of limit-cycle oscillations by exposing all the excitable units to the same electric field. (3) At the lower temperatures, the situation for fluctuating potentials quite resembles the one for the spatially clamped axon at the higher-temperature phase since at the bifurcation from the R_L to O_L state, the specific frequency components of fluctuating potentials increase in amplitude while their phases coincide all over the axon, except in the vicinity of the bifurcation point (see fig. 6). Even in the vicinity of and after the bifurcation point, their phase difference is unstable and fluctuating in their size and sign from time to time. The instability in the R_L state is closely related to the instability of limit-cycle oscillations of action potentials in the O_L state after the bifurcation. These characteristics of their phase are equivalent to those observed for the spatially clamped axon in the R_H state.

As a result, it can be concluded that the stability of the limit-cycle oscillation, spontaneously induced in the O state or forced by external current step in the R state, depends on the phase relationship among the specific frequency components of fluctuating potentials along the axon: the limit-cycle is stabilized and destabilized when the fluctuating potentials are generated out of phase and in phase, respectively. This leads to the question as to why the fluctuating potentials are produced in phase at lower temperatures and out of phase at higher temperatures. For this answer, at the structural base, we note that structural changes of subaxolemmal cytoskeletons take place in a temperature-dependent manner, and that the transition temperature of the structural change for the axon bathed in 40 Ca–ASW (for the axon in the R state) is in good agreement with the bifurcation temperature. At higher temperatures, the organization of the subaxolemmal cytoskeleton can be characterized by two functional domains of microtubule- and microfilament-associated regions [33, 36] (see fig. 8A). At lower temperatures, on the other hand, these domains disappear and the filaments appear to evenly distribute over the axolemma (fig. 8B). The subaxolemmal cytoskeletons are thus structurally differentiated, probably to regulate Na channel distribution over the axolemma in the following ways. (1) Fluorescence photobleaching recovery experiments on rat neurons stained by fluorescent neurotoxin probes specific for the voltage-dependent Na channel show that lateral diffusion of sodium channels in the axon hillock is restricted, suggesting that the sodium channels are localized by direct cytoskeletal attachments or by a selective barrier to channel diffusion [38]. (2) Rat brain Na channel proteins labeled with ^{3}H-saxitoxin are precipitated in the presence of exogenous brain ankyrin by anti-ankyrin antibodies, and ^{125}I-labeled ankyrin binds with high affinity to sodium channels reconstituted into lipid vesicles [39]. These results indicate that brain ankyrin links the voltage-dependent sodium channel to the underlying cytoskeleton. (3) Calmodulin antagonists (W-7, W-5, trifluoperazine, chlorpromazine) are potent blockers of the sodium current of squid giant axons because they inhibit sodium gating [40]. The site of action is the intracellular surface of the axolemma where the Ca^{2+}-calmodulin complex can be formed [40], suggesting that some calmodulin-binding proteins link the sodium channel to the subaxolemmal cytoskeleton.

It is probable that sodium channels in the squid axolemma are homogeneously distributed over the axon immersed in 40 Ca–ASW at temperatures lower than 10°C (corresponding to the R_L state) while they make patches presumably underlying the microfilament-associated domain of the subaxolemmal cytoskeletons above 10°C (corresponding to the R_H state). Patch formation of Na channels is also exemplified for the axons in the nerve fiber layer of the adult rat retina [41]. These specializations of Na channel distribution possibly relate to the difference between fluctuating potential behaviors at higher and lower temperatures. One may say that a small fraction (< 5%) of Na channels labeled "threshold channels" [42] is responsible for the fluctuating potentials at higher and lower temperatures since, in distinction from the normal Na channels, threshold channels open at abnormally negative voltages [42]. However, threshold channels close very slowly [42], leading to the conclusion that the fluctuating potential frequencies should be lower than the macroscopic oscillation of action potentials if threshold channels are responsible for the fluctuation. This is not the case for our experiments.

One major result of this paper is that the membrane state of squid giant axons is subdivided into two phases (higher and lower temperature phases) and that the subdivision is regulated by spatial interactions among Na channels which are different at higher and lower temperatures. The interactions are possibly characterized by Na channel distribution regulated by the subaxolemmal cytoskeleton. These results are consistent with our earlier hypothesis [17] that a specific spatial long-range interaction among Na channels is crucial to induce the Hopf bifurcation in the squid giant axon system, but we do not exclude the possibility of alternative interpretations. Finally, we note that the axon under physiological conditions is situated in the higher-temperature phase since squids prefer to live in seawater of 14–18°C. Thus the higher-temperature phase of the membrane state is actually realized under natural conditions. As a result, the present experiments have given further support to our previous model of nerve excitation [17], in which the generation of action potentials can be grasped as a transition, assisted by externally applied current, from the resting state with a stable focal point to the limit-cycle state. We also note that the spatial clamp, which is generally adopted as a conventional technique for electrophysiological experiments, changes the higher-temperature phase to the lower one. Further, electrophysiological experiments on squid giant axons have usually been performed at lower temperatures below 10°C. These two experimental conditions of the spatial clamping and of < 10°C may lead to the apparent discrepancy between our view and the current conventional view of the spatial properties of the axon. How the spatial properties can be approached by the Hodgkin–Huxley formalism [43] and what kinds of molecular events are related to the temperature-dependent structural alternations of subaxolemmal cytoskeletons are questions currently under investigation.

Acknowledgements

The authors would like to express their hearty thanks to Professors R. Kubo and N. Wakabayashi (Keio University, Department of Physics) for stimulating discussions and critical reading of the manuscripts, and to Dr. Michinori Ichikawa for his valuable technical assistance in the electrophysiological experiments and electron microscopic observation studies. The present paper is dedicated to Professor Akiyoshi Wada in honor of his sixtieth birthday.

References

[1] H.L. Swinney and J.P. Gollub, Phys. Today 31 (8) (1978) 41.
[2] A. Libchaber, C. Laroche and S. Fauve, J. Phys. Lett. 43 (1982) L211.
[3] K. Ikeda, H. Daido and O. Akimoto, Phys. Rev. Lett. 45 (1980) 709.
[4] A.T. Winfree, Science 181 (1973) 937.
[5] J.L. Hudson and J.C. Mankin, J. Chem. Phys. 74 (1981) 6171.
[6] K. Tomita and I. Tsuda, Prog. Theor. Phys. 64 (1980) 1138.
[7] H.L. Swinney, Physica D 7 (1984) 3.

[8] K. Aihara and G. Matsumoto, in: Chaos in Biological Systems, eds. H. Degn, A.V. Holden and L.F. Olsen (Plenum Press, New York, 1987) p. 121.
[9] N. Takahashi, Y. Hanyu, T. Musha, R. Kubo and G. Matsumoto, Physica D 43 (1990) 318.
[10] H. Hayashi, S. Ishizuka and K. Hirakawa, J. Phys. Soc. Japan 55 (1986) 3272.
[11] M.R. Guevara, L. Glass and A. Shrier, Science 214 (1980) 1350.
[12] L. Glass, M.R. Guevara and A. Shrier, Ann. NY Acad. Sci. 584 (1987) 168.
[13] A.T. Winfree, The Geometry of Biological Time (Springer, Berlin, 1980).
[14] S. Kabashima, M. Itsumi, T. Kawakubo and T. Nagashima, J. Phys. Soc. Japan 39 (1975) 1183.
[15] G. Matsumoto, K. Kim, T. Uehara and J. Shimada, J. Phys. Soc. Japan 49 (1980) 906.
[16] G. Matsumoto, I. Tasaki and I. Inoue, J. Phys. Soc. Japan 44 (1978) 351.
[17] G. Matsumoto, Long-range spatial interactions and a dissipative structure in squid giant axons and a proposed physical model of nerve excitation, in: Nerve Membrane, Biochemistry and Function of Channel Proteins, eds. G. Matsumoto and M. Kotani (Univ. Tokyo Press, Tokyo, 1981) p. 203.
[18] B. Hille, Ionic Channels of Excitable Membranes (Sinauer, Sunderland, MA, 1984).
[19] G. Matsumoto, M. Ichikawa, A. Tasaki, H. Murofushi and H. Sakai, J. Membrane Biol. 77 (1984) 77.
[20] A.F. Huxley, Ann. NY Acad. Sci. 81 (1959) 221.
[21] R. Guttman, Biophys. J. 9 (1969) 269.
[22] R. Guttman and R. Barnhill, J. Gen. Physiol. 55 (1970) 104.
[23] R. Guttman, S. Lewis and J. Rinzel, J. Physiol. 305 (1980) 377.
[24] I. Tasaki, Physiology and Electrochemistry of Nerve Fibers (Academic Press, London, 1982).
[25] G. Matsumoto and H. Shimizu, J. Theor. Neurobiol. 2 (1983) 29.
[26] B. Frankenhaeuser and A.L. Hodgkin, J. Physiol. (London) 137 (1957) 218.
[27] G. Matsumoto and W. Stühmer, J. Phys. Soc. Japan 45 (1978) 1069.
[28] G. Matsumoto and H. Shimizu, J. Phys. Soc. Japan 44 (1978) 1399.
[29] G. Matsumoto and T. Kunisawa, J. Phys. Soc. Japan 44 (1978) 1047.
[30] N.J. Abbott, E.M. Lieberman, Y. Pichon, S. Hassan and Y. Larmet, Biophys. J. 53 (1988) 275.
[31] M.L. Astion, J.A. Coles, R.K. Orkand and N.J. Abbott, Biophys. J. 53 (1988) 281.
[32] W.J. Adelman, Jr. and R. FitzHugh, Fed. Proc. 34 (1975) 1322.
[33] S. Tsukita, S. Tsukita, T. Kobayashi and G. Matsumoto, J. Cell Biol. 102 (1986) 1710.
[34] T. Kobayashi, S. Tsukita, S. Tsukita, Y. Yamamoto and G. Matsumoto, J. Cell Biol. 102 (1986) 1699.

[35] G. Matsumoto, S. Tsukita and T. Arai, Organization of the axonal cytoskeleton. Differentiation of the microtubule and actin filament arrays, in: Cell Movement, Vol. 2. Kinesin, Dynein, and Microtubule Dynamics, eds. F.D. Warner and J.R. McIntosh (Liss, New York, 1989) pp. 335–356.

[36] T. Arai and G. Matsumoto, J. Neurochem. 51 (1988) 1825.

[37] S.G. Waxman and J.M. Ritchie, Science 228 (1985) 1502.

[38] K.J. Angelides, L.W. Elmer, D. Loftus and E. Elson, J. Cell. Biol. (1988) 1911.

[39] Y. Srinivasan, L. Elmer, J. Davis, V. Bennett and K. Angelides, Nature 333 (1988) 177.

[40] M. Ichikawa and G. Matsumoto, J. Membr. Biol., in press.

[41] C. Hildebrand and S.G. Waxman, Brain Research 258 (1983) 23.

[42] W.F. Gilly and C.M. Armstrong, Nature 309 (1984) 448.

[43] A.L. Hodgkin and A.F. Huxley, J. Physiol. 117 (1952) 500.

Physica D 49 (1991) 214–223
North-Holland

Sequential events in bacterial colony morphogenesis

James A. Shapiro and David Trubatch

Department of Biochemistry and Molecular Biology, University of Chicago, 920 E. 58th Street, Chicago, IL 60637, USA

Bacterial colonies are organized, differentiated multicellular communities expressing genetically controlled patterns. These patterns can be seen in mature colonies by staining for differential gene expression, by visualization of surface textures, and by microscopic examination of cellular morphologies and multicellular arrays. Colony morphogenesis involves many sequential processes of cellular growth, differentiation and movement which are regulated, at least in part, by cell–cell interactions and communication between groups of cells. These morphogenetic processes can be followed by periodic microscopic examination of developing colonies and by time-lapse video recordings. Since the final colony structure is the integrated product of many steps, pattern formation cannot realistically be explained by assumptions about autonomous cell behaviors. Instead, colony growth is best viewed as a developmental process in which the cells interact and adjust their individual and collective behaviors as morphogenesis proceeds.

1. Introduction

One of the main questions to be addressed at this symposium is the relationship between patterns observed in living and non-living systems. In some cases, such as the similarities between spiral waves in Belousov–Zhabotinsky reactions and in *Dictyostellium discoideum* aggregations, the geometrical similarities are quite striking. Do these morphological similarities mean that the same pattern-generating principles are at work in chemical, physical and biological systems? In order to find an answer, it is necessary to have a detailed knowledge of how patterns arise in a number of different specific situations, such as the growth of microbial populations on laboratory medium.

Understanding biological pattern formation requires an appreciation of the basic properties of living organisms. Among the most fundamental of these is the great structural, metabolic and behavorial complexity needed to carry out life processes. In other words, living organisms are extraordinarily information-rich, and contempo-

rary biological research has deepened our appreciation of the role of information processing. We have learned about how growth and biosynthesis are elaborately regulated and how homeostasis is maintained in all organisms through continual responsiveness to changes in external and internal conditions. Some of the most spectacular progress has been made in deciphering the sophisticated genomic systems that provide for the reliable transmission and precise but flexible expression of hereditary information through cell lineages. Indeed, we now have so much information about structure, function and control in all kinds of organisms, ranging from the smallest microbes to the largest plants and animals, that traditional concepts based on mechanical and statistical models no longer provide satisfactory solutions to major problems. Accordingly, this paper will present some aspects of bacterial colony morphogenesis on laboratory media which illustrate unanswered questions in biological pattern formation [9]. The emphasis will be on the sequential nature of morphogenesis, on the role of changing bacterial populations, and on colony responses to perturbations of the developmental process.

2. Observations with *E. coli*

2.1. Basic organization of E. coli colonies

Patterns in the colonies of standard laboratory bacteria such as *E. coli* can be visualized by histochemical staining for differential biochemical activity and by careful examination of colony surface structure. When *E. coli* strains are engineered to produce the enzyme β-galactosidase under a variety of different genetic control systems, the regulation of enzyme synthesis can be visualized by growth on agar medium containing an indicator dye for enzyme activity. The nondiffusing dye marks those cells which contain the enzyme, and the intensity of staining is proportional to the amount of enzyme produced. Low-magnification views of colonies produced under such conditions reveal reproducible patterns which give each colony a flower-like appearance [5, 6]. These patterns contain radially oriented sectorial elements which reflect hereditary changes in the control of growth and of gene expression affecting the descendants of a single ancestral cell (collectively denoted a clone). The patterns also contain concentric rings which reflect periodic changes in the control of gene expression affecting cells descended from many different ancestors at a common stage of the developmental process (fig. 1). Reflected light photography and scanning electron microscopy (SEM) of colonies show similar sectorial and concentric patterns in surface structure, indicating that multicellular aggregation and synthesis of extracellular components of colonies are also subject to clonal and non-clonal regulatory changes [7, 8]. Higher magnification SEM analysis reveals further that phenotypically distinct macroscopic

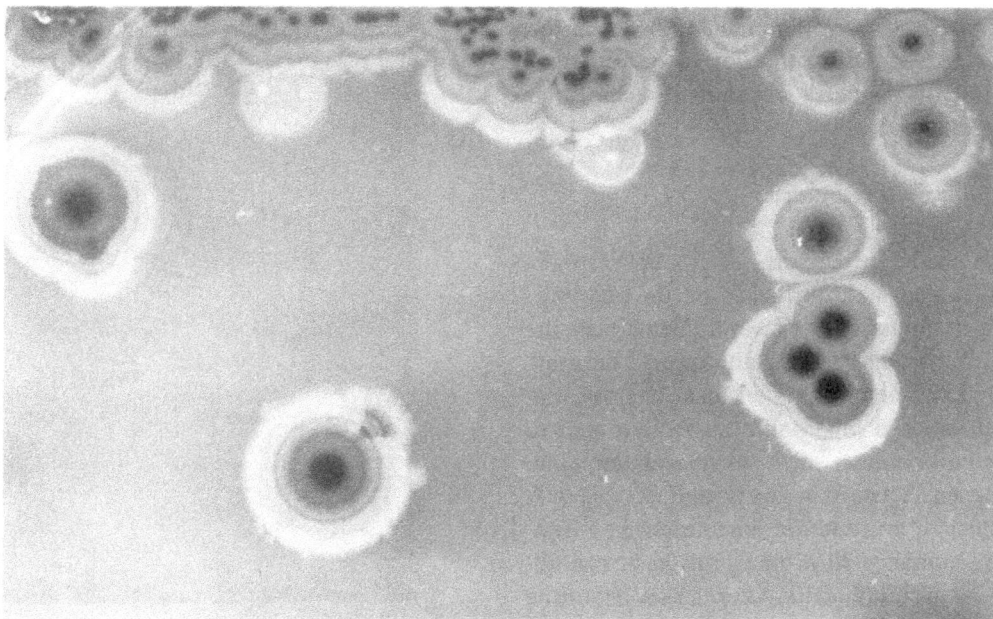

Fig. 1. A field of genetically engineered *E. coli* colonies (strain MS1891 [10]) stained for differential β-galactosidase activity. Note the concentric ring pattern. The bottom colony displays two adjacent sectors with different enzyme levels and growth phenotypes. The outline of the colony at the left was distorted by growth extending along a glass fiber on the agar surface. These colonies grew from individual bacteria. The sectored colony was 5 mm in diameter.

zones can be resolved microscopically into zones consisting of distinct cell morphologies and multicellular arrays.

2.2. Dynamics of E. coli colony development

One way to investigate the origin of macroscopic patterns is to examine the dynamics of *E. coli* colony development. Observing distinct microscopic zones by SEM at different times following inoculation shows that patterns of aggregation and cell morphogenesis are continually changing; for example, the cells at the colony edge have different shapes, sizes and arrangements at various periods of colony development [8]. Time-lapse video recording of the earliest stages of *E. coli* colony development shows that bacterial cell elongation and division occur in a non-random manner and are influenced by the presence of neighboring cells to generate regular multicellular arrays [10]. It is particularly important to note that *E. coli* cells grow in such a way as to form very intimate side-by-side alignments. That is, they maximize cell-to-cell contact rather than individual access to substrate. These observations indicate that the bacteria respond to each other and grow cooperatively.

2.3. Encounters with obstacles

Another useful procedure is to study how *E. coli* colonies deal with obstacles on the agar surface, such as fibers of glass wool. Growth in the presence of obstacles is actually more "natural" than growth on clean agar because bacteria encounter all kinds of fibers and particulate matter in real environments, such as the mammalian digestive tract. Thus, it is likely that *E. coli* will have evolved a routine for encountering obstacles. Upon contact with a fiber and its surrounding meniscus of liquid, cells are released from the expanding colony perimeter, swim around the fiber and establish a multicellular population coating the obstacle. Following this "coating" process, there are several possible consequences for colony pattern formation. One possibility fre-

quently seen with relatively small fibers is that the fiber is incorporated into the colony without any obvious effect on macroscopic pattern. Since contact with the fiber perturbs the usual symmetry of colony expansion, this result indicates a capacity for pattern regulation analogous to that observed with higher organisms, where various experimental manipulations on embryos heal and do not prevent the formation of morphologically normal adults.

Another common possibility is that the colony shape is deformed by the fiber and that growth proceeds outwards from the population coating it (fig. 2). In this case, it is extremely interesting to note that the periodic ring patterns are reproduced quite faithfully extending outwards from

Fig. 2. Two MS1891 colonies which ran into clusters of glass fibers placed on the agar just outside the colony edge. Note the continuation of the concentric ring pattern in the growth originating along the fibers. These colonies grew from spots containing about 10^5 bacteria. The colonies were 9 mm in diameter.

the new growth origins starting from a stage some time after the colony encountered the fiber(s). This result shows that the bacteria coating the fibers can reorganize themselves and continue the colony's normal developmental sequence of changing gene expression. Since this sequence is expressed from regions of the agar gel which have not been subject to the same nutrient depletion and metabolite accumulation as the edge of the expanding colony, the result further suggests that factors other than responses to local modifications of the substrate are responsible for the observed changes in gene expression.

2.4. Mutants with altered developmental patterns

There are several ways to obtain *E. coli* mutants which display altered colony development because of changes in their DNA [10]. One is to pick bacteria from sectors which display novel growth patterns. Another is to isolate bacteria which have altered developmental properties because of extra DNA incorporated into their genome at a position where it disrupts normal expression of genetic information. Two mutants have interesting phenotypes that display complementary aberrations of the normal morphogenetic process. One mutant (carrying the *dev*2187 change in its DNA) was isolated from a spreading sector; it displays quite normal microcolony development, and only at a later stage in colony morphogenesis does it assume an abnormal, more extensive growth pattern. The other mutant (carrying the *dev*2099 change) was isolated because it altered the β-galactosidase expression pattern in a certain strain (i.e. *dev*2099 caused a change in the concentric rings seen by biochemical staining). At the start of development, this second mutant displays clearly abnormal microcolonies containing many elongated cells; however, after three days' incubation, the overall structure of its colonies cannot be distinguished from those of the parent strain. Thus, it can be seen that the effects of particular genetic alterations can be expressed at different stages of the developmental process, and the *dev*2099 phe-

notype also shows that alterations of the control of cell growth at particular stages of morphogenesis do not automatically lead to changes in the macroscopic structure of the final multicellular community.

3. Observations with *Proteus mirabilis*

3.1. Swarm colony morphogenesis

Morphogenetic analysis with *E. coli* is valuable because it illustrates the generality among bacteria of phenomena like cellular differentiation, cell–cell interaction, and coordinated multicellular regulation. Nonetheless, *E. coli* colony development does not proceed as rapidly or dramatically as morphogenesis by highly motile organisms like the Myxobacteria [4] and *Proteus mirabilis* [11–13]. These migrating bacteria form spectacular large swarm colonies on laboratory media; those of *P. mirabilis* can cover an entire 9 cm petri dish in less than a day and display a highly organized terraced appearance (fig. 3).

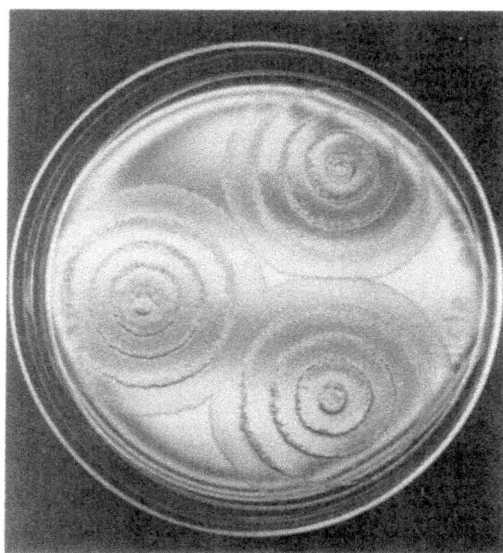

Fig. 3. Three terraced *P. mirabilis* swarm colonies after just one day's growth at 32°C. These colonies were produced by a wild strain (PRM 48 000) isolated from a patient.

Real-time and time-lapse video recordings of *P. mirabilis* swarm colony development using reflected-light microscopy make it possible to see some of the intricacies of bacterial morphogenesis. Of special interest are the rhythmic and sequential nature of alternating cycles of growth and migration [3, 11–13].

3.2. Sequential events in swarm colony morphogenesis

The most convenient way to inoculate *P. mirabilis* for observing swarm colony development is with a small drop of about 1 μl containing on the order of 10^5 bacteria. The bacteria in this

drop are short rods (up to 4 μm in length), and each has a few flagellae needed for swimming motility. As these bacteria multiply on agar, the inoculated zone fills in, structure develops (fig. 4A) and very long multinucleate hyperflagellated swarmer cells appear and migrate to the colony edge [1, 2]. At the edge, the swarmer cells move tangentially, become increasingly active, associate in groups and begin to synchronize their flagellar movements. At a certain point of growth (which depends on the strain and the medium), "rafts" of swarmer cells begin to migrate outwards from all around the colony perimeter (fig. 4B). These rafts contain tens of cells and are encapsulated in polysaccharide-containing polymers [13]. Migra-

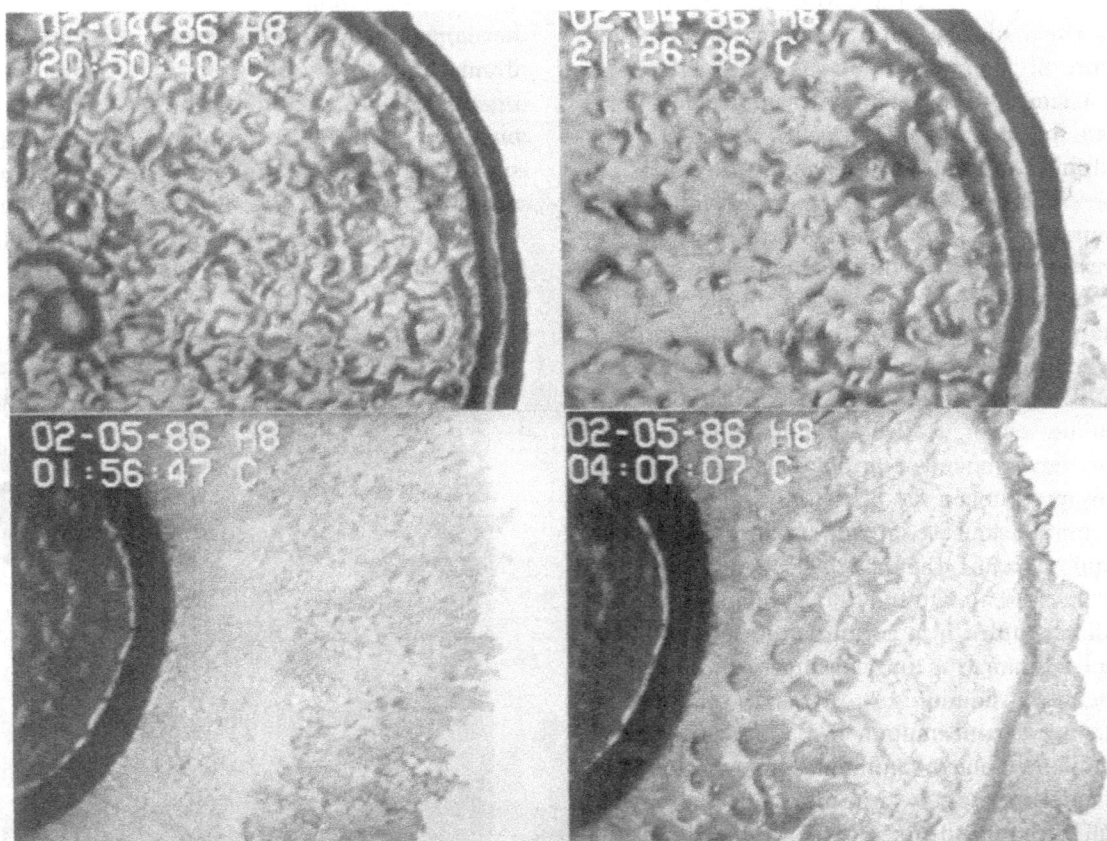

Fig. 4. Frames from a time-lapse video sequence of a wild *P. mirabilis* strain (PRM 1) during swarm colony development. The first two frames (2/4/86 20:50:40 (A) and 21:26:36 (B) are magnified ≈ 50 × ; the second two frames (2/5/86 1:56:47 (C) and 4:07:07 (D) are magnified ≈ 20 × . The colony was inoculated at 11:00 on 2/4/86.

tion of each raft appears to be a cooperative phenomenon: if a single cell is detached from a raft, it stays in place until incorporated into another passing raft. Migration continues outwards until the rafts gradually slow down all around the colony. With the particular strain illustrated here, the rafts in the initial swarm phase are relatively independent and, consequently, the border of the colonized zone is ragged (fig. 4C). As the swarm cells complete their first phase of migration, a wave of cell division and growth spreads from the central inoculation zone and covers the colonized surface; as this wave reaches the colony edge, a second swarm phase begins, this time with more highly organized fronts of migrating swarm cells (fig. 4D). Note that the final outline of the terrace formed by this cycle is not determined by the initial swarm cells but by the subsequent multiplication phase. When the second swarm cycle stops, the whole colony surface becomes still, and there is no wave of multiplication spreading from the center. Instead, the newly colonized area thickens, then movement begins synchronously around the colony just inside the perimeter, and a new swarm phase starts migrating. It is important to observe how the second and later swarm phases begin with a number of separate migration fronts which then align themselves and coalesce into a single unified swarm front (fig. 5).

3.3. Environmental effects on swarm colony morphogenesis

These repetitive cycles of swarming, stopping, multiplication and swarming can proceed quite regularly, and graphs of the movement of the colony edge display a step-function [3]. Terrace periodicity is quite regular for a particular strain under a given set of conditions, but the process is sensitive to environmental parameters such as medium composition and temperature. Terraces are formed more rapidly on richer medium and at 37°C as compared to 32°C. Thus, it appears that the swarming cycle depends upon the ability of

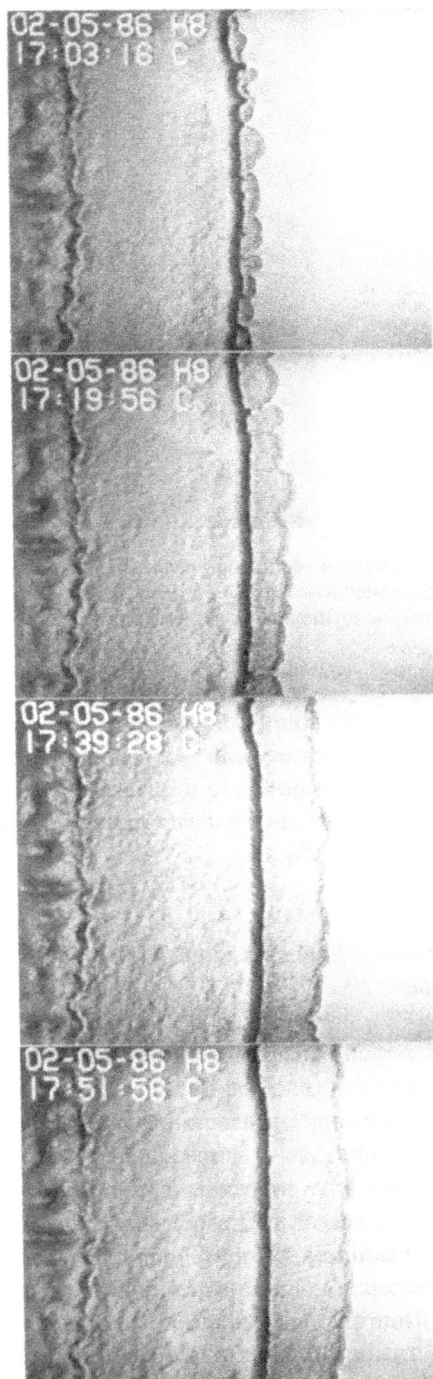

Fig. 5. Frames from the same video sequence as in fig. 4. ≈ 20 ×.

Fig. 6. Temperature-shift experiment. A mutant strain (PRM16) was inoculated and incubated at 32°C for 16 h. The plate was then refrigerated at 4°C for 10.5 h and then returned to 32°C for an additional 24 h incubation. At the time of refrigeration, the terrace marked by the arrow was being formed. Note that the next terrace was shortened, even though it developed at 32°C.

the bacteria to complete a series of physiological events. Temperature-shift experiments indicate that metabolism during the formation of one terrace plays a role in the development of the succeeding terrace (fig. 6).

3.4. Swarm colony development on modified substrates

One popular class of explanation for *Proteus'* cyclic swarming patterns is that cells at the edge exhaust nutrients (or accumulate metabolic waste products) and, as a consequence, periodically differentiate into swarmers and migrate chemotactically in search of fresher substrate. This idea can be examined by modifying the structure of the substrate. A simple experiment is to cut out a trench from the agar substrate, thereby creating a barrier to the diffusion of molecules through the gel. When swarm colonies form on such agar, the trench can have a dramatic effect on the terracing pattern. With certain mutant strains, no migration takes place in a zone of the agar surface behind the trench; in effect, this empty zone is "shadowed" by the trench (fig. 7). Such a result cannot readily be explained by local physiology at the colony edge and implies that the cyclic swarming behavior of the migrating bacteria requires some signal coming from the colony center. The signal could be a diffusible substance moving through the agar. Experiments with crystals of nigrosine dye deposited on agar plates containing trenches similar those in fig. 7 show that the pattern of dye diffusion around the trenches mimics the pattern of PRM16 and PRM2005 swarming. Since the colony center can be excised without altering swarming after two or more terraces have been formed, it is possible that the postulated diffusible signal is emitted only during the early stages of colony development. With the non-mutant parent strains, trenches do not have the same "shadowing" effect. One or more terraces may be modified after the colony front has passed the trench, but then normal swarming and terrace formation recover behind the obstacle. It thus appears that the mutants lack some compensatory mechanism.

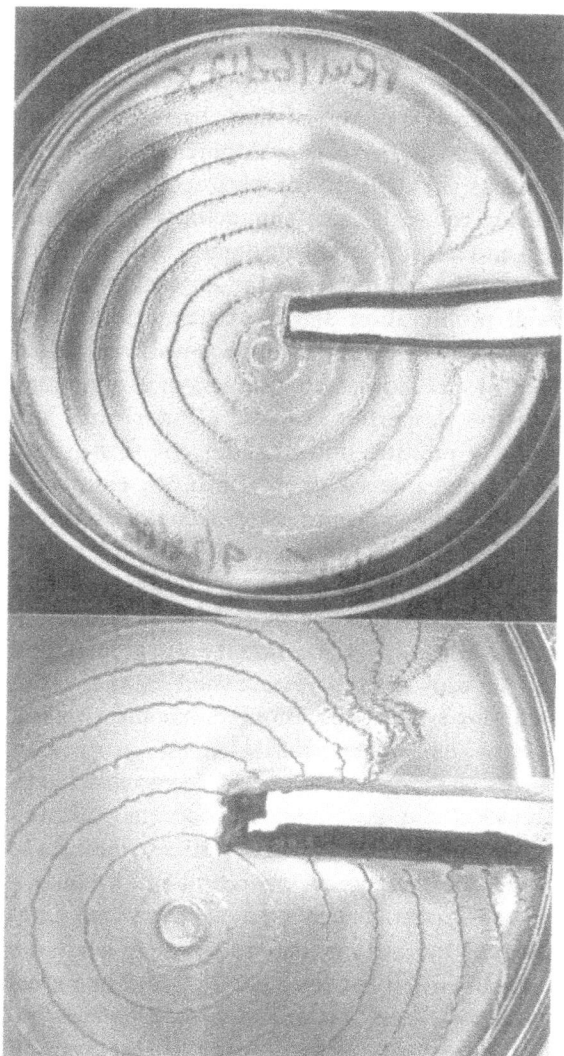

Fig. 7. Shadowing experiment. Mutant *P. mirabilis* strains were inoculated on petri dishes with a trench cut out of the agar. After several days incubation, a portion of the agar behind the trenches remained uncolonized. Top: PRM16 (a mutant of PRM1). Bottom: PRM2005 (a mutant of PRM2).

3.5. A mutant with altered growth control

Mutant isolation is also very useful in disentangling which aspects of colony morphogenesis are due to experimental conditions and which to biological functions of the bacteria themselves. One instructive mutant (PRM2007) was isolated in two

steps from a clinical isolate of *P. mirabilis*. PRM2007 is severely crippled in its swarming capacity and displays grossly abnormal cell morphology during growth in liquid medium. On solid medium, PRM2007 has lost the circular symmetry normally seen in bacterial growth and displays, instead, a branching fractal-like growth pattern on agar plates (fig. 8, top panel). Interestingly, PRM2007 is very sensitive to irregularities in the growth medium, and its initial growth extensions tend to track stress lines normally present in the agar substrate (fig. 8, middle panel). This tracking behavior can be directed by inoculating PRM2007 on medium containing thin depressions in the agar surface made with a sterile platinum bead (fig. 8, bottom panel). Thus, it appears that the normal circular growth of colonies produced by *P. mirabilis* and other bacteria is not the inevitable consequence of many small objects proliferating on an artificially smooth surface. The stress line effect indicates that the substrate is far from uniform at the scale of the bacterial cells, and the existence of a mutant with branching, non-circular colonies shows that the geometry of growth is subject to genetic control. PRM2007 illustrates how genetics can reveal unsuspected control processes in two ways: (a) by removing normal regulatory systems, such as the one that allows the bacteria to ignore stress lines in the agar, and (b) by uncovering otherwise invisible patterns, such as branching growth.

4. Discussion

There are basically two points to be made about the observations reported here. The first is that bacteria are excellent experimental material for studying very fundamental problems of biological pattern formation. Patterns are readily observed with inexpensive equipment, and the system is easy to manipulate experimentally. Many parameters can be altered, and the internal properties of the pattern-forming system can be modified by genetic methods. It is worth emphasizing

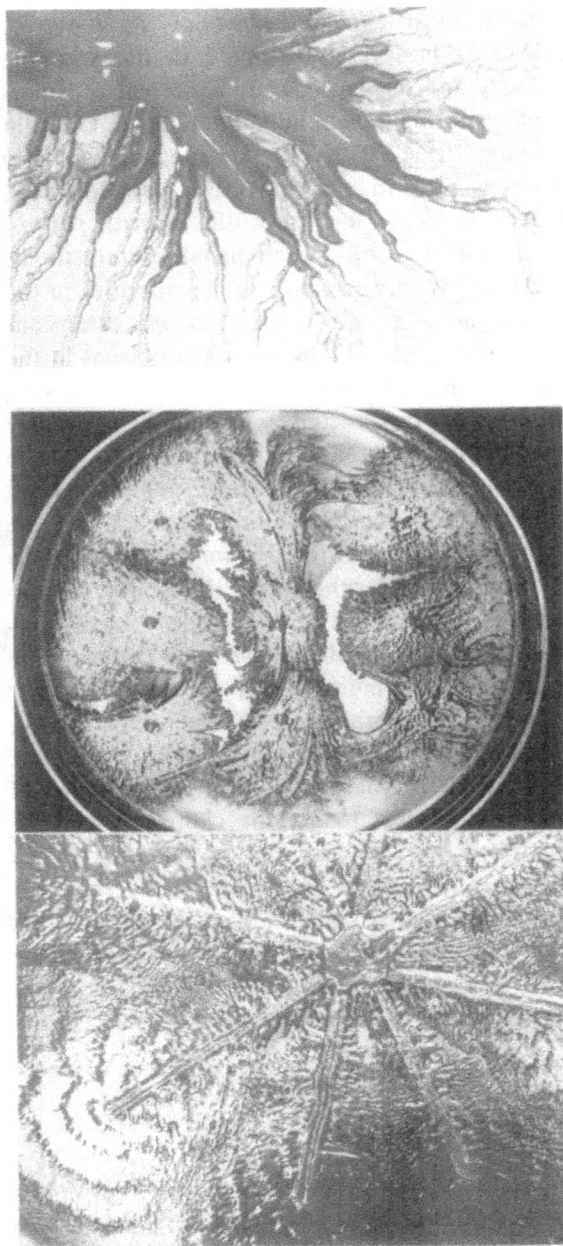

Fig. 8. Aberrant growth of strain PRM2007 (a mutant of PRM2005, hence a double mutant of PRM2). Top: Part of a branching colony growing from a stab inoculation site. Middle: Several stab colonies whose initial growth tracked stress lines in the agar surface. Bottom: The central portion of a colony whose initial growth extensions had tracked shallow indentations placed in the agar with a platinum bead.

that bacterial colonies display the same complexity, regularity and robustness which characterize pattern formation in higher organisms.

The second point is that bacterial colony morphogenesis, like development in higher organisms, is a highly complex sequential affair involving cell–cell signalling, cellular differentiations, and multicellular coordinations that continually change and influence the ongoing processes of cell proliferation and movement. Because of the intricacy of colony morphogenesis, we find it difficult to see how small numbers of assumptions about the properties of basic cellular units will provide a realistic understanding of this or any other complex developmental process. Instead of predicting the behavior of the composite from rules that pertain to the component parts, we consider it likely that successful explanatory models will be based on the more biological concept that the behavior of individual cells is regulated to fit in with patterns established by the entire developing system. For example, our thinking will need to encompass the idea of progressive changes in the properties of the component cells as multicellular development proceeds. In addition, a number of the observations indicate a capacity for overcoming environmental irregularities, adjustment to new circumstances and compensation for genetic defects. This is what the classical embryologists called "developmental regulation". Thus, we will also need to incorporate the concepts of monitoring, evaluation and rectification into theories of biological morphogenesis. Neither the nature of these theories nor the proper experimental approaches for testing them are yet clear. Molecular genetic analysis can tell us about the individual components of a particular morphogenetic system, and this approach has been extremely successful with higher organisms. Nonetheless, identification of particular routines for controlling gene expression, cell–cell recognition, intracellular signal transduction and establishment of spatial coordinates does not explain the overall integration of the developmental process. Thus, a major challenge for the future is to identify the physical basis of central information processing in biological pattern formation.

Acknowledgements

We thank Nancy Cole for technical assistance. This research was supported by grants from the National Science Foundation (DCB-8416998, DCB-8816274, DMB-840140 and DMB-8715935).

References

[1] J.F. Hoeniger, Can. J. Microbiol. 10 (1964) 1.
[2] J.F. Hoeniger, J. Gen. Microbiol. 40 (1965) 29.
[3] J. Kvittingen, Acta Path. Microbiol. Scandinavica 26 (1949) 24.
[4] E. Rosenberg, Myxobacteria: Development and Cell Interactions (Springer, Berlin, 1984).
[5] J.A. Shapiro, Symp. Soc. Gen. Microbiol. 36 (2) (1984) 169.
[6] J.A. Shapiro, J. Gen. Microbiol. 130 (1984) 1169.
[7] J.A. Shapiro, J. Bacteriol. 164 (185) 1171.
[8] J.A. Shapiro, J. Bacteriol. 169 (1987) 142.
[9] J.A. Shapiro, Sci. Am. 258 (1988) 82.
[10] J.A. Shapiro and C. Hsu, J. Bacteriol. 171 (1989) 5963.
[11] S.A. Sturdza, Zbl. Bakt. Hyg. I. Abt. Orig. A 233 (1975) 505.
[12] S.A. Sturdza, Zbl. Bakt. Hyg. I. Abt. Orig. A 238 (1977) 444.
[13] F.D. Williams and R.H. Schwarzhoff, Ann. Rev. Microbiol. 32 (1978) 101.

Physica D 49 (1991) 224–232
North-Holland

Analysis of optical density wave propagation and cell movement in the cellular slime mould *Dictyostelium discoideum*

Florian Siegert and Cornelis J. Weijer[1]

Zoological Institute, University of Munich, Luisenstrasse 14, W-8000 Munich 2, Germany

We have studied optical density wave propagation during aggregation of the cellular slime mould *Dictyostelium discoideum* in a quantitative manner by digital image analysis. The waves are mostly single ended spiral waves starting from an aggregation center. We can measure a variety of parameters such as oscillation frequency, wave propagation velocity and wave shape. This allows the construction of dispersion curves under a variety of experimental conditions. During later development where the optical density waves are no longer visible we have started to measure movement of fluorescently labelled cells. Our main conclusions from these measurements are that the cells continue to move chemotactically to periodic signals both in aggregates and in slugs. There is a dramatic difference in the movement pattern of prestalk and prespore cells: Prestalk cells move perpendicular to the long axis of the slug, they are most likely organized by a scroll wave. Prespore cells seem to move almost perpendicular to the prestalk cells, in the direction of the tip. This behaviour is explained on the basis of different relay properties of prespore and prestalk cells.

1. Introduction

During development of the cellular slime mould *Dictyostelium discoideum* single cells, dispersed over a substratum, aggregate towards an aggregation center. The cells collect in multicellular aggregates (10^3–10^5 cells), which transform into a conditional motile stage, the slug. During slug formation the cells start to differentiate into at least two cell types, prespore and prestalk cells. The slug finally transforms into a fruiting body which is composed of two cell types, stalk cells and spore cells [1, 2]. The aggregation process is brought about by three cellular competences: (1) Periodic production and secretion of cyclic adenosine monophosphate (cAMP) by the cells in the aggregation center. (2) Detection of this signal by cell surface cAMP receptors followed by amplification of the cAMP signal via the activation of adenylate cyclase (relay response). This response is subjected to adaptation which ensures

outward propagation of cyclic AMP waves. (3) Chemotaxis towards increasing cAMP concentrations. Cells respond to temporal increases in cAMP and therefore cells aggregate unidirectionally in the direction of the aggregation center [3]. Many of the biochemical reactions underlying the cAMP relay response are now known since the dynamics of these reactions can be studied effectively in synchronized cell populations in suspension [2–4]. Based on these data several models have been proposed for the cAMP oscillator [5–7], of which the three-variable model of Martiel and Goldbeter [7] seems to be able to explain the basic features of the oscillator. cAMP wave propagation can be seen as an optical density wave propagation with low-power darkfield optics. The wave impression is caused by a change in the shape and coherence of the cells during their chemotactic response. Chemotactically moving cells are seen as white (light scattering) bands [8]. The optical density waves can be seen to propagate outward from the aggregation center. The patterns seen are either concentric waves or spirals [9, 10]. The quantitative study of aggregation

---

[1]To whom all correspondence should be sent.

dynamics in vivo has recently become possible by the use of digital image analysis [11].

During later development all morphogenetic movements are organized by the tip, a distinct morphological structure, which is present from late aggregation onwards. The tip behaves as an organizer because it will induce a secondary axis and take over part of the tissue behind it to form a new secondary slug when it is grafted into the side of another slug [12]. The tip also inhibits the formation of new tips [13]. Many of the properties of the tip can be explained by assuming that it is a pacemaker for cAMP signals. The test of this hypothesis is however not as straightforward as it might seem, since in these later structures darkfield waves are no longer visible. The darkfield waves can only be seen for about 20 periods during early development and disappear when the cells start to make aggregation streams [9–11]. The goal of the work described below is to compare quantitative measurements of optical density wave propagation (spreading of the signal) and of cell movement (the cell response). Analysis of cell movement is currently the only way to deduce the pattern of signal propagation in aggregates and slugs.

2. Results

2.1. Analysis of optical density wave propagation during early aggregation

We have studied several parameters associated with the propagation of optical density waves during the initial stages of aggregation. In general there are around 20–25 waves emanating from an aggregation center. The waves can be concentric, but with our strains and experimental conditions more than 95% of the waves form spirals (fig. 1A). Almost all spirals have only one arm (fig. 1A), but on rare occasions unstable double armed spirals can be observed (figs. 1B, 1C). We have developed a technique based on digital image analysis with which we can easily measure oscilla-

tion frequency, wave propagation speed, wave shape and amplitude of propagating optical density waves [11]. In wild type cells it is seen that the oscillation frequency increases during development (period length decreases from 6 to 3 min), while the propagation speed decreases from 600 to 300 μm/min [10, 11]. From these data one can construct a dispersion curve (relationship between wave propagation velocity and period length, open circles in fig. 2) that fits well with model calculations based on the Martiel–Goldbeter model [14].

2.2. The relay inhibitor caffeine induces biphasic waves

By disturbing the normal kinetics of the oscillation with the aid of mutations or drugs it will be possible to investigate which variables are important for wave formation and stable propagation. We have investigated the effect of caffeine (an inhibitor of adenylate cyclase activation [15]) on wave formation and propagation and found that inhibition of the cAMP relay response (cAMP signal strength) leads to fewer aggregation centers and slower oscillating and slower propagating waves [11]. This is reflected in a widely different dispersion relation as compared to the control case (fig. 2). Since the form of the dispersion relation is mainly dependent upon the chemical kinetics of the oscillator it will be of interest to see how these findings can be accommodated in models for the cAMP oscillator. Since it is now possible for the first time to obtain experimental data rather easily and with good accuracy it offers the possibility for a fruitfull interaction between experiment and theory [11, 14].

2.3. Biphasic waves

We have found biphasic waves under a variety of experimental conditions. They can be very clearly seen in cells aggregating on the relay

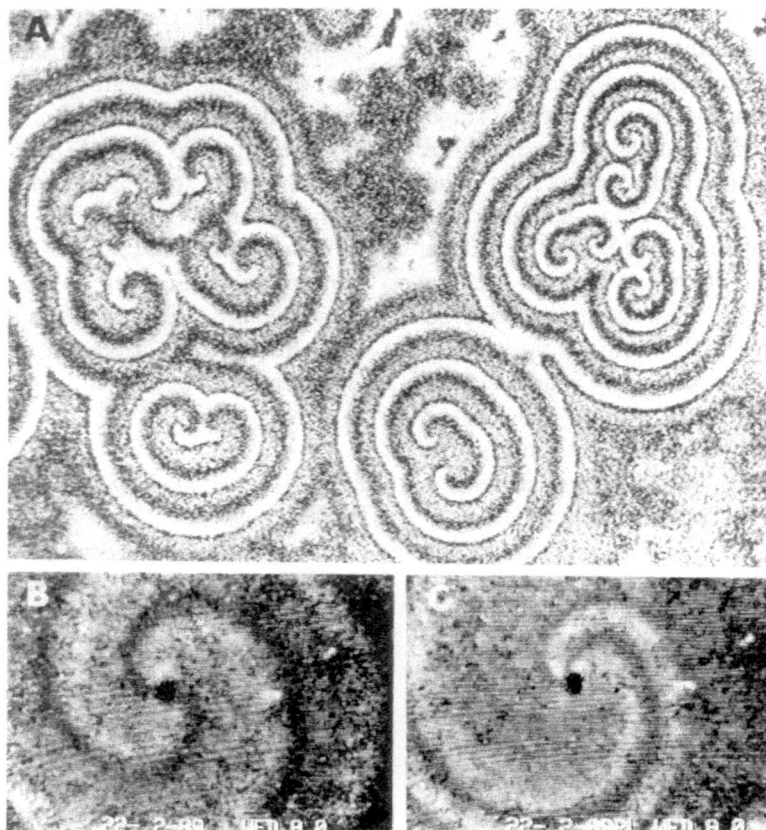

Fig. 1. (A) Low-power view of optical density waves in a field of aggregating AX-2 cells. The appearance of the waves was enhanced by real time image differencing between all incoming waves and an initial reference frame. This procedure detects changes from a reference time point and indicates motion. (B) High-power darkfield photograph of a double armed spiral. (C) High-power photograph of the same spiral shown in (B), 13 min later. The double armed spiral is unstable since one arm overtakes the other, which leads to extinction of both arms.

inhibitor caffeine (fig. 3). Analysis of the power spectrum shows that these signals are composed of two harmonic frequencies [11].

Since the optical density signal consists of the response of many cells we believe that the wave form may reflect a heterogeneity in the response of the cells. The presence of discrete response elements (cells) with variable properties distinguishes this system from more homogeneous chemical systems. Not all cells can make a response every time that they are hit by a signal since they are not yet completely deadapted. This leads to a partial gating of signals coming from the aggregation center, i.e. some cells can re-

spond to every signal while others with an intrinsic longer deadaptation time can only respond to every second pulse. This gating can happen either at the level of the relay response or at the level of the chemotactic response [11, 16]. If gating occurs at the level of the chemotactic response this will lead to a sorting out of fast and slow responding cells. Fast cells will end up in the tip, thereby creating an inhomogenous excitable medium. In order to distinguish between these two alternative hypotheses we will have to measure the rate of movement of individual cells under conditions where the optical density signals are mono- and bi-phasic.

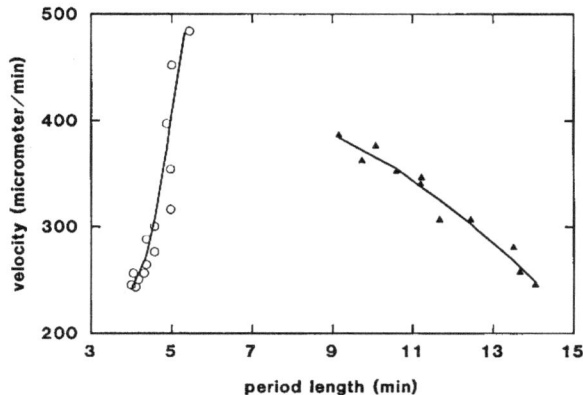

Fig. 2. Relationship between wave propagation velocity and period length (dispersion relation) of successive waves during aggregation. The dispersion relation is shown for wild type AX-2 cells (open circles) and AX-2 cells aggregating on 5 mM caffeine (solid triangles). The data shown are the mean values derived from more than 40 aggregation centers in 15 independent experiments. Wave propagation velocity and period length were determined by digital image processing as described in ref. [11].

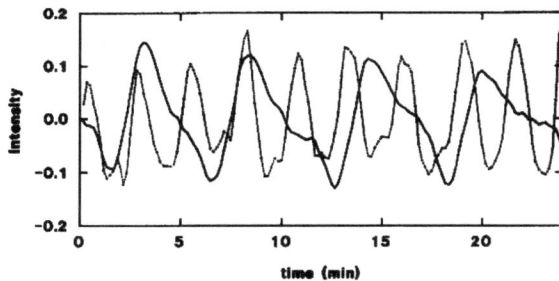

Fig. 3. (A) Optical density oscillation of AX-2 cells showing regular oscillations (dotted line) and of cells aggregating on 5 mM caffeine showing biphasic oscillations (solid line). For methods see ref. [11].

2.4. Analysis of movement of single cells

As a first step we have measured the movement of individual cells aggregating to an aggregation stream. We have chosen fluorescent tagging of the cells since this allows us to follow individual cells in multicellular stages of development, such as in aggregation streams, moulds and slugs. We can detect and follow several labelled cells every few seconds anywhere in a video frame

and save the grey level information of all the pixels of the cell and a small surrounding square. This allows us to follow the cell shape and fluorescence intensity changes while it is moving. Simultaneously we can calculate several parameters like rate and direction of movement and changes in cell shape. This is a procedure which will also allow us in the near future to correlate changes in physiological variables during chemotaxis using the newly developed fluorescent ion specific membrane potential sensitive dyes.

As can be seen the rate of movement of these cells is highly periodic and can be easily measured (fig. 4).

2.5. Movement of cells in aggregation streams

To test whether it is possible to measure periodic cell movement in streams we have measured movement rates of individual cells in aggregation streams of streamer mutants. Streamer mutants form long aggregation streams in which one can see the propagation of optical density waves (figs. 5A, 5B) [11, 16, 17]. These waves arise by the local accumulation and depletion of cells and reflect differences in stream diameter. As can be seen in fig. 5C individual cells do move periodically although the periodicity is not as clear as in the darkfield waves. This must be attributed to the fact that we are now looking to the response of an individual cell and not to a population response as in the case of darkfield waves. However, power spectrum analysis indicates a clear periodicity of cell movement of approximately 2.5 min (not shown).

2.6. Movement of cells in slugs

We have performed similar measurements in slugs and it appears that the movement of single cells in slugs is at least as periodic as that of cells in aggregation streams (fig. 6B). Furthermore it can be seen that the cells change their shape in a characteristic fashion during the periodic move-

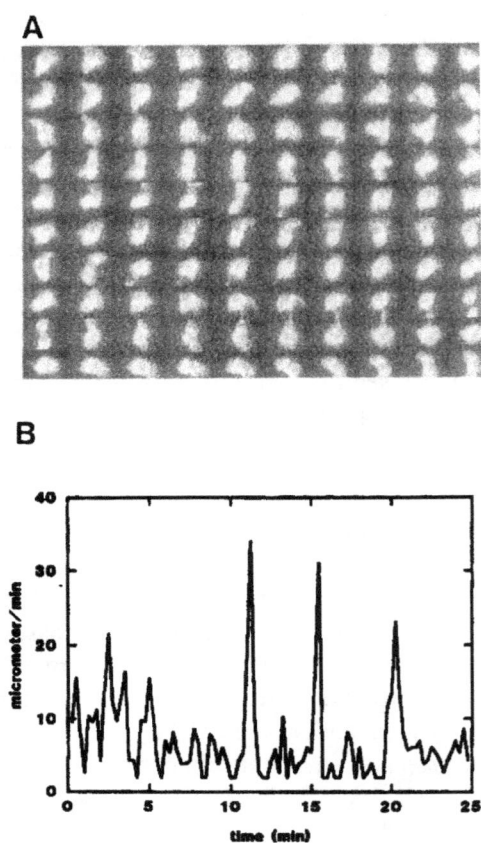

Fig. 4. Movement of a single cell towards an aggregation center. (A) Successive frames showing the shape changes of a single cell aggregating towards an aggregation center. The cell was labelled with a rhodamine dextran (MW 10 000 D) conjugate that was brought in the cell by filter loading. The cells to be loaded were pressed through a 7 μm Nythal filter (Schweiz. Seidengazefabrik AG Thal) in the presence of rhodamine dextran. Under the conditions employed about 1% of the cells were labelled and retained the label for more than 24 h. Fluorescence was detected by a Hamamatsu silicon intensified target (SIT) camera (C-2400-8) and the resulting image digitized with 512 × 512 pixel resolution. Cells were detected by recording intensity values over a certain threshold and the center of mass was calculated. A window of 50 × 50 pixels around the center of the cell was saved on a hard disk. This procedure was repeated every 10 or 15 s. From these data a new picture was constructed by placing all consecutive images in rows behind each other. The top left image is the first image and the bottom right image is the last. (B) Rate of movement of the cell seen in (A). The movement was calculated on the basis of the displacement of the center of the cell as described in (A). Every data point corresponds to a picture in (A).

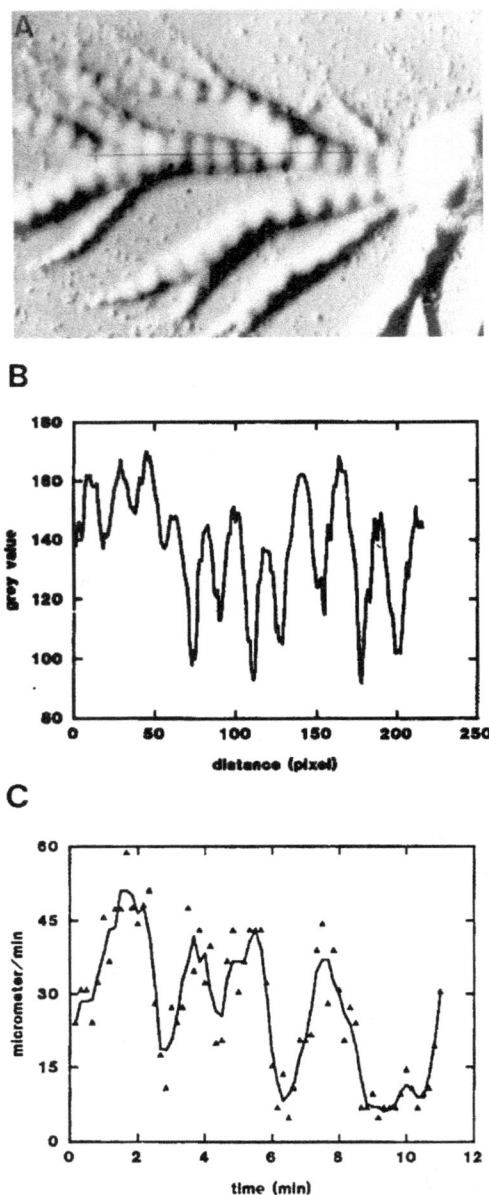

Fig. 5. Optical density wave propagation and cell movement in aggregation streams of streamer mutant NP377. (A) Digitized image of aggregation streams of the streamer mutant NP377. (B) Optical density waves measured along the black line through the aggregation stream shown in (A). (C) Rate of movement of a fluorescently labelled cell in such an aggregation stream. The rate of movement was determined as described in the legend of fig. 4. The solid line was calculated by moving average smoothing over three neighbours. The original data are shown as solid triangles.

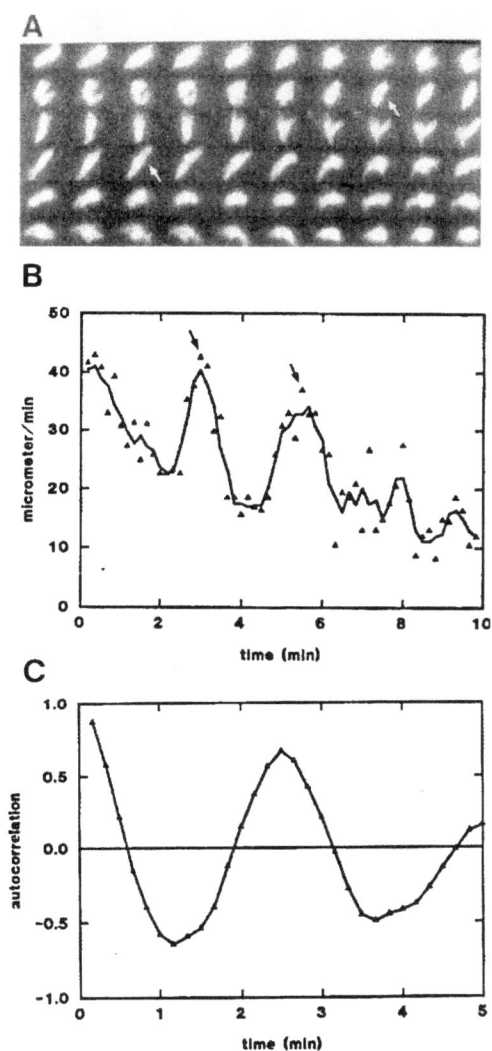

Fig. 6. Movement of a fluorescently labelled cell in the pre-spore zone of a slug (AX-2). (A) A time series of images showing the successive shape changes of the cell. Notice the periodic change in cell shape. If the cell is moving fast it is elongated, when it is moving slow it extends pseudopods. Two successive time points where the cell moves fast are indicated by white arrows in the second and the fourth row of the time series. (B) Rate of movement of the cell shown in (A). Two time points where the cell is elongated (two white arrows in (A)) are indicated by two black arrows. This shows that elongated cells move fast. Methods as described in fig. 4. The solid line was calculated by moving average smoothing over three neighbours. The original data are shown as solid triangles. (C) Autocorrelation of data in (B) shows that the cell movement is clearly periodic, with a for slugs typical period length of 2.5 min.

ment. At the times of slow movement they stick out pseudopods in different directions before they decide to continue in one or to proceed in the going direction (fig. 6A). This is a behaviour that is well known from cells making chemotactic movement and shows that cells in slugs move in a chemotactic fashion.

2.7. Global analysis of movement in aggregates and slugs

Since a slug consists of 10^3-10^5 cells it is rather difficult to extract data about the behaviour of the cell population by the above mentioned methods. Therefore we have resorted to a technique that allows us to map the movement of many cells in a slug simultaneously. In a slug that contains many labelled cells, we can detect which cells have a fluorescence above a certain threshold and keep track of the times that pixels are above threshold. If this information is displayed one obtains tracks where cells have moved. Additionally we keep track how many times a pixel at a certain location is above threshhold. This is displayed in a 16-valued red scale from black to bright red whereby bright red indicates that a pixel has been above threshhold 16 times or more. Therefore if a cell moves slowly the resulting track appears bright red at this location and if the cell is moving fast the track appears dark red or black (plates I and II). This method allows us to track fluorescently labelled moving cells in real time and to display this information in an easily interpretable way.

To test the method we have followed several labelled cells in an AX-2 aggregate. As can be seen (plate IA) cells in streams move periodically in the direction of the aggregation center, while the cells in the body of the aggregate move in a spiral fashion around a central core. The movement of the cells is indicated in the lower graphical representation of the results (plate IB).

We have observed spiral movement in aggregates and mounds in many experiments (results will be published in detail elsewhere). Now the

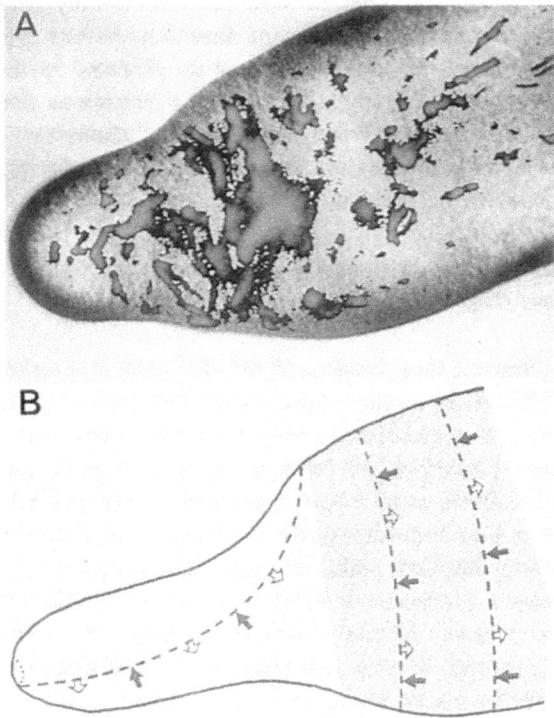

Plate I.

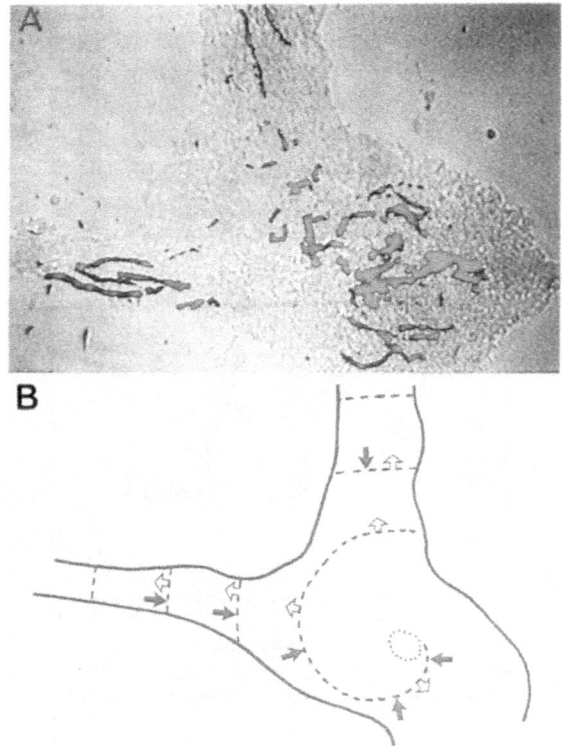

Plate II.

Plate I. Direction and rate of movement of several cells in an aggregate. (A) Cells were fluorescently labelled as described in fig. 4. The image of the aggregate was digitized and a threshold was set such that all the intensity values of fluorescent cells were above the threshold. The position of all pixels above threshold was saved in the memory of our video digitizer board and given the value 1. This procedure was repeated for successive times and when the grey level of a pixel above threshold had been above threshold previously the value of this pixel was incremented by 1. This procedure leads to the formation of tracks which show where the cells have moved. The tracks are shown superimposed on the original picture. The value of a given pixel in the track is color coded in a 16-valued red scale. Black means occupied only once (fast movement) and bright red means occupied 16 times or more (slow or no movement). It can clearly be seen that the cells in the aggregation stream move periodically in the direction of the center while the cells in the aggregation center move more continuously in spirals. (B) Schematic diagram of observed cell movement (black arrows) and the thereof derived propagation of waves. The wave fronts are indicated as broken lines, the direction of propagation is indicated by the open arrows.

Plate II. Movement of cells in a slug. (A) Several fluorescently labelled cells both in the prespore and prestalk zone of a slug are followed with the method described in plate I. Cells in the prespore zone move in the direction of the tip, while cells in the prestalk zone move perpendicular to the slug tips main axis. (B) Schematic diagram depicting cell movement in slugs and the derived propagation of the waves. The direction of cell movement is indicated by the black arrows, the wave fronts are indicated by the broken lines and the direction of wave propagation by the open arrows.

interesting question is what happens in slugs? As can be seen in plate IIA cells in the prestalk zone of slugs move in a direction almost perpendicular to the long axis of the slugs, while cells in the back (prespore zone) of the slug move in a direction parallel to the long axis of the slug. Since the chemotactic movement in the prestalk zone is perpendicular to the propagating wave front, this implies that the wave front must extend along the long axis of the slug tip. Therefore the slug tip is a scroll wave which decomposes in waves traveling almost perpendicular to the long axis of the slug in the prespore zone. This situation is depicted schematically in plate IIB. These observations are not compatible with the inverse fountain flow model for slug movement as proposed recently [18], but are compatible with the observations of movement of neutral red stained anterior-like cells in the prespore zone of slugs [19, 20].

This change in the geometry of the propagating waves is in our opinion most likely caused by a difference in the oscillatory properties of the cells in the prestalk and in the prespore zone. It is well documented that prespore and prestalk cells sort out from each other when mixed experimentally and that this sorting out is due to chemotaxis [21, 22]. This opens in principle the possibility that one collects cells with different oscillatory properties in the prestalk and prespore zone of the slug. We have previously shown that one can separate aggregation stage cells that will sort to the tip of the slug from those that will sort to the back of the slug. The cells that will sort to the front have higher intrinsic oscillation frequencies than cells that will sort to the back of a slug [23]. Furthermore cells in the prespore zone can break-down the external cAMP less efficiently as the cells in the prestalk zone since they have less phosphodiesterase [24]. This would leave them always being partly adapted. Since they are partly adapted they will oscillate slower and the amplitude of the released signal will be smaller than that from the prestalk cells. The result of these differences will be that cells that are in the tip are those cells that will be able to oscillate and chemotax rapidly.

They will move around in a slightly tilted scroll wave [25] along the tip long axis. The other cells outside the tip cannot propagate the cAMP signal anymore around the tip axis, which will be the minimal refractory period attainable by a spiral wave. The end of the scroll wave however will be a continuous source of cAMP which can trigger new waves in the prespore zone which can then move along the slug main axis. Waves will be able to propagate either along the slug main axis or as slowly winding spirals depending on the diameter of the prespore piece.

It is clear that there has to be a supercellular control mechanism to coordinate the behaviour of a mass of single cells in the slime mould. On the basis of the experiments described above, we believe that the tissue is an excitable three-dimensional medium, and that morphogenesis must be based on slight local changes in signal propagation and chemotactic responses. An extra complication in the description of this system is that the responding units (cells) can change their neighbours and rearrange themselves in new patterns, contributing to spatial heterogeneity. This introduces an extra feedback element into the system that will make it even more interesting to study. We are looking forward to a comparison of this intuitive explanation with three-dimensional model calculations for example based on cellular automata [26].

Acknowledgements

We like to thank Dr. P.C. Newell (Oxford) for his gift of the streamer F mutants and Dr. H.K. MacWilliams for critically reading the manuscript. This work was supported by a grant from the Deutsche Forschungsgemeinschaft to C.J. Weijer (We 1127).

References

[1] W.F. Loomis, Dictyostelium discoideum, A Developmental System (Academic Press, New York, 1975).

[2] R.H. Kessin, Microbiol. Rev. 52 (1988) 29.

[3] P.N. Devreotes, Adv. Cycl. Nucl. Res. 15 (1983) 55.

[4] P.C. Newell, G.N. Europe-Finner and N.V. Small, Microbiol. Sci. 4 (1987) 5.

[5] H. Meinhardt, Differentiation 24 (1983) 191.

[6] L.A. Segel, A. Goldbeter, P.N. Devreotes and B.E. Knox, J. Theor. Biol. 120 (1986) 152.

[7] J.L. Martiel and A. Goldbeter, Nature 313 (1985) 590.

[8] F. Alcantara and M. Monk, J. Gen. Microbiol. 85 (1974) 321.

[9] A.J. Durston, Dev. Biol. 37 (1974) 225.

[10] J. Gross, M. Peacy and D. Trevan, J. Cell Sci. 22 (1976) 645.

[11] F. Siegert and C.J. Weijer, J. Cell Sci. 93 (1989) 325.

[12] J. Rubin and A. Robertson, J. Embryol. Exp. Morphol. 33 (1975) 227.

[13] H.K. MacWilliams, Symp. Soc. Devel. Biol. 40 (1984) 463.

[14] J.J. Tyson and J.D. Murray, Development 106 (1989) 421.

[15] M. Brenner and S. Thoms, Dev. Biol. 101 (1984) 136.

[16] K. Gottmann and C.J. Weijer, J. Cell Biol. 102 (1986) 1623.

[17] F.M. Ross and P.C. Newell, J. Gen. Microbiol. 127 (1981) 339.

[18] G.M. Odell and J.T Bonner, Phil. Trans. R. Soc. London BM12 (1986) 487.

[19] A.J. Durston and F. Vork, J. Cell Sci. 36 (1979) 261.

[20] R.L. Clark and T.L. Steck, Science 204 (1979) 1163.

[21] S. Matsukuma and A.J. Durston, J. Embryol. Exp. Morphol. 50 (1979) 243.

[22] J. Sternfeld and C.N. David, Differentiation 20 (1982) 10.

[23] C.J. Weijer, S.A. McDonald and A.J. Durston, Differentiation 28 (1984) 9.

[24] A.P. Otte, M.J.E. Plomp, J.C. Arents, P.M.W. Janssens and R. van Driel, Differentiation 32 (1986) 185.

[25] A.T. Winfree, When Time Breaks Down (Princeton Univ. Press, Princeton, NJ, 1987).

[26] M. Gerhardt, H. Schuster and J.J. Tyson, Science 247 (1990) 1563.

Physica D 49 (1991) 233-239
North-Holland

Quantitative analysis of periodic chemotaxis in aggregation patterns of *Dictyostelium discoideum*

Oliver Steinbock, Hajime Hashimoto[1] and Stefan C. Müller

Max-Planck-Institut für Ernährungsphysiologie, Rheinlanddamm 201, W-4600 Dortmund 1, Germany

Wave patterns in the cellular slime mold Dictyostelium discoideum are investigated quantitatively by determining the chemotactic motion of the amoebae cells towards their aggregation center. The velocity of moving cells is analyzed by a pixel-based correlation program applied to digital microscopic video images of approximately 0.4×0.3 mm^2 area. The average velocity component in the direction of the center is clearly periodic with periods of 6-9 min, a maximum velocity of 20 to 30 μm/min and a minimum velocity close to zero. Details concerning the asymmetric shape of the velocity function are observed. The new technique allows the detection of oscillating behaviour in chemotactic motion, even after the macroscopic patterns observed by dark-field techniques have disappeared.

1. Introduction

During the developmental cycle of *Dictyostelium discoideum* amoebae from an assembly of single cells to a multicellular fruiting body, an aggregation phase of single cells is observed when cells start to undergo coherent motion towards a later aggregation center [1]. This motion is stimulated by intercellular wave-like propagation of the biochemical transmitter molecule cyclic AMP establishing communication between the cells [2, 3]. During aggregation the majority of the population rests in an excitable state but there are assemblies of cells serving as pacemakers in that they emit the cAMP pulses which excite neighboring cells. The geometries of the resulting waves are, in general, concentric circles (target patterns) or rotating spirals.

The aggregation patterns formed in this biological excitable system have been investigated over a long time as a model case for biological self-organization. There is a remarkable similarity to the trigger waves found in the Belousov–

Zhabotinsky reaction, the chemical model system for spatiotemporal structure formation under far-from-equilibrium conditions [4, 5].

This similarity was recently corroborated by quantitative methods [6]. The aggregation patterns are usually detected by dark-field illumination of macroscopic dimension, which takes advantage of the difference between the light scattering behaviour of moving, slightly elongated cells and that of resting circular cells. Thus, the regions of moving cells appear as bright bands (circles or spirals) (fig. 1). By combining this detection optics with digital recording of video images and computerized evaluation techniques, it was shown that the spiral shape is that of an Archimedian spiral or of an involute of a circle (not distinguishable within the range of experimental error). Overlay of image sequences reveal that there exists a core region in spiral patterns of approximately 300 μm radius around which the dark-field waves rotate. The relationship between normal propagation velocity and local curvature of a wave front fulfills a linear law and yields the diffusion coefficient of the autocatalytic species cAMP (0.66×10^{-5} cm^2 s^{-1}).

Together with the dispersion relation of aggregation waves, as reported in ref. [7], these find-

[1]Permanent address: Department of Information Science, Oshima National College of Maritime Technology, Komatsu 1091-1, Oshima-cho, Oshima-gun, 742-21, Japan.

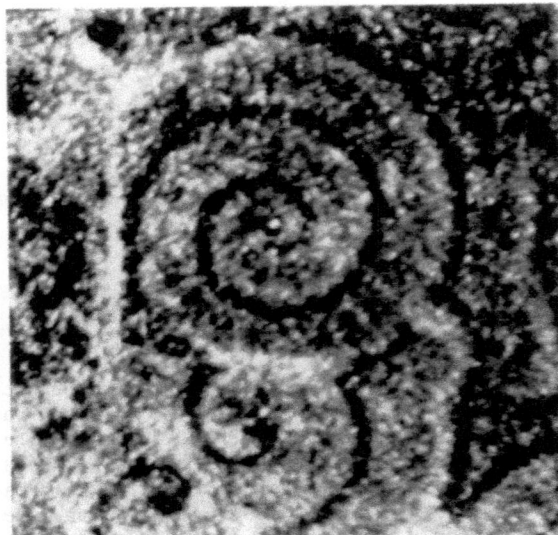

Fig. 1. Spiral wave pattern in *Dictyostelium discoideum* AX-2 aggregating on an agar surface, detected by dark-field photography. The brightness differences are correlated with their moving behaviour and regulated by differences in the local cAMP concentration.

ings verify predictions made on the basis of a reaction–diffusion model for excitable systems presented in refs. [8, 9]. This model contains two reaction–diffusion equations describing the dynamic interactions of extracellular cAMP and the active membrane receptor to which the secreted cAMP is bound, thus stimulating further cAMP production. The reaction kinetics are derived from the Martiel–Goldbeter model [10] and a diffusion term is implemented only for extracellular cAMP. The basic structure of this reaction–diffusion scheme is analogous to a model proposed for trigger waves in the BZ reaction, the predictions of which have been verified experimentally in a similar fashion [11].

In the case of the slime mold aggregation patterns the direct comparison between theory and experiment is complicated because of the following reasons: (1) the theory predicts the dynamic behaviour of the autocatalytic species cAMP, the experimental detection of which is very difficult. Only a small amount of data, obtained with isotope dilution-fluorography, is available for

frozen-in snapshots of the cAMP-structure [12]. The commonly used dark-field data supply only indirect evidence for comparison with theory. (2) The chemotactic motion of the cells during pattern evolution has to be taken into account in realistic models of the aggregation process. The global dynamic features of this motion have not yet been characterized quantitatively.

In this work we introduce a quantitative method for measuring the velocity of chemotactic cell motion. The method is based on mutual correlation analysis of temporal sequences of digital images with microscopic resolution recorded during the phases of dynamic pattern formation of the cell population [13]. This quantification of chemotaxis provides information about relevant parameters concerning the process of self-organization in this excitable system.

After briefly specifying the preparation of the *Dictyostelium discoideum* system and the principle of the applied correlation method, we present results on the velocity field analysis of chemotaxis in a small section of an aggregation pattern through which waves of chemotactic activity propagate periodically. The more complex features of chemotactic vortex-like cell motion around the core of a spiral-shaped aggregation pattern are discussed qualitatively.

2. Experimental

The cells of *Dictyostelium discoideum*, axenic strain AX-2, were cultivated on nutrient medium and harvested at a density of 5×10^6 cells/ml, washed three times with buffer and spread uniformly on an agar surface (containing 2 mM caffeine) in a petri dish at a density of approximately 4×10^5 cells/cm^2. The dishes were stored in the dark at 21°C for 4–6 h.

A petri dish was mounted on the stage of an inverse microscope (Zeiss IM 35). Pictures obtained by bright field microscopy were recorded with a video camera (Hamamatsu C2400) connected to a time-lapse video recorder (Sony EVT-

801CE). With this recorder quick-motion pictures (factor 25) were taken. The area of the recorded pictures was 0.39×0.32 mm^2.

The electronic equipment for image evaluation consists of a personal computer with a hard disk (PC, Siemens PCD-3TS) and a Convex-C201 computer. An optional image acquisition card (Data Translation, DT-2851, 512×512 pixels, 8 bits grey level, 3:4 aspect ratio and 50 Hz sampling speed) and a 6 MByte expanded memory card are installed in the PC.

With this equipment it is possible to store sequential image data with arbitrary image size, image position, and sampling speed. These data are taken into the PC extended buffer memory through the image acquisition card and then sent to the Convex computer to which the PC is connected by a local area network (LAN).

The information of the cell motion is calculated from the sequential image data by an algorithm described in the following section.

3. Mutual correlation method

Velocity analysis was done with a pixel-based correlation program [13, 14]. The main idea of this program is as follows: The movement of the objects causes a temporal change of the grey level at each pixel. The local velocity at a certain pixel site is estimated by analyzing the mutual correlation between the temporal brightness change of the central pixel and that of neighboring pixels. The distance between the reference site and each of the eight neighbors was chosen to be three pixels.

The mutual correlation function $M_0^k(\tau)$ for the central pixel 0 and the neighboring pixel k is described for the case of continuous intensity functions $A_0(t)$ and $A_k(t)$ by the equation

$$M_0^k(\tau) = \frac{1}{T} \int_{-T}^{T} [A_0(t) - \overline{A}_0][A_k(t+\tau) - \overline{A}_k]\, dt$$

where $\overline{A}_0$ and $\overline{A}_k$ denote the mean values of

$A_0(t)$ and $A_k(t)$. For digitized time series the function $M_0^k(\tau)$ can be reduced to an equation containing only summation terms, which is given in ref. [13].

The direction of motion is determined by that mutual correlation function that has the highest maximal value. The speed is calculated from the lag time of the peak of that function and the distance to the central pixel.

Since velocity information is available only at the discrete pixel sites, an interpolation algorithm is implemented in the calculation that results in a realistic estimate of the actual speed and direction of motion [14]. Furthermore, we add several techniques to reduce the calculation errors. Our program evaluates the amplitude of temporal data of the target pixel and rejects the velocity information of this point if the amplitude of intensity variation is lower than a special value (here grey level changes of 30 are necessary). We require the maximum mutual-correlation value to reach at least 0.7. Some methods using additional criteria to suppress unreliable velocity data are described in ref. [14].

Due to the necessary restrictions to be met, a considerable number of pixels will be omitted in the calculations. Therefore we commonly use spatial averages in appropriately selected areas.

4. Results

With this quantitative method we investigated the chemotactic movement of the amoebae in regions far from the center of the patterns, where the curvature of the wave fronts is very small. A typical microscopic image recorded with the video camera is shown in fig. 2. It depicts a dense assembly of a few hundred amoebae. An erratic motion and change of cell shape is constantly observed. Looking at the quick-motion pictures (factor 25), we can see periodically occurring pulses of more coherent motion on the TV screen. In our experiments the small areas of the patterns under investigation are oriented such that

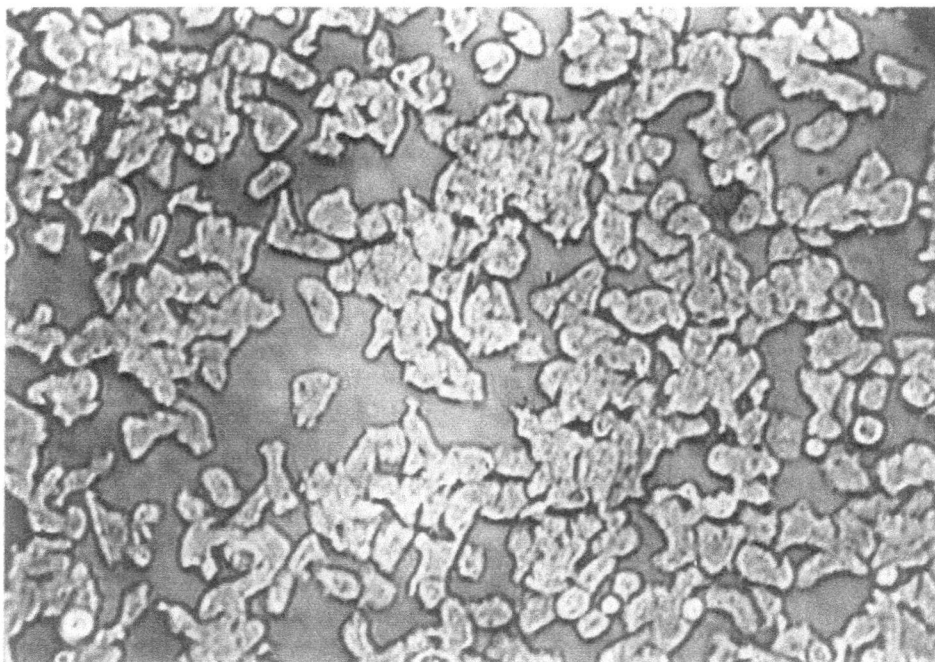

Fig. 2. *Dictyostelium discoideum* amoebae during aggregation. The microscopic photography covers an area of 0.39×0.32 mm^2. Sequences of such images were analyzed with respect to the velocity of cell motion.

these pulses propagate in horizontal direction of the recorded pictures, that is along their x-axis.

For the velocity analysis sequential image data are taken from the quick-motion video movies at equidistant time intervals. The complete data set consists of 30 frames and has a total duration of 100 s. The area of each single frame is chosen to be 0.07×0.31 mm^2. The longer side of this rectangular frame is oriented parallel to the wave front moving in x-direction. Spatial averaging over all estimated velocities of the 85×500 pixel area results in the velocity information at a given point in time. As found in many preparations, the time course of v_x, which is the velocity component in the direction of the aggregation center, varies in a periodic fashion, having periods of 6–9 min. $v_x(t)$ reaches maximum values of approximately 20–30 μm/min and its minimum values are close to zero. Within the experimental error the data sug-

gest that the shape of the velocity functions $v_x(t)$ is non-symmetric. Each maximum is followed by a shoulder or a small second peak, while the leading front is very steep.

A typical example of the temporal evolution of the velocity $v_x(t)$ is shown in the experiment of fig. 3. It has a period of $T = 429$ s and a maximum for v_x of 26 μm/min. Fig. 4 shows the angular distribution of cell velocities for the same image data at the time $t_0 = 220$ s. Its sharp peak at 173° with respect to the x-axis indicates that the motion in the investigated region is approximately unidirectional.

Furthermore, we calculated the velocity field for this experiment at the time $t_0 = 220$ s (fig. 5). Here the used image area was 0.33×0.27 mm^2, corresponding to 432×432 pixels in the digitized data. Each velocity vector is calculated from the velocity information of 54×54 pixels. One can

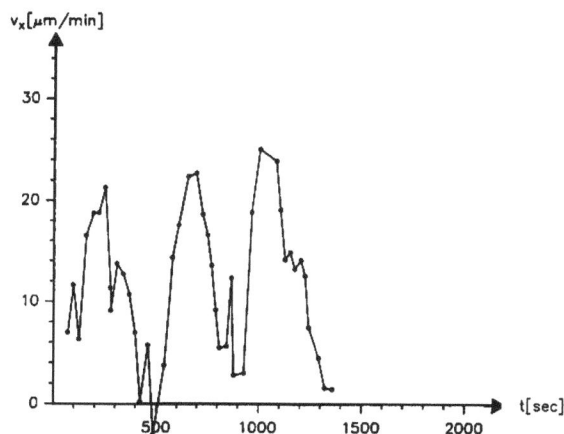

Fig. 3. Velocity component v_x in the direction of the aggregation center as a function of time, obtained for an area of 0.07×0.31 mm². Three waves of cAMP propagate through the investigated area and cause periodic movement of the amoebae.

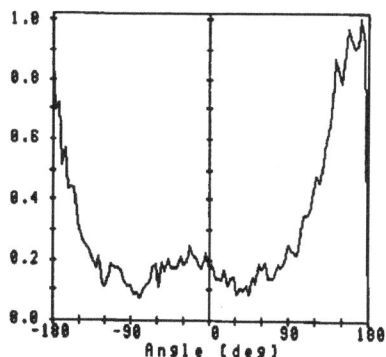

Fig. 4. Angular distribution of the velocities detected in fig. 3. This histogram corresponds to the time $t_0 = 220$ s. 173° is the angle towards the aggregation center (x-direction).

see that a component of cell motion in x-direction predominates, indeed, over the whole observation territory.

We also estimated the speed of the chemically induced wave of excitation by analyzing the cellular movement in different sections of the image sequence. For this purpose we divided each picture into several stripes perpendicular to the propagation direction. In each stripe the chemotactic response reaches its maximum at a different time, which allows the measurement of the

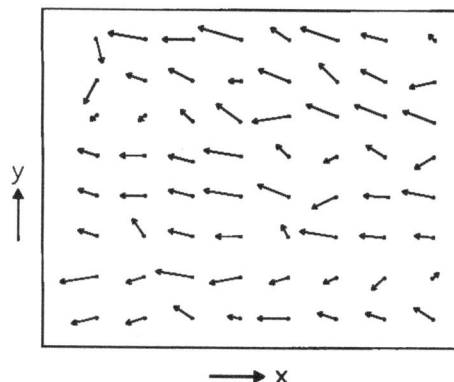

Fig. 5. Velocity field of cellular motion, corresponding to the time $t_0 = 220$ s in fig. 3. Each velocity vector is estimated from 54×54 pixels. The investigated area is 0.33×0.27 mm².

retardation Δt_{ij} between two functions $v_{x_{i,j}}(t)$. Division of the distance Δx_{ij} between two stripes i and j by the retardation Δt_{ij} yields directly the wave speed. The estimated values are in good agreement with the wave speeds detected by dark-field photography [7].

An experimental example for this calculation is given in fig. 6. The retardation between these two functions $v_x(t)$ was roughly estimated as $\Delta t = 102$ s and the distance between the respective two stripes was 319 μm. Thus, we found a wave speed of approximately 190 μm/min. The periods of these functions were calculated as $T = 491$ s. After completing this experiment no wave pattern could be detected any more by dark-field visualization, but apparently the periodic chemotactic motion continued, as proven by our results.

We also performed preliminary studies of the core region of spiral patterns. In the center of these patterns our time-lapse video movies reveal a more or less circular region of low cell density, around which a wave of excitation circulates. This results in a circulating sector of homogeneous cell motion – the tip of the spiral – directed tangentially to the low-density region. In different experiments the diameter of this region varies by a factor of approximately four. The period of spiral rotation increases with the diameter. Fig. 7 illus-

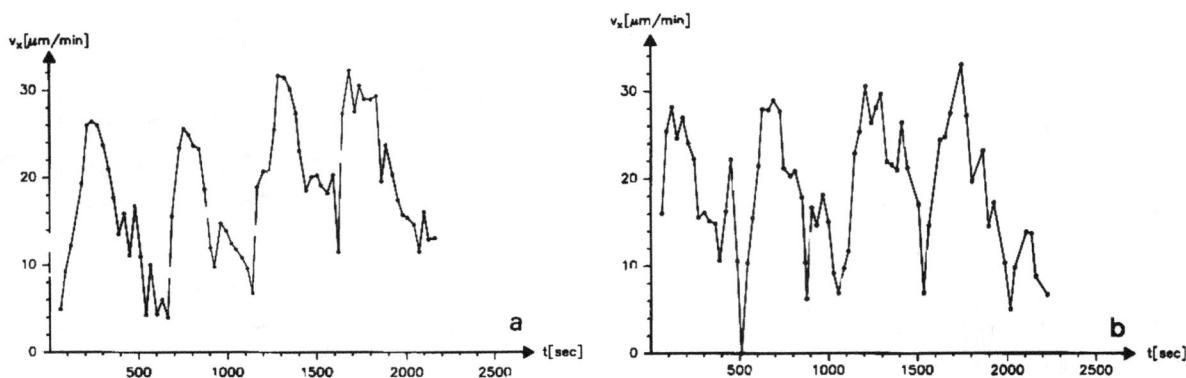

Fig. 6. Velocity component v_x taken in two different areas of the same image sequence. From the retardation $\Delta t = 102$ s of these two functions $v_x(t)$ the speed of the excitation wave is estimated as approximately $c = 190$ μm/min.

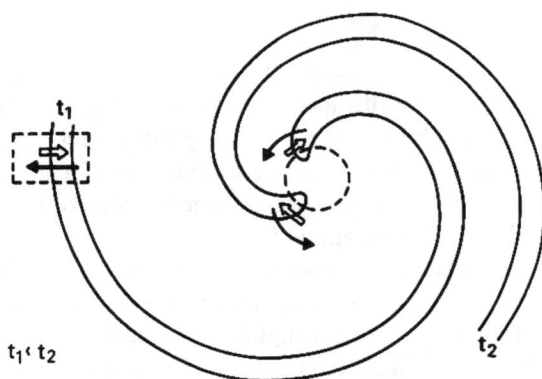

Fig. 7. Schematic illustration of an isoconcentration level of a spiral pattern at two times t_1 and t_2. The tip of the spiral circulates around the core region. The arrows indicate the velocities of the outward propagating chemical wave (thin) and chemotaxis oriented towards the aggregation center (open). The rectangle shows a typical observation area (cf. fig. 2).

trates our finding that the cells move tangentially to the spiral core close to the spiral center, and radially in the periphery of the spiral pattern.

5. Discussion

Our results present a quantitative analysis of only microscopically detectable properties ($v_{\text{cell}}^{\text{max}}$, angular distribution of velocity, time course of v_x) in *Dictyostelium discoideum*, which are not

measurable with the conventional dark-field methods. For comparison it is possible to calculate from these experiments well known parameters like the wave period T and the wave velocity c.

The new dynamic image processing technique, which is based on microscopic video imaging, adds a powerful tool of investigation to the conventional dark-field method. As an advantage, we are able to detect oscillating behaviour when macroscopic patterns have already disappeared. Thus, the time evolution of important parameters can be observed over much longer time intervals. In fact, we detect oscillatory chemotactic behaviour even when a large portion of the cells have already accumulated in the core region (approximately 1 h after disappearance of the dark-field patterns).

The observed periodicity of v_x shows a high correspondence to results obtained by analyzing dark-field intensities in a fixed location of the patterns [7]. It should be noted that both methods result in a non-symmetric shape of the velocity, respectively the intensity oscillations. In their experiments Siegert et al. showed that the non-symmetric shape is caused by addition of caffeine, which we also used in our preparations. In ref. [7] it was suggested that the population of amoebae consists of two groups having different chemotactic response behaviour with respect to the travelling cAMP signals. We did not find any direct evidence yet in our analysis which supports this view. The asymmetric shape in the velocity curves could also be caused by non-linearities in the regulation of chemotaxis or be the result of a non-symmetric shape of the cAMP wave. It is a common feature in the BZ reaction or in glycolysis that non-symmetric shapes of waves and oscillations are a result of the non-linearity of the underlying mechanism.

Our results on chemotaxis close to the core of a *Dictyostelium* spiral are only preliminary, but they clearly show the preferentially tangential direction of cell motion along the periphery of a spot with relatively low cell density. We have not yet obtained conclusive data about how far this tangential motion extends into the spot. Further outside the main component of chemotaxis has radial direction. In accordance with theoretical assumptions, we have drawn schematically in fig. 7, how this change of direction can be correlated with the spiral-shaped level of cAMP concentration, where the gradient is particularly high. This correlation should be further elucidated by appropriate models including chemotaxis.

Acknowledgements

We thank Professor B. Hess for fruitful discussions and Dr. P. Foerster for advice in microbiological preparations. This work was supported by the Stiftung Volkswagenwerk, Hannover, Germany.

References

[1] G. Gerisch, Periodische Signale steuern die Musterbildung in Zellverbänden, Naturwissenschaften 58 (1971) 430–438.

[2] W.F. Loomis, Development of *Dictyostelium* (Academic Press. New York. 1982).

[3] P. Devreotes, A model system for cell–cell interactions in development, Science (1989) 1054–1058.

[4] R.J. Field and M. Burger, eds., Oscillations and Travelling Waves in Chemical Systems (Wiley, New York, 1985).

[5] J. Ross, S.C. Müller and C. Vidal, Chemical waves, Science 240 (1988) 460–465.

[6] P. Foerster, S.C. Müller and B. Hess, Curvature and spiral geometry in aggregation patterns of *Dictyostelium discoideum*, Development 109 (1990) 11–16.

[7] F. Siegert and C. Weijer, Digital image processing of optical density wave propagation in *Dictyostelium discoideum* and analysis of the effects of caffeine and ammonia, J. Cell Sci. 93 (1989) 315–335.

[8] J.J. Tyson and J.P. Keener, Singular perturbation theory of travelling waves in excitable media (a review), Physica D 32 (1988) 327–361.

[9] J.J. Tyson, K.A. Alexander, V.S. Manoranjan and J.D. Murray, Spiral waves of cyclic AMP in a model of slime mould aggregation, Physica D 34 (1989) 193–207.

[10] J.-L. Martiel and A. Goldbeter, A model based on receptor desensitization for cyclic AMP signalling in *Dictyostelium* cells, Biophys. J. 52 (1987) 807–828.

[11] P. Foerster, S.C. Müller and B. Hess, Critical size and curvature of wave formation in an excitable chemical medium, Proc. Natl. Acad. Sci. USA 86 (1989) 6831–6834.

[12] K.J. Tomchik and P.N. Devreotes, Adenosine 3',5'-monophosphate waves in *Dictyostelium discoideum*: a demonstration by isotope dilution-fluorography, Science 212 (1981) 443–446.

[13] H. Miike, Y. Kurihara, H. Hashimoto and K. Koga, Velocity-field measurement by pixel-based temporal mutual-correlation analysis of dynamic image, Trans. IEICE Japan E 69 (1986) 877–882.

[14] H. Hashimoto, H. Miike, K. Koga and S.C. Müller, manuscript in preparation.

Physica D 49 (1991) 240–246
North-Holland

Chapter 5. Image processing

Can excitable media be considered as computational systems?

A.V. Holden[a], J.V. Tucker[b] and B.C. Thompson[b]

[a]*Centre for Nonlinear Studies, University of Leeds, Leeds LS2 9JT, UK*
[b]*Department of Mathematics and Computer Science, University College, Swansea SA2 8PP, Wales, UK*

Actual excitable media, and discrete models of excitable media, may be used to process images. Discrete time, discrete space models of excitable media are shown to be examples of synchronous concurrent algorithms and so may be considered as formal computational systems.

1. Introduction

A number of mathematical approaches may be used to model a given excitable system. For an excitable system that is not spatially extensive, for example an isopotential neurone, a single map or a system of *nonlinear ordinary differential equations* may be appropriate. For a spatially extensive excitable medium, for example a sheet of cardiac tissue, a system of *partial differential equations* in which space is assumed continuous may be appropriate. Alternatively, a *cellular automaton*, or a *coupled map lattice*, in which space is assumed discrete might be used. These different types of models are all nonlinear, and are often analytically intractable, and so their behaviour is usually investigated by numerical approximation methods on a digital computer. Thus the problem arises of faithfully simulating excitable media using appropriate algorithms.

In this paper we consider the converse problem: To what extent can an excitable medium be considered and used as a computational device that carries out some set of algorithms, or indeed can be programmed in a general way? In particular, we are interested in the possible use of two- or three-dimensional excitable media (photo-chemical, opto-electronic or neurobiological) as computing devices, in which data in the form of a spatially continuous function (image or field) are processed by the nonlinear dynamics and nonlinear wave properties of the medium.

The idea that nonlinear reaction–diffusion systems, such as actual excitable media, could be used to process patterns is not new. It is implicit in Kuhnert's use of the light sensitive Belousov–Zhabotinsky reaction to process images [1], and is explicit in Kirby and Conrad's [2] distinction between structurally intelligent systems, which perform intelligent actions without the use of symbols, and symbolically intelligent systems.

A spatially extensive excitable medium used as a computing device is an example of what we will call an *analogue-field computer* (AFC). An AFC is simply a spatially extensive physical system that processes spatially extensive data. The study of excitable media as computational systems is of interest for a variety of reasons:

(1) From the point of view of theoretical computer science, such analogue field computers enlarge our understanding of computations by

broadening substantially our notions of machines and algorithms. The excitable medium is a simple form of *infinitely* parallel computer, as it represents the simplest form of infinite parallelism, and so could be used to investigate the asymptotic properties of parallel computer systems.

(2) From the point of view of computer system engineering, such AFCs would find application in novel hybrid computer systems as special purpose elements (say for extracting global properties of a field) or preprocessors (say in image processing).

(3) From the point of view of natural sciences, the study of AFCs should provide new methods for the classification of the behaviour of actual excitable media in nature.

Thus, to study the question "Can excitable media be considered as computational systems?" we *appear* to need a theory of analogue field computation in which computers and computations are *continuous* in space and time. Ideally, such a theory of computation should address basic matters to do with computers, including

(i) design and programming of devices;
(ii) correctness and efficiency of programs;
(iii) existence of universal devices; and
(iv) limits of computation.

However, excitable media can be successfully simulated by models that are *discrete* in space and time, such as certain numerical approximations to partial differential equations, cellular automata, and coupled map lattices. Thus we may begin to study the question using these models of excitable media, together with an appropriate theory of computation that addresses (i)–(iv) above.

In this paper we introduce the theory of *synchronous concurrent algorithms* and show that it provides a framework for investigating the computational properties of discrete space, discrete time approximations to excitable media. A synchronous concurrent algorithm (SCA) is an algorithm that consists of processes or cells that compute and communicate in parallel and are synchronized by a global clock; SCAs process infinite streams of input data, and return infinite

streams of output data. An SCA is a general concept that characterizes the essential features of many types of parallel computer architectures. A general mathematical theory of SCAs is being developed that addresses (i)–(iv) above for parallel deterministic computation by hardware. At the heart of the theory of SCAs simultaneous recursive functions on (many-sorted) algebras represent the algorithm's architecture [3]. The algebras represent the data and basic processing elements of the algorithm, and the mathematical structure the recursion. The application of the theory of SCAs to discretized models of excitable media allows us to examine excitable media as computational systems in terms of (i)–(iv) above.

The structure of the paper is this. In section 2 we recall the representation of an excitable medium as a reaction–diffusion equation. In section 3 the essential features of three types of discrete space, discrete time models are considered, namely: discrete approximations to nonlinear partial differential equations; cellular automata; and coupled map lattice dynamical systems. In section 4 we describe SCAs and, in section 5 and 6, show that the discrete systems are examples of SCAs. In section 7 we review the computational nature of these models of excitable media.

2. Field equations for excitable media

An actual excitable medium (say a thin layer of Belousov–Zhabotinsky reagent) is a concrete object that can be idealised by a partial differential equation. The behaviour of excitable media and their models is surveyed in ref. [4]. A general isotropic excitable medium may be represented by a nonlinear reaction–diffusion equation:

$$\frac{\partial u(x,t)}{\partial t} = K(u(x,t)) + D \nabla^2 u(x,t), \qquad (1)$$

where $u \in \mathbb{R}^n$ are the variables in state space, $x \in \mathbb{R}^m$ $m = 1$, 2, or 3 forms the physical space and $\mathbf{D} \in \mathbb{R}^{n \times n}$ is a diagonal matrix of real-valued, positive diffusion coefficients.

For nerve and muscle all the "diffusion coefficients" D_{ij} are zero except for D_{11}: in this single diffusion case only the voltage interacts diffusively. In chemical excitable media all D_{ii} are positive and often approximately equal, while all D_{ij}, $i \neq j$, are zero.

3. Discretised excitable media

An actual excitable medium, which has a continuous state space, continuous physical space and operates in continuous time, may be modelled by a system in discrete space and discrete time. This may be done by discretising the field equations (1), or by considering models defined on discrete physical space and in discrete time, such as cellular automata [5] or coupled map lattice models [6] for actual excitable media.

3.1. Numerical solutions of partial differential equations

The nonlinear partial differential equations that represent excitable media are almost invariably intractable, and so are usually approached numerically. Although a number of explicit and implicit methods are in current use the simplest finite-difference approximation

$$\delta t \left(\frac{\partial f_r}{\partial t} \right)_{t_0} = f_{r,\, t\, + \delta t} - f_{r,\, t_0},$$

$$f_{r,\, t_0 + \delta t} = f_{r,\, t_0} + \frac{\delta t}{(\delta x)^2} (f_{r-1} - 2f_r + f_{r+1})_{t_0},$$

$$r = 1, 2, \ldots, n - 1$$

for a one-dimensional parabolic partial differential equation illustrates the discretisation of space and time with a fixed space step and time step. Note that this approximation is a simultaneous primitive recursion. For two- and three-dimen-sional systems a regular spatial grid may be used, or more sophisticated multigrid methods used.

As long as appropriate step sizes are used, the numerical solutions of partial differential equations provide accurate approximations to analytical solutions of the partial differential equations when these are known, and have the same asymptotic properties as the solutions for analytically intractable equations.

The essential point is that the partial differential equation is replaced by a regular lattice, and the next value at a node is computed from the preceding values at the same and neighbouring nodes. The numerical solution of partial differential equations, although usually carried out on computers with a conventional, von Neumann architecture, is ideally suited for parallel algorithms such as synchronous concurrent algorithms and parallel machines such as an array processor.

3.2. Cellular automata

Cellular automata (CA) are dynamical systems with discrete time, discrete space and discrete state. CA were invented to model the development and organisation of biological systems [7] and recently many applications of CA have been discovered in physics and computer science [8]. For further information on the modelling of systems by CA the reader is referred to the book by Soulié et al. [5], in which the basic ideas underlying the account of CA given here can be found.

Most generally a (deterministic) cellular automaton is an infinite dynamical system of processing elements or cells that occupy discrete n-dimensional space and compute on a finite state space Q in discrete time $t = 0, 1, 2, \ldots$.

Specifically we consider a cellular automaton on a finite subset $S \subseteq \mathbb{Z}^n$ such that cells are indexed by elements of S, and connections between a cell $x \in S$ and other cells are given by a neighbourhood $N(x) = \{y_1, \ldots, y_k\} \subseteq S^k$ when x has a direct connection from cells $y_1, \ldots, y_k$. In general both the number of neighbours k and the neighbours themselves may vary from cell to cell.

Each cell x has an associated *local transition function* f_x: $Q^k \to Q$ when x has k neighbours and the computation performed by a complete cellular automaton is the parallel execution of these transition functions in the following way:

Initially, that is at time $t = 0$, each cell x is in some given state $q^0(x)$. The system then evolves in discrete time t according to the iteration

$$q^{t+1}(x) = f_x(q^t(y_1), \ldots, q^t(y_k)) \qquad (2)$$

when $N(x) = \{y_1, \ldots, y_k\}$.

Two further properties of CA are as follows. Firstly, it is usually the case that Q involves a distinguished state denoted by 0 called the *quiescent* state and each f_x must satisfy

$$f_x(0, \ldots, 0) = 0.$$

Thus local quiescence cannot generate non-quiescence.

Secondly, we may allow a cell's computation to vary with time. In such a case the local transition function for a cell x has the form f_x: $T \times Q^k \to Q$ where $T = \{0, 1, 2, \ldots\}$ measures time t and the iteration (2) becomes

$$q^{t+1}(x) = f_x(t, q^t(y_1), \ldots, q^t(y_k))$$

when $N(x) = \{y_1, \ldots, y_k\}$. CA with such cells are called *flexible* CA [9].

Example For any $m \geq 1$ consider the subset S of $\mathbb{Z}^2$ defined by

$$x = (i, j) \in S \Leftrightarrow 0 \leq i \leq m - 1 \text{ and } 0 \leq j \leq m - 1.$$

This S is simply a square comprising m^2 cells lying in the first quadrant of the plane with one corner of the square at the origin $(0, 0)$.

We can make a cellular automaton on S by defining

$$N(x) = \{(i, j-1), (i, j+1), (i-1, j), (i+1, j)\}$$

for every cell $x = (i, j)$. To ensure $N(x) \subseteq S$ we can assume the arithmetic on indices i and j is taken modulo m, and this imposes the topology of a torus on the automaton.

Alternatively we can define

$$N(x) = \{(i, j-1), (i, j+1)\}$$

if $i = 0$ or $i = m - 1$,

$$N(x) = \{(i, j-1), (i, j+1), (i-1, j), (i+1, j)\}$$

otherwise

for each $x = (i, j)$ and this defines a cylindrical structure (again taking arithmetic modulo m where necessary).

For a specific cellular automaton we choose the automaton of D. Griffeath described in ref. [10]. We begin by choosing

$$N(x) = \{(i, j), (i, j-1), (i, j+1),$$
$$(i-1, j), (i+1, j)\},$$

which gives a toroidal topology in which each cell is directly connected to itself in addition to its four orthogonal neighbours.

Next we take $Q = \{0, 1, \ldots, n - 1\}$ for some $n \geq 1$ and define f_x: $T \times Q^5 \to Q$ by

$$f_x(t, q, q_1, \ldots, q_5) = q + 1$$
$$\text{if } q_1 = q + 1 \text{ or } q_2 = q + 1$$
$$\text{or } q_3 = q + 1 \text{ or } q_4 = q + 1,$$
$$f_x(t, q, q_1, \ldots, q_5) = q$$

otherwise (3)

and thus a cell changes state if at least one of its orthogonal neighbours has state one greater than the current state of x. (Here the arithmetic is considered modulo n so that $q = 0$ is considered one greater than $n - 1$.)

If we visualize the behaviour of this automaton and execute the automaton on initial random states the system evolves in approximately 200 time steps into a spatially coherent system composed of a number of spirals.

3.3. Coupled map lattices

A coupled map lattice (CML) is a dynamical system with discrete time, discrete space and continuous state. At each lattice site i ($i = 1, \ldots, N$, N = number of elements in the lattice), the activity $x(i)$ evolves as a nonlinear mapping f of its preceding value and some function of the preceding activities at different sites $x(j)$. If there are no interactions between the activities at different sites the whole system is simply N independent mappings f; if there are interactions, these can be local, global, or in between these extremes and specified by some specific connectivity matrix.

If the connections are local, an element i interacts with some elements in its vicinity; these could be the nearest neighbours, or a larger vicinity. For a one-dimensional lattice, the nearest neighbours of i are simply $i \pm 1$; in two- and higher-dimensional lattices the neighbourhood needs to be specified (e.g. a von Neumann 4- or Moore 8-neighbour neighbourhood for a rectangular lattice in the plane). The connections can be unidirectional or symmetrical, and different weights can be associated with the different connections. An example of a one-dimensional, diffusively coupled CML is

$$x_{n+1}(i) = (1 - \varepsilon)f(x_n(i))$$
$$+ \tfrac{1}{2}\varepsilon\left[f(x_n(i+1)) + f(x_n(i-1)) \right] \qquad (4)$$

where n is a discrete time step. The behaviours of locally coupled map lattices are described in ref. [11].

4. Synchronous concurrent algorithms

A *synchronous concurrent algorithm* (SCA) is an algorithm based on a network of modules, channels, sources and sinks that compute and communicate in parallel and are synchronized by a global clock $T = \{0, 1, 2, \ldots\}$. The SCA pro-

cesses *data* taken from a set A as a sequence $a(0), a(1), a(2), \ldots$ of clocked data: this sequence is a function $a: T \to A$. Each *module* is a unitary computational device that executes a (time-dependent) operation that is specified by a function $f_m: T \times A^n \to A$, when the module m has n input channels. Communication between modules occurs along *channels* that can transmit only a single datum $a \in A$ at any time; channels can branch but not merge. A *source* reads data into the network: it has no input channels, and a single output. A network with n sources will process n streams $a_1, a_2, \ldots, a_n$ that form the vector-valued stream $a: T \to A^n$. A sink has a single input channel and no output channel; data are read out of the network. The *architecture* consists of a finite network of modules connected by channels. Two modules are neighbours if the output channel of one is an input channel of the other.

Let N be an SCA over a data set A with a clock T, with $n > 0$ sources. The input to N is a stream $a: T \to A^n$. If N has $k > 0$ modules the initial state of the network is a vector $x = (x_1, \ldots, x_k) \in A^k$, where x_i denotes the value output from the ith module at time zero. At each time $t \in T$ there is a single value output that can be determined from the t, a, and x. The *value functions* of the network $V_i(t, a, x)$, for $i = 1, \ldots, k$ are total functions

$$V_i: T \times [T \to A^n] \times A^k \to A$$

that denote the value output from the ith module at time t when the network is executed on input a and initial data x. The state of the channels is then given by *the* value function for the network

$$V_N(t, a, x) = \{V_1(t, a, x), \ldots, V_k(t, a, x)\}$$

for each $t \in A$, $x \in A^k$.

To obtain the value functions V_i for each $t \in T$ we first need $V_i(0, a, x)$, and then obtain

$V_i(t + 1, a, x)$ from $V_i(t, a, x)$. In the case $t = 0$:

$$V_i(0, a, x) = x_i.$$

If the ith module m_i has $n(i) > 0$ inputs and a functional specification f_i, then if at time t the input is $b_1, \ldots, b_{n(i)}$ the value output at time $t + 1$ is $f_i(t, b_1, \ldots, b_{n(i)})$. However, for $j = 1, \ldots, n(i)$, the jth input channel is the output of either a source $\lambda \in \{1, \ldots, n\}$, and so

$$b_j = a_\lambda(t), \tag{5a}$$

or a module $\mu \in \{1, \ldots, k\}$, and so

$$b_j = V_\mu(t, a, x). \tag{5b}$$

Thus the value functions of the network are

$$V_i(t + 1, a, x) = f_i(t, b_1, \ldots, b_{n(i)}). \tag{6}$$

The indices λ and μ are independent of t, a and x and are determined by the architecture; in programming terms they are syntactic quantities. For a specified architecture eqs. (5) and (6) collapse into a single equation.

Example 2. Consider a linear array of n identical modules with nearest neighbour connections. For $t = 0$

$$V_i(0, a, x) = x_i,$$

and for $t > 0$

$$V_1(t + 1, a, x) = f_1(t, V_1(t, a, x), V_2(t, a, x)). \tag{7}$$

For $i = 2, \ldots, n - 1$

$$V_i(t + 1, a, x) = f_i(t, V_{i-1}(t, a, x), V_i(t, a, x),$$
$$V_{i+1}(t, a, x)),$$

and

$$V_n(t + 1, a, x) = f_n(t, V_{n-1}(t, a, x), V_n(t, a, x)),$$

where for $i = 1$ and $i = n$, $f_i: T \times A^2 \to A$, and for $i = 2, \ldots, n - 1$, $f_i: T \times A^3 \to A$ are any given operations. (If $f_1, \ldots, f_n$ are all to have the same functionality then we need a ring structure.)

Thus we have a general formalisation that allows us to specify the output from a network of computing modules and any architecture if we know the architecture, functionalities of the modules and the initial inputs.

5. CA as SCAs

Clearly the terminology of CA maps easily onto that of SCAs: a cellular automaton is an SCA over $A = Q$ whose modules are the cells of the automaton. For example, value functions for the Griffeath's automaton are

$$V_{00}, \ldots, V_{m-1, m-1}: T \times Q^{(m^2)} \to Q,$$

$$V_{ij}(0, a, x) = x_{ij}, \quad x = (x_{00}, \ldots, x_{m-1, m-1}),$$

$$V_{ij}(t + 1, x) = f_{ij}(t, V_{ij}(t, x), V_{i, j-1}(t, x),$$

$$V_{i, j+1}(t, x), V_{i-1, j}(t, x), V_{i+1, j}(t, x)),$$

where $f_{ij}: T \times Q^5 \to Q$ is given by (3) above.

6. CMLs as SCAs

The terminology of CMLs maps easily onto that of SCAs: a CML is an SCA over $A = \mathbb{R}$ whose modules are the elements of the CML. For example, value functions for the one-dimensional diffusively coupled CML given by (4) are

$$V_1, \ldots, V_N: T \times \mathbb{R}^N \to \mathbb{R},$$

where, for $i = 1, \ldots, N$

$$V_i(0, x) = x_i,$$

$$V_i(t+1, x) = f_i(t, V_{i-1}(t, x), V_i(t, x), V_{i+1}(t, x)),$$

where $f_i: T \times \mathbb{R}^3$ is defined by

$$f_i(t, u, v, w) = (1 - \varepsilon)f(V) + \tfrac{1}{2}\varepsilon[f(u) + f(w)]$$

with $f: \mathbb{R} \to \mathbb{R}$ given, for example, by a piecewise linear map with a threshold [12].

7. Concluding remarks on the theory of SCAs

The theory of synchronous concurrent algorithms has been developed in the context of theoretical computer science, to provide a neutral mathematical formalism for the analysis of architectures of processing elements that compute and communicate in parallel in a deterministic way. However, an SCA is a general notion that encompasses all discrete time, discrete physical space, continuous or discrete space dynamical systems; in fact, an SCA *is* a discrete time, discrete space dynamical system.

The theory of SCAs is developed in ref. [13] and applied to coupled networks of excitable elements (biological and various formal neural networks) in ref. [14]. Its application to discretized representations of excitable media emphasises the computational poverty of excitable media: the architecture of the SCA that is implemented by the representation of the excitable medium is simply nearest neighbour, diffusive coupling. The structure of this architecture is not changed by allowing the diffusion coefficients to be adjustable. The only way to extend the computational capabilities of an excitable medium is by

radical anisotropy, in which different parts of the medium comprise different components, and also have different connectivities, involving specific non-local connections, as in a neural network.

References

[1] L. Kuhnert, A new photochemical memory device in a light sensitive active medium, Nature 319 (1986) 393.
[2] K.G. Kirby and M. Conrad, Intraneuronal dynamics as a substrate for evolutionary learning, Physica D 22 (1986) 205–215.
[3] B.C. Thompson, A mathematical theory of synchronous concurrent algorithms, Ph.D. Thesis, School of Computing Studies, University of Leeds (1987).
[4] V.S. Zykov, Simulation of wave processes in excitable media (Manchester Univ. Press, Manchester, 1987).
[5] F. Fogelman-Soulié, Y. Robert and M. Tchuente, eds., Automata Networks in Computer Science (Manchester Univ. Press, Manchester, 1987).
[6] J.P. Crutchfield and K. Kaneko, Phenomenology of spatio-temporal chaos, in: Directions in Chaos, ed. Hao Bai-Lin (World Scientific, 1988, Singapore).
[7] J. von Neumann, in: Theory of Self-reproducing Automata, ed. A.W. Burks (University of Illinois Press, Urbana, IL, 1966).
[8] D. Farmer, T. Toffoli and S. Wolfram, eds., Cellular Automata, Physica D 10 (1984) 1–248.
[9] M. Tchuente, Computation in Automata Networks, in: Automata Networks in Computer Science, eds. F. Fogelman-Soulié, Y. Robert and M. Tchuente (Manchester Univ. Press, Manchester, 1987).
[10] A.K. Dewdney, Computer recreations, Sci. Am. (1989) 88–91.
[11] K. Kaneko, Pattern dynamics in spatio-temporal chaos, Physica D 34 (1989) 1–41.
[12] E. Labos, Spike generating dynamical systems and networks, in: Dynamical Systems, Proceedings of IIASA Workshop on Mathematics of Dynamical Processes, eds. A.B. Kurzhanski and K. Sigmund, Lecture Notes in Economics and Mathematical Systems (Springer, Berlin, 1987).
[13] B.C. Thompson and J.V. Tucker, Synchronous concurrent algorithms, Computer Science Division, University College of Swansea Research Report, in preparation.
[14] A.V. Holden, J.V. Tucker and B.C. Thompson, The computational structure of neural systems, in: Neurocomputers and Attention. I. Neurobiology, Synchronisation and Chaos, eds. A.V. Holden and V.I. Kryukov. (Manchester Univ. Press, Manchester, 1990) pp. 223–240.

Physica D 49 (1991) 247–253
North-Holland

Autowave principles for parallel image processing

V.I. Krinsky, V.N. Biktashev and I.R. Efimov

Institute of Biological Physics of the USSR Academy of Sciences, 142292, Pushchino, Moscow Region, USSR
and the Computing Research Center, 142292, Pushchino, Moscow Region, USSR

A highly parallel autowave method for pattern analysis and topological feature detection is presented. It is invariant against translations, rotations and scalings of the input pattern. The method yields an increase in computational speed of 3 to 6 orders of magnitude in comparison with a sequential (von Neumann) computer. The method can be realized in principle using only one chip with simple uniform connections of elements.

1. Introduction

Modern computers, even highly parallel ones, are relatively ineffective in pattern analysis when compared to the visual systems of animals, especially man. Therefore, more attention is now paid to the investigation of formal neural networks.

A considerable advance of the last decade was the discovery of analogies in the mathematical descriptions of the spin glass behavior in physics and of the formal neural network dynamics [1, 2]. The analogies have provided a theoretical background for designing neural networks with the required properties, at least for networks with symmetrical connections.

On the other side, the capability of the existent systems for image analysis based on neural networks is a sorry sight. There has been the opinion for many years that they cannot compete with computers and specialized technical systems. Here we demonstrate that some results that are likely to be competitive can be obtained with so-called autowave neural networks for primary feature analysis (for topological feature detection).

In the Hopfield networks each neuron is coupled with every other and the corresponding inte-

grated circuit (VLSI) is not technologically feasable. In contrast, the elements in an autowave network are coupled only to the nearest neighbors. The autowave VLSI has a simple structure, close to that of the VLSI of dynamic memory. It can detect a large number of simple topological features. It has high reliability, as is the case with neural networks. This means that the devices can run even though some percentage of elements are damaged.

2. Autowaves

Autowaves represent a particular class of nonlinear waves, which spread in active excitable media at the expense of the energy stored in the medium [3, 4]. Auto-oscillations are the logical predecessor of autowaves. It is well known that, mathematically, auto-oscillations are simply a movement along a limit cycle, and the oscillation amplitude after a relaxation time can assume only two values, 0 or 1. The value 1 is assigned to the movement along the limit cycle, while the value 0 is reached when the oscillations are absent. Waves can propagate in a distributed medium consisting of such oscillators. Under some conditions the steady state amplitude of such a wave has essentially only two values: 0 or 1 (almost everywhere in the medium). Propagating waves possessing these properties are called *autowaves* [3, 4]. The differential equation describing autowaves is:

$$\frac{\partial u}{\partial t} = D\frac{\partial^2 u}{\partial X^2} + f(u),$$

where, for a chemical autowave medium, $\partial u / \partial t$ is the rate of change of the reagent concentrations u and is induced by the kinetics of chemical reaction $f(u)$ plus the diffusional term $D\partial^2 u /\partial X^2$. A wave front – a trigger wave of transitions from state 0 to state 1 – can propagate in the system. If u is a vector, impulses of finite length can spread.

The fundamental properties of autowaves differ basically from those of classical waves in conservative systems, including nonlinear waves. Two waves spreading in opposite directions do not pass through each other, but annihilate. This is obvious for the one-dimensional case, for example, for nerve impulses spreading in an axon to meet one another, or for two combustion waves.

A two-dimensional example is shown in fig. 1. Fig. 1a shows a classical picture of the interference of waves from two sources in a conservative medium, and fig. 1b shows the interaction of autowaves from two sources in an active medium (the Belousov–Zhabotinsky reaction). There is no interference of the autowave, and a specific

Table 1
Properties of waves and autowaves.

Properties	Waves	Autowaves
1. Conservation of energy	+	–
2. Conservation of amplitude and waveform	–	+
3. Reflection	+	–
4. Annihilation	–	+
5. Interference	+	–
6. Diffraction	+	+

pattern appears due to annihilation of colliding autowaves.

The fundamental properties of classical waves and autowaves are summarized in table 1. They differ with respect to interference, annihilation, and reflection from obstacles or from the medium boundaries. The only common feature of the two types of waves is their ability to bypass obstacles (diffraction) because the propagation is described by the Huygens principle in both cases. These properties of autowaves make them a useful tool for image processing.

3. Image analysis

Two different problems are involved here: pattern recognition and primary image processing. Pattern recognition is the ultimate goal of all the efforts in this direction. The formulation of the problem is as follows: a small set of numbers (features) is at the input, and one number (identifier of the object) is at the output. Modern computers are rather effective in solving this problem. However, to get input data for pattern recognition one should first extract these features from the image. That is the aim of primary image processing. In this case, there is an image raster at the input, and the input data consist of a lot of numbers. For example, if the raster is $10^3 \times 10^3$ pixels, then the input array contains 10^6 numbers. The architecture of modern computers is not adequate for such a problem. Some artificial approaches must be used to account for many inter-

Fig. 1. Interaction of waves from two local sources. (a) Classical conservative waves. Interference of waves is observed. (b) Autowaves. There is no interference; colliding waves annihilate.

nal symmetries, compressing considerably the information, such as translational invariance (images which are different in location are equivalent), rotational or scaling invariance, and division into equivalence classes according to the most important topological characteristics.

Primary image processing is the very problem that we bring to a focus.

4. Image processing using autowaves

As was mentioned, autowaves are not reflected by obstacles and boundaries and do not interfere (see table 1). Thus autowaves are very promising tools for investigating images. The diffraction of autowaves by the image does not create an excess complexity that would appear due to reflections and interference.

If an image is projected on a photosensitive autowave chemical medium, the elements illuminated below some threshold become excited [5]. An autowave induced by the image starts to propagate in the medium. Diffraction of the autowave permits an easy distinction between closed and unclosed curves, and the finding of the shortest path in the labyrinth. By changing the external parameters that control the wave propagation, it is possible to pursue many image processing operations, including contrast regulation, restoration of a broken contour, edge detection, and highly parallel extraction of some topological and geometrical characteristics ([5–7]; see also section 5 below).

One of the important advantages of autowave media for image processing is that all elements of the medium evolve their states simultaneously, resulting in a very high degree of parallelism. For example, it is possible to have about 10^8 excitable elements in a photosensitive autowave chemical medium. But the switching frequency of an element of the chemical autowave medium [5] is very low: the characteristic time is about 1 min.

Autowave image processing can be executed not only by analogous autowave media but also by

digital computers. A computer can be used for modelling the autowave propagation along the image raster. In this case, the switching frequency is satisfactory but the low degree of parallelism makes the image processing ineffective. For example, for rasters of $10^3 \times 10^3$ pixels, the wave propagation for only one time step requires the calculation of new states for all 10^6 pixels, and wave propagation along the whole raster (from the left side to the right) requires the calculation of 10^9 states.

That is why the advantages of autowave methods for image processing can only be realized by using a special highly parallel hardware (autowave VLSI) that performs an independent evolution of the state for each pixel as it occurs in an autowave chemical medium.

5. Autowave algorithms

5.1. Detection of closed curves

The first parallel computer specialized for the detection of closed curves was CLOPAN (CLOsed Pattern ANalyser). To determine whether a curve is closed, it searches for the points that are electrically isolated from the raster boundary. A system of switches is located at every point of the raster, and the keys in lightened points are turned off. Therefore, points inside a closed curve are electrically isolated from the raster boundary.

In fig. 2 the process of detecting a curve's closedness using the autowave method is shown [6]. The medium is constructed in such a way that the points belonging to the curve cannot conduct the excitation. At the initial moment, the wave triggering from one stable state (white) to another (black) is initiated in the left bottom corner. Since the trigger wave cannot overcome the projection of the curve on the medium, it will bypass the curve. If the curve is not closed, then the whole medium becomes black (figs. 2a, 2b, 2c, and 2g). In contrast, if the curve is closed, then the trigger wave cannot reach the inside region

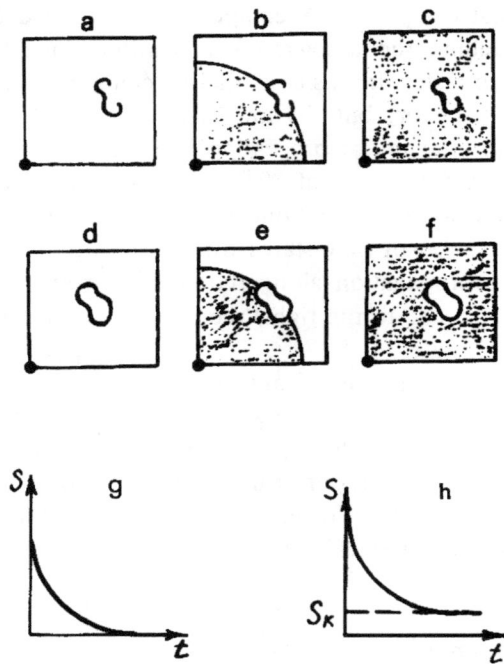

Fig. 2. Detection of closed curves with the help of autowaves. (a, b, c) Propagation of a trigger wave when the obstacle is an unclosed curve. (d, e, f) The same for a closed curve. (g, h) Graphs of the total area of the white region S versus time t; (h) shows that for a closed curve the area S does not vanish with time but reaches some finite value $S(k)$ which is equal to the area enclosed by the curve.

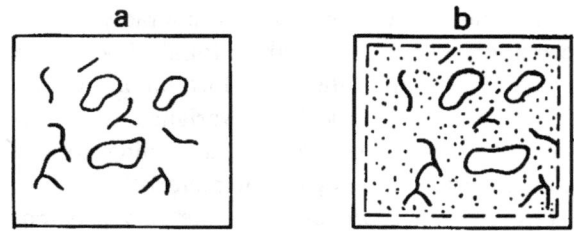

Fig. 3. Detection of a closed curve in a noisy image. Due to diffraction, the trigger wave bypasses all the noise obstacles during its propagation (see fig. 2). The obstacles do not affect the final result.

bounded by the curve (figs. 2d, 2e, 2f, and 2h), and a part of the medium will remain white. This criterion can distinguish a closed curve from an unclosed one.

With this simple example, it is easy to demonstrate the basic advantages and disadvantages of autowave image processing. If a sufficiently large raster is processed, say, of the order of $10^3 \times 10^3$ pixels, then the standard procedure of closedness detection with a usual computer implies (i) scanning the raster (in a random or systematic way) until at least one point on the curve is found, and (ii) following the curve by the method of trial and error, with the scanning pointer permanently losing the curve and returning to it. The time needed for such a procedure is proportional to the number of pixels on the raster, i.e., not less than the order of 10^6 CPU cycles. The time needed for the

corresponding autowave processing is only of the order of 10^3 cycles, since as an autowave propagates through the medium, a large number of elements (about 10^3) operate simultaneously.

The autowave method of curve closedness detection is invariant against translations, rotations and scaling, and is not sensitive to hindrances in the form of unclosed curves (see fig. 3). When a wave propagates, hindrances blend with the dark background and do not affect the final result.

The degree of parallelism is not a constant value and varies depending on the problem. There exist problems where all the 10^6 elements can change their state simultaneously, and the parallelism is of the order of 10^6. An example of such a problem is the detection of self-intersection points on a curve [7]. The time needed for solving this problem with an autowave processor is independent of the raster size.

Autowave methods can also be used for solving such problems as the detection of simply connected regions, the segmentation of images, and noise filtering.

5.2. Restoration of a broken contour

Suppose an image has some breaks because of hindrances during the transfer, along with those initially present in it (see fig. 4a). It is necessary that the breaks of the first type be connected and those of the second type be allowed to remain as they are. Formally, let breaks of length $L < L_0$ be the hindrance breaks and all those with $L > L_0$

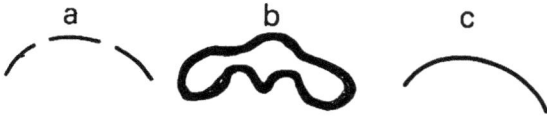

Fig. 4. Restoration of a broken line using the autowave method. (a) Initial image. (b) After propagation of the wave for time *t*. (c) After propagation in the back direction for time *t*.

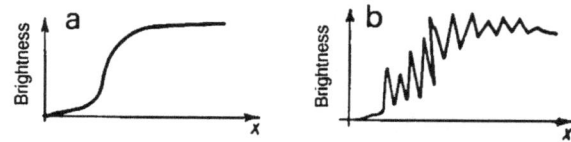

Fig. 5. Brightness distribution near the edge of an image: (a) an idealized case, and (b) a real case.

contain useful information. How can the initial image be restored?

We excite, by turning to the state 1, all elements belonging to the image received. Thereafter we let an autowave propagate through the medium. In the course of time, the broken edges can be seen to become "spliced" (fig. 4b), starting from the narrowest breaks, and after a time $t = L_0/2v$, where v is the velocity of wave propagation, all breaks of width $L < L_0$ are spliced. However, unfortunately, all the lines are blurred and have a width equal to L_0. To restore the contrast (to make the lines thinner), we let the wave propagate backward for the same time $t = L_0/2v$, and stop the process. Now the initial width of the lines and the sizes of all the breaks of width $L > L_0$ are restored, and the breaks of width $L < L_0$ turn out to be spliced (fig. 4c).

Thus to restore the image the following three operations are required: (1) Propagation of a wave for time *t*. (2) Stop. (3) Backward propagation for time *t*. An analogous approach has been used in mathematical morphology [8], but the computation was not a parallel one.

5.3. Edge detection

For the detection of an edge of an image, differential operators are usually employed (see ref. [9] and references therein). However, real brightness data are noisy (see fig. 5), and differentiation will not lead to the detection of an acceptable edge line. Preliminary smoothing of the image is used in this case, which results in blurring of weak edges and in time consuming

computation. If, for instance, a sliding window of size n^2 is used for smoothing, it requires a large number of cycles of a sequential computer. For a raster of 10^6 pixels, approximately $n^2 \times 10^6$ CPU cycles are necessary.

How can one accelerate this operation with the help of autowaves? It is known that the initial shape of an autowave is not conserved during autowave propagation. The wave form is smoothing due to annihilation of the autowave fronts under collisions and due to other effects [3, 4]. This property of propagating autowaves can be used for designing a highly parallel procedure of an edge-preserving smoothing.

The basic idea is as follows. The image is projected on a photosensitive autowave medium, and, as mentioned in section 4, the elements illuminated above some threshold acquire value 0, while others acquire value 1. An autowave (the trigger wave of transitions from 0 to 1) induced by the image starts propagating through the medium. The geometry of the wavefront is close to the edge line for an ideal image but not for the real one. If the distortions are induced by noise only, the situation is not hopeless. Autowave propagation can be used for smoothing the defects of a different kind. An obvious shortcoming is that the image becomes enlarged during autowave propagation. This can be avoided as before by reversing the direction of the autowave propagation.

Smoothing with the 3×3 sliding window on a sequential processor takes about 10^7 CPU cycles. As computer experiments have shown (see fig. 6), autowave smoothing of approximately the same quality requires only several cycles of the autowave processor.

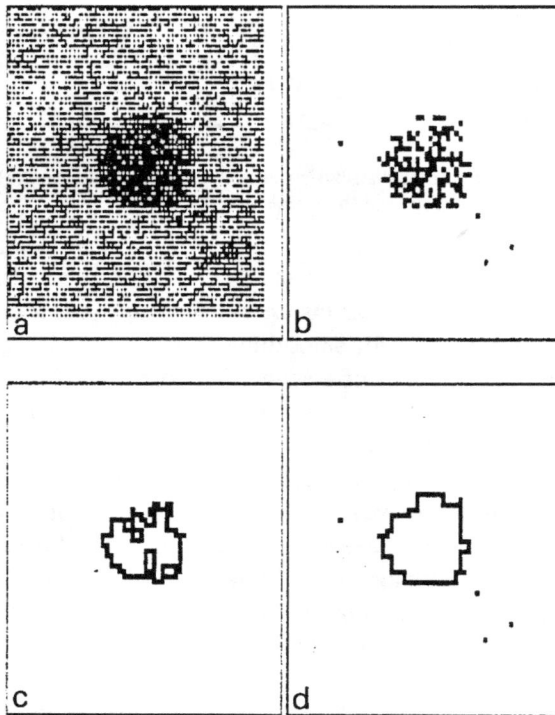

Fig. 6. Edge detection. Comparison of autowave smoothing and smoothing with a sliding window: results of a numerical experiment. (a) The image to be processed; raster 128×128 pixels, 128 grey levels. (b) The edge detected by threshold binarization without preprocessing. (c) The edge detected after the smoothing with a sliding window of 3×3 pixels with equal weights. It requires about $128 \times 128 \times 10$ CPU cycles of the universal computer used. (d) The edge detected after autowave smoothing with 1-step forward and 1-step backward autowave propagation. It requires only two cycles of the simulated autowave processor; the increase in speed is about 2^{16}.

Two algorithms were compared:

Algorithm 1 (sliding averaging):

(a) Averaging: Each point of the new image acquires a brightness equal to the arithmetic average of the brightness of the old image in the 3×3 square with the center at this given point.

(b) Binarization: The brightness of a point of the new image is equal to 1 if the brightness of the corresponding point of the old image exceeds a prescribed threshold. Otherwise, the new brightness is equal to 0.

(c) Edge detection: If a point with brightness equal to 1 in the old image has no neighboring points with brightness equal to 0 in the 3×3 square, the brightness of the point will be equal to 0. Other points remain without any changes.

Algorithm 2 (autowave smoothing):

(a) Binarization (the same as 1b).

(b) Direct propagation of autowave (dilation): The point of the new image acquires brightness equal to 1 if there is at least one point with brightness equal to 1 in the 3×3 square; otherwise the brightness will be equal to 0.

(c) Backward (reverse) propagation of autowave (erosion): The point of the new image acquires brightness equal to 0 if there is at least one point with brightness equal to 0 in the 3×3 square, otherwise the brightness will be equal to 1.

(d) Edge detection (the same as 1c).

As one can see from fig. 6, the results obtained with autowave smoothing are comparable to those obtained by smoothing with the sliding window, but the time consumed is much less (the gain in processing time is of the order of 10^6).

Algorithms for intersection point detection, skeletonization, filtration of hindrances of different types, following a moving object, and others, have been developed. As we have shown above, autowave methods can provide an enormous acceleration in image processing time, particularly when large rasters are used. The autowave algorithms described here can be applied directly only to binary images, but autowaves can process grey-level images as well [5].

6. Autowave processor and neural networks

The basic operation of an autowave processor is the autowave propagation along the raster, that is, the propagation of the trigger wave of transitions from 0 to 1. The simplest autowave processor is the highly parallel VLSI where neighboring cells are connected so that if at the initial moment some cells contain 1, then at the next moment all the neighboring cells will also contain 1 and all the cells evolve their states simultaneously.

The autowave process is similar to technical realizations of neural networks but differs from them in the following aspects:

(1) In a neural network, an element has connections spreading over almost the whole network. In an autowave processor, only a few (e.g. 4, 6, or 8) neighboring elements are connected.

(2) The connection strengths in a neural network have continuous values, and can differ for different pairs of neurons. In the autowave processor, they can be either 0 or 1.

Thus an autowave medium is a simplified caricature of a neural network. We can see, however, that some interesting results can be obtained when analyzing topological features, and invariance under translations, rotations, and scaling or deformation is provided. The speed can be accelerated by several orders of magnitude compared with modern specialized technical systems. Meaningful results can be obtained in spite of the over-simplification.

Summarizing, we can say that:

(1) The natural parallelism during autowave propagation across an autowave processor provides a considerable acceleration (by 3–6 orders of magnitude, when detecting topological features).

(2) The corresponding VLSI is realizable by modern technology.

(3) Due to the diffraction of autowaves, the autowave VLSI is reliable with respect to damages of some elements of the VLSI.

The rejection of a number of the properties of neural networks is certainly caused not by considerations of principal scientific but only by technical reasons. The autowave VLSI is more suitable for modern technology. No doubt, the future development of the computational technology in this field will tend to approach the possibilities of neural networks, which were designed by nature for image processing.

References

[1] J.J. Hopfield, Neural networks and physical systems with emergent collective computational abilities, Proc. Natl. Acad. Sci. USA 79 (1982) 2554.
[2] H. Haken, ed., Computational Systems – Natural and Artificial, Synergetics, Vol. 38 (Springer, Berlin, 1987).
[3] V.I. Krinsky, ed., Self-Organization. Autowaves and Structures Far from Equilibrium, Synergetics, Vol. 28 (Springer, Berlin, 1984).
[4] G.R. Ivanitsky, V.I. Krinsky, A.N. Zaikin and A.M. Zhabotinsky, Autowave processes and their role in disturbing the stability of distributed excitable systems, Sov. Sci. Rev. 1 (1981) 79.
[5] L. Kuhnert, K.I. Agladze and V.I. Krinsky, Image processing using light-sensitive chemical waves, Nature 337 (1989) 244.
[6] K.I. Agladze, F.E. Ilyasov, V.I. Krinsky and O.A. Mornev, Patent of the USSR 4280381/24-112818 (1987).
[7] F.E. Ilyasov, Ya.B. Kazanovich and V.I. Krinsky, Patent of the USSR 4424588/24-74229 (1989).
[8] R.M. Haralick, S.R. Sternberg and X. Zhuang, Image analysis using mathematical morphology, IEEE Trans. Patt. Anal. Mach. Intel. PAMI 9 (1987) 532.
[9] A. Kundu, Robust edge detection, Pattern Recognition 23 (1990) 423.

Physica D 49 (1991) 254
North-Holland

List of contributors

Physica D 49 (1991) 255–256
North-Holland

Analytic subject index